# Measurement Science for Engineers

**P P L Regtien**

*University of Twente, The Netherlands*

and **F van der Heijden, M J Korsten and W Olthuis**

*Department of Electrical Engineering, University of Twente*

*Foreword by*
**Ludwik Finkelstein**, *Senior Research Fellow*
*and Professor Emeritus, City University, London, UK*

London and Sterling, VA

**Publisher's note**
Every possible effort has been made to ensure that the information contained in this book is accurate at the time of going to press, and the publishers and authors cannot accept responsibility for any errors or omissions, however caused. No responsibility for loss or damage occasioned to any person acting, or refraining from action, as a result of the material in this publication can be accepted by the publisher or any of the authors.

First published in Great Britain and the United States in 2004 by Kogan Page Science, an imprint of Kogan Page Ltd
Reprinted 2004
Apart from any fair dealing for the purposes of research or private study, or criticism or review, as permitted under the Copyright, Designs and Patents Act 1988, this publication may only be reproduced, stored or transmitted, in any form or by any means, with the prior permission in writing of the publishers, or in the case of reprographic reproduction in accordance with the terms and licences issued by the CLA. Enquiries concerning reproduction outside these terms should be sent to the publishers at the undermentioned addresses:

| | |
|---|---|
| 120 Pentonville Road | 22883 Quicksilver Drive |
| London N1 9JN | Sterling VA 20166-2012 |
| United Kingdom | USA |
| www.koganpagescience.com | |

© P P L Regtien, 2004

The right of P P L Regtien to be identified as the author of this work has been asserted by him in accordance with the Copyright, Designs and Patents Act 1988.

ISBN 1 9039 9658 9

---

**British Library Cataloguing-in-Publication Data**

A CIP record for this book is available from the British Library.

**Library of Congress Cataloging-in-Publication Data**

Regtien, P. P. L.
  Measurement science for engineers / Paul P.L. Regtien.
     p. cm.
  Includes bibliographical references.
  ISBN 1-903996-58-9
  1. Mensuration. I. Title.
T50.R45 2004
620'.0044--dc22
                                                    2004007394

---

Typeset by Newgen Imaging Systems (P) Ltd., Chennai, India

Transferred to digital print 2008

Printed and bound in Great Britain by CPI Antony Rowe, Eastbourne

# Contents

| | | |
|---|---|---|
| | **Preface** | vi |
| | **Foreword** | viii |
| **1** | **Introduction** | **1** |
| | 1.1. Views on measurement science | 1 |
| | 1.2. Trends in measurement science and technology | 5 |
| | 1.3. General architecture of a measurement system | 8 |
| | 1.4. Further reading | 11 |
| | 1.5. Exercises | 12 |
| **2** | **Basics of Measurement** | **15** |
| | 2.1. System of units | 15 |
| | 2.2. Standards | 18 |
| | 2.3. Quantities and properties | 21 |
| | 2.4. Transducers | 36 |
| | 2.5. Further reading | 40 |
| | 2.6. Exercises | 40 |
| **3** | **Measurement Errors and Uncertainty** | **43** |
| | 3.1. True values, errors and uncertainty | 43 |
| | 3.2. Measurement error and probability theory | 44 |
| | 3.3. Statistical inference | 50 |
| | 3.4. Uncertainty analysis | 55 |
| | 3.5. Characterization of errors | 63 |
| | 3.6. Error reduction techniques | 74 |
| | 3.7. Further reading | 83 |
| | 3.8. Exercises | 84 |
| **4** | **Analogue Signal Conditioning** | **87** |
| | 4.1. Analogue amplifiers | 87 |
| | 4.2. Analogue filters | 97 |
| | 4.3. Modulation and detection | 105 |
| | 4.4. Further reading | 113 |
| | 4.5. Exercises | 114 |

## 5 Digital Signal Conditioning — 117
- 5.1. Digital (binary) signals and codes — 117
- 5.2. Sampling — 119
- 5.3. Sampling devices and multiplexers — 128
- 5.4. Digital functions — 133
- 5.5. Hardware for digital signal conditioning — 135
- 5.6. Further reading — 138
- 5.7. Exercises — 139

## 6 Analogue to Digital and Digital to Analogue Conversion — 141
- 6.1. Conversion errors — 141
- 6.2. Parallel DA-converters — 145
- 6.3. Serial DA-converters — 148
- 6.4. Parallel AD-converters — 149
- 6.5. Direct AD-converter — 152
- 6.6. Integrating AD-converters — 154
- 6.7. Sigma-delta AD-converters — 157
- 6.8. Further reading — 160
- 6.9. Exercises — 161

## 7 Measurement of Electrical, Thermal and Optical Quantities — 163
- 7.1. Measurement of electrical quantities and parameters — 163
- 7.2. Measurement of time and time-related quantities — 173
- 7.3. Measurement of magnetic quantities — 174
- 7.4. Measurement of thermal quantities — 179
- 7.5. Measurement of optical quantities — 188
- 7.6. Further reading — 193
- 7.7. Exercises — 195

## 8 Measurement of Mechanical Quantities — 197
- 8.1. Potentiometric sensors — 197
- 8.2. Strain gauges — 200
- 8.3. Capacitive displacement sensors — 205
- 8.4. Magnetic and inductive displacement sensors — 210
- 8.5. Optical displacement sensors — 217
- 8.6. Piezoelectric force sensors and accelerometers — 231
- 8.7. Acoustic distance measurement — 236
- 8.8. Further reading — 247
- 8.9. Exercises — 248

## 9 Measurement of Chemical Quantities — 251
- 9.1. Some fundamentals and definitions — 251
- 9.2. Potentiometry — 252

|  |      |                                 |     |
|--|------|---------------------------------|-----|
|  | 9.3. | Amperometry                     | 257 |
|  | 9.4. | Electrolyte conductivity        | 267 |
|  | 9.5. | Further reading                 | 275 |
|  | 9.6. | Exercises                       | 275 |

**10 Imaging Instruments** — 277
- 10.1. Imaging principles — 278
- 10.2. Optical imaging — 286
- 10.3. Advanced imaging devices — 295
- 10.4. Digital image processing and computer vision — 301
- 10.5. Further reading — 306
- 10.6. Exercises — 307

**11 Design of Measurement Systems** — 309
- 11.1. General aspects of system design — 309
- 11.2. Computer based measurement systems — 314
- 11.3. Design examples — 318
- 11.4. Further Reading — 329

**Appendices** — 331
- A. Scales — 331
- B. Units and quantities — 334
- C. Expectations, variances and covariance of two random variables — 335

**References** — 339

**Solutions to Exercices** — 345

**Index** — 355

# Preface

Throughout history, measurement has played a vital part in the development of society, and its significance is still growing with the ongoing advances in technology. The outcome of a measurement may have important implications for the resultant actions. The correct execution of a measurement is, therefore, highly important, making heavy demands not only on the executor's skills but also on the designer's competences to create measurement systems with the highest possible performance.

This book on measurement science is adapted from a course book on measurement and instrumentation, used by undergraduate students in electrical engineering at the University of Twente, The Netherlands. It deals with basic concepts of electronic measurement systems, the functionality of their subparts and the interactions between them. The most important issue of all is the quality of the measurement result. How to design a measurement system or instrument with the highest performance under the prevailing conditions. How to get the most of the system in view of its technical limitations. How to minimize interference from the environment. And finally: how to evaluate the measurement result and account for the remaining errors.

A course book cannot give the ultimate answer to such questions. Its intention is merely to offer basic knowledge about the physical and instrumentation aspects of measurement science, the availability and characteristics of the major measurement tools and how to use them properly. More importantly, the book tries to make students aware of the variety of difficulties they may encounter when setting up and using a measurement system and make them conscious of the fact that a measurement result is always error-ridden. Solving such problems is not only a matter of knowledge; experimental skills and experience are even important to arrive at the specified performance. To that end, the course in measurement and instrumentation of which this book is a consequence, is accompanied by a series of practical exercises, directed to the evaluation of sensor characteristics, signal processing and finally the total measurement system.

The organization of the book is as follows. Chapter 1 is an introductory chapter in which concepts of measurement science and the general architecture of a measurement system are described. In Chapter 2 some fundamental aspects of measurement science are reviewed: the international system of units, measurement standards and physical quantities and their relations. Measurement errors and uncertainty are discussed extensively in Chapter 3; it includes the basic elements of probability theory, methods to analyse and evaluate uncertainty and general techniques to minimize measurement errors.

Chapters 4, 5 and 6 deal with signal conditioning and conversion: Chapter 4 is on analogue signal conditioning (amplification, filtering, modulation); Chapter 5 on digital signal conditioning (sampling, multiplexing and digital operations); and Chapter 6 on the conversion from analogue tot digital and vice versa.

For each (physical) quantity a multitude of sensing strategies is available. In the next three chapters the major sensing methods are reviewed: in Chapter 7 we discuss the measurement of electrical, magnetic, thermal and optical quantities; in Chapter 8 mechanical quantities and in Chapter 9 chemical quantities. Measurement of multi-dimensional geometry is performed by imaging: the result is some kind of image in some domain. How these images are acquired and can be interpreted are the topics of Chapter 10. Finally, Chapter 11 covers particular aspects of system design and virtual instrumentation.

The book is written by a team of authors, all from the Department of Electrical Engineering, University of Twente. Chapters 3 (uncertainty) and 10 (imaging) are authored by F van der Heijden, Chapter 5 (digital signal conditioning) by M Korsten, Chapter 9 (chemical quantities) by W Olthuis and the remainder by P Regtien. V Pop contributed to the section on virtual instruments in Chapter 11.

Reading this book requires some basic knowledge of calculus, complex functions, physics and electronics, about equivalent to the first term undergraduate level. As with any course book, this one covers just a selected number of issues from an otherwise wide area. Connection to other relevant literature has been made by adding a short list of books at the end of each chapter, with a short indication of the subject and level. The enthusiastic reader may consult the cited references to the scientific literature for more detailed information on the subjects concerned.

Paul P L Regtien<br>
University of Twente, The Netherlands<br>
April 2004

# Foreword

Measurement, and the instrumentation by which it is implemented, is the basic tool of natural science and a key enabling technology.

Measurement is the essential means by which we acquire scientific and technical knowledge on which modern society and the economy depend. Further, all aspects of technology rely on measurement and could not function effectively without it. Driven by technical progress and by economic and social requirements, measurement and its instrumentation are advancing rapidly in capability and range of application.

A few examples of the main areas of application illustrate the significance of modern measurement technology. Modern manufacture is based to a substantial extent on the application of measurement and measuring systems. They are an essential component of automation, which both raises productivity and enhances quality. Application of measuring instrumentation has extended from the automatic control of continuous processes to applications in the automation of the manufacture of discrete engineering products, and to more general uses of robotics. Transport, in the air, on the sea, and on land, is to an increasing extent relying on the use of advanced measuring systems for navigation and propulsion management. Instrumentation has an ever-increasing role in providing the security of property from fire and theft. Modern medicine increasingly relies on advanced instrumentation for diagnosis and some aspects of therapy such as intensive care. Finally, we must mention the importance of instrumentation in the safeguarding of the environment. Monitoring of pollution, ranging from the analysis of noxious chemicals to the measurement of noise and the remote sensing from space are two leading areas of application. It should also be added that the design and manufacture of measuring equipment and systems is an important economic activity.

Given the central and far-ranging importance of measurement and instrumentation, all engaged in technology require a sound education and training in this field. This is widely recognised.

The structure and content of appropriate education in measurement and instrumentation is the subject of international interest and debate. Professor Paul Regtien, as Chairman of the Technical Committee on Education and Training of the International Measurement Confederation, is spearheading this debate.

Some of the principal problems in designing and delivering education in the field are as follows. Firstly, it is generally accepted that education in measurement and instrumentation should inculcate a well-organised body of basic concepts and principles and that it should also foster a competence to apply that fundamental knowledge in practice through the study of particular measurement equipment and applications. There is continuing discussion concerning the establishment of the nature and scope of the fundamental concepts and principles of the domain and of the framework into which they are organised. There are debates concerning the nature of the discipline; whether it is to be viewed from an information perspective regarding measurement as an information acquisition and handling process, or whether to focus on metrological problems of unit, standards, calibration and uncertainty. There is an issue of the autonomy of the domain and of the extent to which it is related to other disciplines; to physical science in respect of sensors and the sensing process, and to information technology in respect of information handling in measuring systems. There are issues of the nature of the balance between the teaching of abstract principles and the provision of a survey of practice. Finally, there is a growing recognition that education in measurement and instrumentation must be design-orientated, raising the question of how this influences formal presentation of the subject.

The form of education in any technology depends on the nature and level of the competence for which students are being educated. Further it requires a choice from an increasing arsenal of means of teaching: formal presentations, web-based information, laboratories, simulation and teaching machines.

While progress in educational tools is fuelling great interest, it is recognised that core of knowledge must be taught through formal structured presentations.

Regarding the nature of competence for which technical personnel are being educated trained and developed it is necessary to recognise different levels of competence. We may distinguish craftsmen, technicians, technician engineers, and scientific engineers. The names are not agreed in all languages and cultures, but the levels of competence are recognisable everywhere. Craftsmen are generally expert in good practice and in manual skills. Technicians have the practical expertise of craftsmen, but are educated to have an understanding of the basic principles underlying their work. Technician engineers are educated and trained in both the science and practice of their speciality and are able to apply established practice to problems of some size and complexity. Scientific engineers are educated in both science and practice to a level that enables them to advance them, to innovate and to tackle problems that are new and complex. They are formed to provide technical leadership.

In the education and training of all levels of scientific and technical personnel, we may distinguish the basic education in the ability to measure, which must be acquired by all natural scientists and engineers, and is generally provided by laboratory courses. A higher level of attainment in measurement and instrumentation, which we may call ancillary, is required by engineers in a number of specialisations, notably control engineering. The highest level of attainment is required by those who will engage in design and development of measurement instrumentation.

The present book is an introductory presentation of the basic concepts and principles of measurement science. It forms the basis of an introduction to the subject for scientific engineers, but its simplicity and clarity makes it useful at technician engineer level. It is suitable for a wide spectrum of engineering specialisations.

Professor Regtien and his team are internationally known both for their educational work and for their contribution to the science and technology of measurement and instrumentation. Their views, therefore, carry authority.

This book is a very welcome contribution to the educational literature of the field. It is particularly interesting for a good integration of general principles and the description of the realisation of measurement equipment and methods. It achieves a good balance between the information and metrological perspectives on measurement science. It takes into account the state of the art in information technology. A particularly innovative contribution of the book is the inclusion of a chapter on design, recognition of the significant.

The book will be of significant interest to teachers and students of measurement and instrumentation, as a view of the subject by authoritative teachers from an institution of high standing.

Ludwik Finkelstein
Measurement and Instrumentation Centre,
City University, London

# Chapter 1

# Introduction

Any book on measurement science should start with a definition of the word "measurement". Any attempt, however, to fully grasp the notion of measurement within a single, concise sentence is doomed to fail, or would at least cause endless debate. Nevertheless, a few definitions are given in the first section that will throw light on various aspects of measurement.

In sections that follow the general concept of measurement, the physical and mathematical aspects, the impacts of technology, telematics and computer science on practicing and training of measurement are briefly discussed.

This is followed by a section in which the functionality and architecture of a measurement system is considered, and then, in subsequent chapters the various parts of the system are set out in more detail while in the last chapter the measurement system is considered again, but this time from the viewpoint of the designer.

This introductory chapter ends with a list of other recommendable books on measurement and instrumentation, including a short review of each.

## 1.1. Views on measurement science

The history of measurement goes back to the time when communication between people first became apparent. To negotiate on food, clothing or tools, to talk about events having taken place in the past, about the age of tribe members, for all such matters and many others, a measure had to be found to exchange ideas that are associated with distance, time and weight. This very early need for a measure as a kind of reference has survived over time. Such references have been developed since then, stimulated by the expansion of trade, science and production.

Measurement, according to the *Encyclopaedia Britannica*, is the process of associating numbers with physical quantities and phenomena. Measurement is fundamental to the sciences; to engineering, building, and other technical matters; and to much everyday activity. For that reason the elements, conditions, limitations, and theoretical foundations of measurement have been much studied. Measurement theory is the study of how numbers are assigned to objects and phenomena, and its concerns include the kinds of things that can be measured, how different measures relate to each other, and the problem of error in the measurement process [1].

In particular, the aspect of accuracy will be considered extensively in this book. The result of a measurement is not only a number but should include, without any exception, information about the error in that number, as a result of the imperfection of the measurement process. This error can be given implicitly (by the number of significant digits) or explicitly, by a specified error band or in terms of probability.

A definition of measurement that is often cited is the one posed by L.Finkelstein [2]:

*Measurement is the process of empirical, objective assignment of numbers to the attributes of objects and events of the real world, in such a way as to describe them.*

This definition accounts for the fact that measurement applies to *properties* of an object or event, not the object or the event itself. It requires an explicit, unambiguous description of those properties, in terms of mathematical relations and in terms of experimental conditions and procedures. This is what can be called the *model* of the measurement process, which is another issue that is strongly emphasised in this book. The model includes properties of the object to be measured, the measurement system, the environment, and the interactions between them. Such a viewpoint is close to the definition according to DIN 1319 [3], which states: "Measurements are executions of planned actions for a qualitative comparison of a measurement quantity with a unit".

The last phrase brings us to another aspect of measurement: units. Quantitative measurements do not only require a definition of the property to be investigated, but also a reference or standard for that property to determine the unit of it. For instance, the resistance of a resistor is defined as the ratio of voltage and current, and the reference is the international standard for resistance (Section 2.1). As a consequence, measurement is an act of comparison, something that is already included in the definition by the 16th century scientist Christian Freyherr von Wolff: "Measurement means to take a certain quantity for one and to investigate the ratio of others of the same quality to it" [4]. We can state that traceability of a measurement tool to the highest standard is nowadays an important issue for worldwide exchange of information and materials, and for international agreements on matters concerning safety, legislature and environment.

However, for the process of classification (which is considered unquestionably as an act of measurement) such a unit can hardly be found in many applications: another type of representation applies here, for instance a set of references (as the Minsel cards for colour), or a precise description of the property to be considered (for instance the quality of an apple to distinguish between unripe, ripe and overripe). Obviously, in the last example and in many such cases one can argue about the possibility of generating a description of the property to be measured that is sufficiently objective. One approach to account for the limited use of the classical standards is the extension of standards to other classes. Next to the classical, metrological measurement standard, other classes of standards have been proposed leaving scope for subjectivity of perception and for virtual (digital) implementation. This leads to descriptions of properties that are only valid for a restricted group of users of these measurements, where the interpretation of the "unit" for a class is left to that group.

A common idea in all definitions is a *transformation* from (properties of) events or objects to an n-dimensional space containing numbers, symbols, or subspaces. This is not very surprising: it is close to the concept that measurement stands at the basis of communication: we observe an object or event and want to communicate about the perceptions with others who (possibly) did not observe the phenomenon. A purely formal definition of measurement is based on this mapping concept:

$$M : Q \rightarrow Z$$

where $Q$ is the set of properties describing the event, $Z$ the set of symbols and $M$ the transformation rule. Indeed, most measurement aspects discussed before could be included in this relation and from a scientific point of view it is an elegant and very concise definition. However, its practical significance is doubtful because nothing is said about how such a mapping is obtained.

Figure 1.1 illustrates some of the aspects discussed before in a graphical way, with two examples: the measurement of a resistance value and the classification of an animal (a cat).

Obviously, in case of the resistor we have a clear notion of the property "electrical resistance" and there exists a well-defined reference value for this property, a standard which is based on quantum effects. Any resistance measuring instrument is (or should be) calibrated, which means that its internal reference (possibly a precision resistor) has a specified accuracy with respect to the (inter)national standard.

| Object | Physical property | Mathematical model | Transformation/ classification |
|---|---|---|---|
| Resistor | $V/I$ | $V = IR$ | Value in Ω on a scale |
| 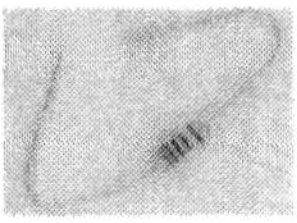 | 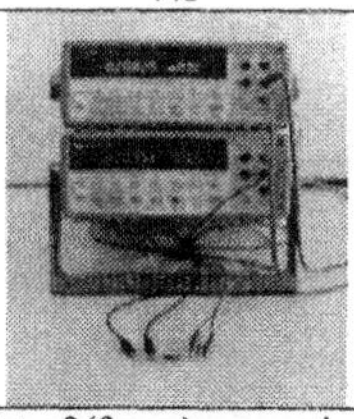 | 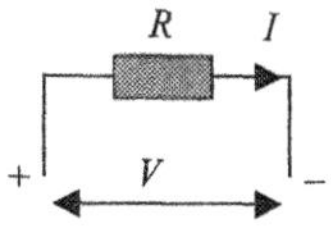 $R \quad I$ | 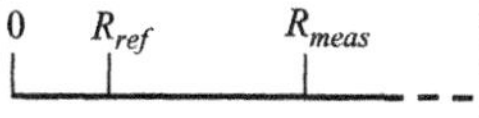 $0 \quad R_{ref} \quad R_{meas}$ |
| Cat on couch | Set of (fuzzy) properties | Subjective model | Zoological scheme |
|  | Animal: about 20 cm high; with 4 legs; long tail; hairy skin; sneaky and limber gait; whiskers; small, upright, pointed ears; …. |  | ANIMALS<br>Vertebrata<br>Mammalia<br>CARNIVORA<br>C.fissipedia<br>Felidae<br>F.maniculata<br>⇒ F. maniculata domestica |

**Figure 1.1** *The information sequence of a measurement*

The classification of the cat, on the contrary, is based on an imprecise model we have built through personal experience and, at least for observations in daily life we do not need the zoological scale for a successful classification. Moreover: even the human observation of part of this animal or a glimpse of it would suffice for a correct classification. Man's recognition ability surpasses by far that of the fastest computer equipped with the most ingenious image processing software. Image processing based measurement systems are still an intriguing challenge, and therefore a special chapter on this subject has been included in this book.

Classification is closely related to identification. In object identification the goal of the measurement is to classify a unique object (each class just contains one particular object). An example is the identification of a person based on unique properties such as a fingerprint, features, iris and DNA. System identification concerns the determination of particular system parameters, for instance the number of poles and zeros or the order of the system. Such measurements reveal particular properties about the structure of a system, which are useful for instance in control applications.

Measurements allow us to distinguish between or order similar objects or processes according to a particular property. The symbols of the measurement set can be positioned along a line in a random or specified order, which line is called the measurement scale. For instance, along the resistance scale in Figure 1.1 resistors can be ordered according to their resistance value. Examples of measurement scales are (Appendix A):

- Mohs scale for hardness
- Richter scale for seismic activity
- Beaufort scale for wind force.

The following types of scales can be distinguished:

*Nominal scale*: the most simple scale, just for classification; the order of this scale is not defined. The zoological system in Figure 1.1 is an example of a nominal scale.

*Ordinal scale*: the symbols or numbers of this scale are arranged in a specified order, but the intervals between the numbers are meaningless. On the basis of this scale we can compare properties in terms of less, equal or larger. Examples are the scales of Mohs and Beaufort. Ordinal scales are frequently used in, for instance, social and live sciences, to represent statistics on health, prosperity, safety, intelligence.

*Interval scale*: as the ordinal scale, but now the intervals between the numbers are quantified. An example is the temperature scale in celsius. The choice for the zero of this scale is arbitrary, and so ratios and summations of the values are meaningless.

*Ratio scale*: as the interval scale, but with a unique zero point. Examples: the temperature scale in Kelvin, scales for length, voltage, etc. Most physical and technical quantities can be measured using the ratio scale. The consequence is that these quantities can be compared numerically with a particular reference value, the unit.

*Absolute scale*: a scale defined by the nature of the measurement process. Example: measuring by counting (of objects).

## 1.2. Trends in measurement science and technology

Although measurement is almost as old as mankind, measurement science as a discipline is quite new. Indeed, in the ancient world the purpose and value of having standards was already recognized, but these standards had only a very limited accuracy, and their use was restricted to the areas governed by local rulers. Early scientists were aware of the need for measurement instruments to explore and understand physical processes. "Numeric relations are the key to the understanding of the whole mystery of nature", Plato stated [6]. But how to verify these relations experimentally? It was only in the Renaissance that measurement instruments were developed with sufficient reproducibility and accuracy, enabling the evaluation of relations between physical quantities, resulting in the formulation and verification of physical laws.

We have already mentioned the global need for standards, serving as a reference for some basic physical quantities. The major aim of establishing the SI system has been to get rid of the large variety of local units and standards [7], hence to facilitate and expand trade. This was only the starting point. Since France founded the Metre Convention in 1875, other nations gradually started to establish metrological institutes, to which the responsibility for the realization and maintenance of national standards was allocated. With the growth of trade and industrial and technological advances the role of these institutes became more and more important. The first standards for physical quantities were just objects. Later, stimulated by the need for ever-increasing stability and reproducibility, such standards gradually were replaced by well defined physical processes based on quantum effects (Section 2.2). The quality of standards (primary as well as lower standards) has reached a very high level, enabling, for instance, the manufacturing of machine parts in various countries to be spread over the world, and to assemble them at one location, building up a complete high-precision instrument.

Although the process of standardization and unification has set in since the 18th century, even nowadays (2003) the need for conversion tables seems still to be present. On the Internet many sites can be found with such conversion tables. A recent work provides thousands of conversion factors in over 800 densely printed pages [8].

Measurement is not only a vital activity of designers, engineers and scientists. It is also deeply rooted in our modern society, and even more: it has almost become an essential element of life. An average household counts dozens of measuring instruments: clocks (to display time; to serve as an alarm or a kitchen timer), thermometers (to measure room and body temperature; in fire alarms; and embedded in all kind of thermally controlled implements), pressure (in central heating system), electric power (to monitor electricity consumption), flow (to monitor gas and water consumption), speed (for speed control in video and audio equipment), level (in a washing machine), light and sound (in alarm and security systems), tape measures and rulers for domestic use, distance sensors in printers, scanners and other equipment, and many others. In a modern car the number of measurements easily exceeds 100. People are often unconscious of performing measurements. There are many hidden measurement instruments, embedded in appliances, tools and cars. Without these built-in measurement devices such mechanisms would have a much poorer performance.

The role of measurement is crucial in almost any field: medical and health care, transport and trade, production and factory management, communication and navigation, space exploration and particle physics, just to mention a few. Incorrect measurement results can have a tremendous economic and societal impact. Correct measurements (which implies correct interpretation of the results as well!) are of paramount significance.

Current development in measurement science aims at the improvement of the performance of measurement equipment according to particular requirements. There is a continuous demand for measurement instruments that perform better with respect to:

- speed (faster)
- dimensions (smaller)
- applicability (wider range, higher environmental demands)
- automation (to be embedded in automated systems)
- quality (higher accuracy; higher sensitivity; lower noise levels)
- energy (minimizing power consumption)
- reliability (more robust, long life time)
- cost (cheaper).

Furthermore, the development time of new (measurement) systems is subject to increased pressure. Designing a sensor system or a complete measurement instrument requires overall knowledge about the many aspects that are involved in the execution of a measurement: the translation from the properties of interest to measurable physical quantities, the transformation into electrical quantities, the conversion into digital signals, the processing of both analogue and digital measurement signals, the properties and limitations of available sensors, signal processing devices and software packages, the consequences of particular signal transformations on the quality of a measurement signal and its information content, the origin and impact of interfering signals, etc.

Most measurement systems are used by non-experts in measurement science; many measurement systems are designed or assembled by non-experts. Nevertheless, this will not necessarily result in a poor measurement, provided that the design and the execution are completed in the approved manner. In a separate chapter we will discuss some general design rules, to help the non-expert create his/her own measurement system in a proper way.

Rapid growth in computers and computer science has contributed to an expansion of software tools for the processing of measurement signals. The computer is gradually taking over more and more instrumental functions. For instance, the visualisation of time-varying measurement signals, traditionally performed by an oscilloscope, can now be carried out by a computer and a data acquisition card, almost in real time. The development of such "virtual instrumentation" is still expanding, driving back the usage of conventional real instruments. Obviously, virtual instruments have a lot of advantages, but will never take over the measurement itself. The advance of virtual instruments simplifies the set-up of a measurement; however it also tends to mask the essences of a measurement: the transformation from quantifiable properties to an electric signal and the associated measurement errors. A measurement can never be instrumented by software alone: "All measurements must be done by hardware" [9].

Any measurement is an experiment, and therefore closely associated with the real world. This real world is not ideal: theory does never completely describe the physical behaviour of a real process or system. There always exists a difference between the theoretical model that describes the system under consideration, and the reality as becomes apparent by measurement. Two major causes of this discrepancy are:

(a) the model is inadequate: it does not comprise all relevant properties that we are interested in or it is based on simplified assumptions;
(b) the measurement contains errors, due to for instance unforeseen loading, instrument uncertainties and erroneous operation.

All of these errors might occur simultaneously, hindering a clear analysis of the situation. An important and primary competence for anyone performing a measurement is the ability to be conscious of all kinds of errors and to identify and quantify them. This competence can be mastered by a thorough insight in the subsequent processes that make up the measurement, and a clear methodology for fault isolation and detection. Moreover, a fair assessment of the various errors should reveal the main causes of the observed deviations from the expected behaviour, leaving other possible causes to be negligible in a specific situation. So, the act of measurement is not only a matter of proper equipment and technical skills, but requires a critical attitude as well as sufficient prior knowledge towards the process that is subject to measurement. The apprentice must learn to always be aware of measurement errors and to quantify them in an appropriate way. The need and importance of these additional abilities should be made explicit and requires training.

## 1.3. General architecture of a measurement system

A measurement system can be interpreted as a channel for transporting information from a measurement object to a target object. In a measurement-and-control system (part of) the target object is identical to the measurement object. Three main functions can be distinguished: data acquisition, data processing and data distribution (Figure 1.2):

- *Data acquisition*: to obtain information about the measurement object and convert it into an electrical signal. The multiple input in Figure 1.2 indicates the possibility for the simultaneous measurement of several parameters. The single output of the data acquisition element indicates that, in general, all data is transferred via a single connection to the data processing element.
- *Data processing*: includes processing, selecting or otherwise manipulating measurement data in a prescribed manner. This task is performed by a microprocessor, a micro-controller or a computer.
- *Data distribution*: to supply processed data to the target object. The multiple output indicates the possible presence of several target instruments such as a series of control valves in a process installation.

The data acquisition and data distribution elements are subdivided into smaller functional units. Since most physical quantities are non-electric, they should first be converted into an electric signal to allow electronic processing. This conversion is called *transduction* or sensing and is performed by a transducer or sensor (Figure 1.3).

Many sensors or transducers produce an analogue signal. However, since processors can only handle digital signals, the analogue signal must be converted into a digital one. This process is called analogue-to-digital conversion or *AD-conversion*. It comprises three steps: *sampling* (at discrete time intervals samples are taken from the analogue signal), *quantization* (the rounding off of the sampled value to the nearest fixed discrete value) and *conversion* to a binary code. Both sampling and quantization may introduce loss of information. Under specific conditions, however, this loss can be limited to an acceptable minimum.

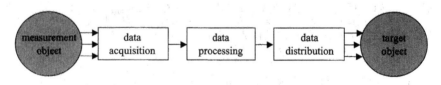

**Figure 1.2** *Three main functions of a measurement system*

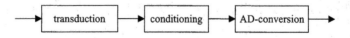

**Figure 1.3** *Three main functions of data acquisition*

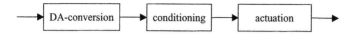

**Figure 1.4** *Three main functions of data distribution*

Generally, the output signal of a transducer is not suitable for conversion by an AD-converter: it needs to be processed (conditioned) first in order to optimize signal transfer. The major processing steps necessary to achieve the required conditions are explained briefly below:

- *amplification*: to increase the signal's magnitude or its power content;
- *filtering*: to remove non-relevant signal components;
- *modulation*: modification of the signal shape in order to enable signal transport over a long distance or to reduce the sensitivity to interference during transport;
- *demodulation*: the reverse operation of modulation.

After being processed by the (digital) processor, the measurement data experiences the reverse operation (Figure 1.4). The digital signal is converted into an analogue signal by a digital-to-analogue converter or *DA-converter*. It is supplied to an *actuator* (or effector, excitator, output transducer), to perform the required action. If the actuator cannot be connected directly to the DA-converter, the signal is conditioned first. This *conditioning* usually consists of signal amplification and some type of filtering.

Depending on the goal of the measurement, the actuation function could be:

- *indication* (for instance by a digital display);
- *registration* or *storage* (on paper using a printer or plotter, or on a magnetic or optical disk);
- *control*, e.g. driving a valve, a heating element or an electric motor.

For the (quasi)simultaneous processing of more than one variable, the outputs of the AD-converters are alternately connected to the processor (Figure 1.5).

The *multiplexer* acts as an electronically controlled multi-stage switch, controlled by the processor. This type of multiplexing is called *space division multiplexing* (SDM). Obviously, the number of signal conditioners and AD-converters equals the number of sensor channels. When this number becomes large, as for instance in temperature profile measurements or in seismic measurements, another system architecture should be considered. Figure 1.6 shows such an approach, where the multiplexing takes place prior to AD-conversion. Now the AD-converter is shared with all sensors. In some cases it is possible to share also the signal conditioning. Such a concept is called *time division multiplexing* or TDM. Evidently, a similar approach can be applied for the data distribution part of the measurement system: the processed data either are demultiplexed and individually converted to the analogue domain, or first converted sequentially, and next demultiplexed to the actuators.

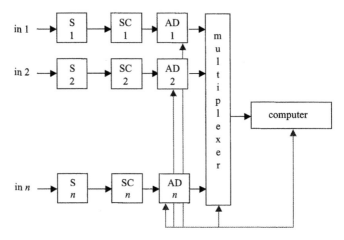

**Figure 1.5** *Space division multiplex, with digital multiplexer; $S$ = sensor; $SC$ = signal conditioner; $AD$ = analogue-to-digital converter*

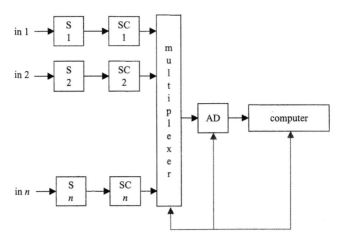

**Figure 1.6** *Time division multiplex, with analogue multiplexer and shared AD-converter; $S$ = sensor; $SC$ = signal conditioner; $AD$ = analogue-to-digital converter*

Note that the requirements for the converters and the multiplexers differ substantially for these two architectures. In the SDM architecture digital signals are multiplexed, while in TDM the multiplexer handles analogue signals, which makes the latter more susceptible to errors and the demands on the multiplexer much greater. Further, the AD-converter in TDM should be very fast in order to maintain a high sampling rate per channel, whereas the AD-converters in the SDM architecture may have larger conversion times up to the total cycle time of the multiplexer sequence.

At first sight, the concept of time multiplexing (either SDM or TDM) has the disadvantage that only the data from the selected channel is processed while the information from the non-selected channels is blocked. It can be shown that, when the time between two successive selections for a particular channel is made sufficiently short, the information loss is negligible.

**1.4. Further reading**

We list here a selection of books on measurement and instrumentation, published during the last decade, and recommended for undergraduate students to learn more about measurement systems in general, or just to broaden their view on measurement science. Books on specific topics dealt with in other chapters are listed at the end of those chapters.

J. McGhee, W. Kulesza, M.J. Korczyñski, I.A. Henderson, *Measurement Data Handling – Hardware Technique*; Technical University of Łodz, 2001, ISBN 83-7283-008-8.
This book is the second volume of *Measurement Data Handling*. The first volume considers theoretical aspects of data handling; this volume focuses on the hardware implementation of the signal processing part of a measurement system. It contains chapters on microprocessors and micro-controllers, data communication systems (networks), analogue signal conditioners, AD converters, analogue and digital filters and system reliability. A highly readable book, with many practical recommendations for the designer, and with useful end-of-chapter tutorials and exercises.

J.G. Webster, *The Measurement, Instrumentation, and Sensors Handbook*, Boca Raton, FL [etc.]: CRC Press [etc.] (1999); ISBN: 3-540-64830-5.
From the introduction: "The Handbook describes the use of instruments and techniques for practical measurements required in engineering, physics, chemistry, and the life science. It includes sensors, techniques, hardware, and software. ... Articles include descriptive information for professionals, students, and workers interested in measurement."
The handbook is structured according to the topics: measurement characteristics, measurement of spatial variables; time and frequency; mechanical variables (solid; fluid; thermal); electromagnetic and optical variables; radiation; chemical variables; biomedical variables; signal processing; displays; control. Each article starts with a brief introduction to the subject and ends with references to original work.

T.P. Morrison, *The Art of Computerized Measurement*, Oxford [etc.]: Oxford University Press (1997); ISBN 0-19-856542-9, 0-19-856541-0.
A book on handling digital measurement data. It provides much practical information on topics like data formats, file formats, digital hardware (microprocessors), compilers and I/O software. Suitable for users who like to write their own software for the processing of (digitized) measurement data. Examples are based on C++. The text is mainly on computerizing, not so much on measurement.

A.F.P. van Putten, *Electronic Measurement Systems*, Bristol: Institute of Physics Publishing (1996) (2nd ed.); ISBN 0-7503-0340-9.

Apart from those topics usually discussed in a text book on instrumentation and measurement, this book contains various chapters on some particular subjects: reliability, guarding and shielding, and noise. The chapter on transducers mainly concerns silicon microsensors. The last chapter is on ergonomics and human engineering. Every chapter ends with examples and problems; unfortunately the answers to the problems are not included in the book. Nevertheless, a valuable compilation of the many aspects that are involved in the design and use of a measurement system.

J.P. Bentley, *Principles of Measurement Systems,* Burnt Mill [etc.]: Longman Scientific & Technical (3rd ed. 1995); ISBN 0-582-23779-3.
This book is divided into three parts. The first part deals with general aspects of measurement systems (static and dynamic characteristics, noise, reliability and loading). The second part discusses various elements of a measurement system: sensors, signal conditioners, AD-converters, microprocessors and displays. In the third part some particular measurement systems are reviewed: flow measurement systems, systems based on thermal, optical and ultrasonic principles. In addition, there are special sections on gas chromatography and data-acquisition. Theory and practice are well balanced throughout the book. End-of-chapter exercises and solutions at the back make the book suitable for home study.

L. Finkelstein, K.T.V. Grattan, *Concise Encyclopedia of Measurement and Instrumentation,* Pergamon Press, 1994; ISBN 0-08-036212-5.
A typical reference book, to get a quick answer to particular questions on measurement science. Unlike traditional encyclopedia, being structured as a continuous list of entries in alphabetical order, this work has entries at a more general level. An extensive table of contents helps the reader to quickly find the proper section.

W. Bolton, *Electrical and Electronic Measurement and Testing,* Longman: Harlow (1992); ISBN 0-582-08967-0.
An introductory text covering a wide spectrum of topics: error sources, reliability, oscilloscopes, analogue meters and recorders, signal sources, test procedures. Measurement bridges are discussed in great detail. Suitable for undergraduates.

## 1.5. Exercises

1. What scale is used when presenting the results of the following processes:

   a. ranking of submitted papers for a conference based on scientific relevance
   b. sorting of stamps according to country
   c. determination of the amount of money in a purse
   d. deforestation rate in Brazil (in $km^2$ per year)
   e. observing the time on a watch
   f. determination of body weight.

2. Compare the demands on the multiplexer and the analogue-to-digital converter(s) for a space division multiplex measurement system and a time division multiplex system.

3. Identify the main functions (according to Figures 1.3 and 1.4) of the following instruments:
   a. liquid in glass thermometer
   b. electronic balance
   c. car cruise control system.

# Chapter 2

# Basics of Measurement

In this chapter we discuss various basic aspects of measurement science. First, the system of units is presented, and how this system has been developed from its introduction in 1795 to the present time with basically seven standard units. The materialization of a unit quantity is the next issue to be discussed. It will be shown that at present all standards (except for the kilogram) are related to fundamental physical constants. Obviously, there are many more quantities than those seven basic quantities. In the third section we give an overview of the most important quantities and properties, used in various physical domains: the geometric, electrical, thermal, mechanical, and optical domain. Relations between quantities from different domains that are fundamental to the measurement of non-electrical quantities and parameters are also given. Finally, some general aspects of sensors, the devices that convert information from one domain to another, are discussed.

## 2.1. System of units

A unit is a particular physical quantity, defined and adopted by convention, with which other particular quantities of the same kind are compared to express their value. The value of a physical quantity is the quantitative expression of a particular physical quantity as the product of a number and a unit, the number being its numerical value.

For example, the circumference of the earth around the equator is given by:

$$c_e = 40{,}074 \cdot 10^3 \text{ m} \qquad (2.1)$$

where $c_e$ is the physical quantity and the number $40{,}074 \cdot 10^3$ is the numerical value of this quantity expressed in the unit "meter". Obviously, the numerical value of a particular physical quantity depends on the unit in which it is expressed. Therefore, to any measurement result the unit in which it is expressed should be added explicitly. To avoid misinterpretations the use of the international system of units (Système International d'unités, SI) is highly recommended when presenting measurement results.

The foundation for the SI system was laid during the French Revolution, with the creation of a metric decimal system of units and of two platinum standards representing the meter and the kilogram (1799). The unit of time was based on the astronomical second. Since then the SI has been gradually improved and extended to the present system of 7 base units: meter, kilogram, second, ampere, Kelvin, mole and candela.

An important property of the system is its coherence: derived units follow from products or quotients of the base units, with a multiplication factor equal to 1. Examples: $Q = I \cdot t$ (defining the unit of charge from the base units ampere and second), $Q = C \cdot V$ (defining the unit of electric capacity), etc.

The choice for a basic electrical quantity has long been a disputable matter. First there were two parallel systems: the electrostatic and the electromagnetic subsystem. In the electrostatic system the fourth unit was $\varepsilon_0$ and in the electromagnetic system $\mu_0$. Hertz combined the electrostatic and electromagnetic units into a single system, through the relation $c^2 \varepsilon_0 \mu_0 = 1$, with $c$ the speed of light in vacuum.

In 1938 it was proposed to take $\mu_0$ as the fourth unit, with the value $10^{-7}$, although any other electrical quantity could have been accepted. Finally, in 1960 the 11th Conférence Générale des Poids et Mesures adopted the ampere (and not $\mu_0$ or $\varepsilon_0$) as the basic unit for electric current, with such a value that the permeability in vacuum $\mu_0$ is just $4\pi \cdot 10^{-7}$ H/m (the "rationalized" mks system). Moreover, two other basic units have been added to the system, for temperature (Kelvin) and light (candela), and in 1971 another for substance (mole). The seven standard units are defined as follows:

The **meter** is the length of the path travelled by light in vacuum during a time interval of 1/299 792 458 of a second [17th CGPM (1983), Res.1].

The **kilogram** is the unit of mass; it is equal to the mass of the international prototype of the kilogram [1st CGPM (1889)].

The **second** is the duration of 9 192 631 770 periods of the radiation corresponding to the transition between the two hyperfine levels of the ground state of the cesium 133 atom [13th CGPM (1967), Res.1].

The **ampere** is that constant current which, if maintained in two straight parallel conductors of infinite length, of negligible circular cross-section, and placed 1 meter apart in vacuum, would produce between these conductors a force equal to $2 \times 10^{-7}$ newton per meter of length [9th CGPM (1948)].

Note that the effect of this definition is to fix the permeability of vacuum at exactly $4\pi \cdot 10^{-7}$ H $\cdot$ m$^{-1}$.

The **Kelvin**, unit of thermodynamic temperature, is the fraction 1/273.16 of the thermodynamic temperature of the triple point of water [13th CGPM (1967), Res.4].

The **mole** is the amount of substance of a system which contains as many elementary entities as there are atoms in 0.012 kilogram of carbon 12 [14th CGPM (1971), Res.3]. When the mole is used, the elementary entities must be specified and may

**Table 2.1** Accuracy of the base units

| Unit | Accuracy |
|---|---|
| Meter | $10^{-12}$ |
| Kilogram | – |
| Second | $3 \cdot 10^{-15}$ |
| Ampere | $4 \cdot 10^{-8}$ |
| Kelvin | $3 \cdot 10^{-7}$ |
| Mole | $8 \cdot 10^{-8}$ |
| Candela | $10^{-4}$ |

be atoms, molecules, ions, electrons, other particles, or specified groups of such particles.

The **candela** is the luminous intensity, in a given direction, of a source that emits monochromatic radiation of frequency $540 \cdot 10^{12}$ hertz and that has a radiant intensity in that direction of 1/683 watt per steradian [16th CGPM (1979), Res.3].

Evidently, the meter, the second and the Kelvin are defined by fundamental physical phenomena; the kilogram is (still) based on a material artefact, and the ampere, the mole and the candela are related indirectly to the kilogram. Table 2.1 shows the accuracy with which the base units have been determined [1].

Other quantities than the base quantities are called *derived quantities*. They are defined in terms of the seven base quantities via a system of quantity equations. The *SI derived units* for the derived quantities are obtained from these equations and the seven SI base units.

Examples are:

| Quantity | Unit name | Unit |
|---|---|---|
| Volume | Cubic meter | $m^3$ |
| Current density | Ampere per square meter | $A/m^2$ |
| Magnetic field strength | Ampere per meter | $A/m$ |

Some derived quantities have got special names and symbols, for convenience. Examples are:

| Quantity | Name | Unit symbol | Derived unit | Expressed in SI-base units |
|---|---|---|---|---|
| Frequency | Hertz | Hz | – | $s^{-1}$ |
| Electric potential difference | Volt | V | W/A | $m^2 \cdot kg \cdot s^{-3} \cdot A^{-1}$ |
| Capacitance | Farad | F | C/V | $m^{-2} \cdot kg^{-1} \cdot s^4 \cdot A^2$ |

All 22 derived quantities with official SI names are listed in Appendix B1.

To express multiples or fractions of the units a notation system of prefixes is established, based on decimal fractions. Decimal multiples have capital prefixes, decimal fractions are written in lower case. The only basic unit with a prefix in its name is the kilogram (kg). It is important to apply these prefixes carefully, to exclude any confusion. For instance, a MHz and a mHz are quite different units, and pressure is expressed in (for instance) Pa $\cdot$ m$^{-2}$ and not PA/M$^2$. A full list of standardized prefixes is given in Appendix B2.

## 2.2. Standards

For each of the basic units a physical materialization should be available that enables a comparison of the associated quantity with other instruments: *a standard*. The value of the quantity that is represented by such a standard is exactly 1 by convention. For instance the unit of the meter was the distance between two small scratches on a particular platinum bar that was kept under the supervision of the Bureau International des Poids et Mesures (BIPM) in Sèvres, France. The national standards of other countries are derived from and continuously compared with this international standard (Figure 2.1) [3].

For other units too (primary) standards have been constructed, and they have been used for a long time to compare the unit value with other standards, called *secondary standards*. The unit value for the quantities mass, length and time have been chosen with a view to practical applicability. For instance, the circumference of the earth could be a proper standard but is highly impractical; therefore the meter was originally defined (1791) as one ten millionth of a quarter of the earth's meridian passing

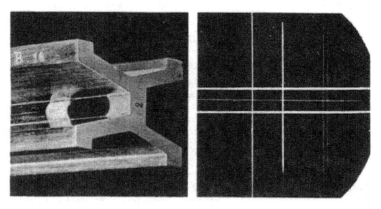

**Figure 2.1** *A former standard meter; left: end view of the British copy of the International Meter; right: rulings on the polished facet; the two thick vertical lines indicate the length at 0°C and 20°C, taking into account the thermal expansion [National Physical Laboratory, Teddington, Middlesex]*

**Figure 2.2** *Length of a foot, defined as the average of the feet of 12 men [3]*

through Paris, which is a much more practical measure for daily life usage. Also the units for time (second) and weight (kilogram) were defined likewise.

Evidently, these physical standards were much more stable and reproducible than those defined earlier on the basis of human properties (Figure 2.2).

A measurement standard should be based on some highly stable and reproducible physical phenomenon representing a particular quantity. This quantity could be a basic unit, like the platinum meter as shown in Figure 2.1, but this is not strictly necessary. Due to the growing need for higher stability and reproducibility of standards, other physical phenomena are being used for the realization of standards. All modern standards are related to fundamental physical constants (such as Planck's constant[1] $h$, Boltzmann's constant[2] $k$ and the charge of an electron $e$), except for the kilogram [2, 4]: an alloy cylinder (90% Pt, 10% Ir) of equal height and diameter, held at the Bureau International des Poids et Mesures, Sèvres, France. Two examples of modern standards will be discussed here: the standards for volt and ohm.

The standard for *electrical potential difference (volt)* is realized using the *Josephson effect*[3]. This is a tunnelling effect that occurs in a structure of two superconductors separated by a very thin isolating layer (about 1 nm). When irradiated

---

[1] Max Planck (1858–1947), German physicist; Nobel Prize for physics in 1918.
[2] Ludwig Edward Boltzman (1844–1906), Austrian physicist.
[3] Brian David Josephson (1940), British physicist; described in 1962 the Josephson effect when he was a student.

with microwave radiation of frequency $f$, the current-voltage characteristic is a staircase function, in which the voltage increases discontinuously at discrete voltage intervals equal to

$$V_j(N) = \frac{N \cdot f}{K_j} \qquad (2.2)$$

where $K_j$ is the Josephson constant, equal to $2e/h$, $f$ the frequency of the radiation and $N$ indicates a specific flat plateau in the current-voltage characteristic. As a consequence, within the specified range (plateau) the voltage is determined by the frequency and two fundamental physical constants. The frequency can be measured with very high accuracy. On the 1st of January 1990 the CIPM set the value for $K_j$ to exactly [5]:

$$K_{j-90} = 483597.9 \text{ GHz/V} \qquad (2.3)$$

thereby having replaced the former standard cells based on saturated salts. The subscript 90 denotes that this value has taken effect in 1990, not earlier, when the value was about 9 ppm (parts per million) lower. The Josephson voltage is quite small: in the order of some $10\,\mu\text{V}$. However, the international reproducibility is better than $5 \cdot 10^{-8}$ relative.

The voltage from a single Josephson junction is too small (a few mV at most) for practical use. Therefore, devices have been developed containing many such junctions in series (over 20,000), to arrive at about 1 V and even 10 V [6]. Using micro-technology, the dimension of such a multiple Josephson junction can be limited to several cm. Although the Josephson voltage is a DC voltage, it is also the basis for AC voltage standards [7].

For *electrical impedance* a standard is used based on a particular capacitor configuration. With such a calculable capacitor the capacitance per unit length (F/m) is related to the permittivity of vacuum. The best construction represents a capacitance with an accuracy of about $10^{-8}$.

Another, and more accurate, electrical standard is based on the *quantum Hall effect*[4]. This effect occurs in for instance a junction made from GaAsAl$_x$Ga$_{1-x}$As or in silicon MOSFETs. In a strong magnetic field (>10 T) and at low temperatures (<1.2 K) the electron gas is completely quantized: the device shows regions of constant voltage (Hall plateaus) in the $V$–$I$ characteristic [8]. The resistance of the $i$th plateau is given by the equation

$$R_H(i) = \frac{V_H(i)}{I} = \frac{R_k}{i} \qquad (2.4)$$

---

[4] Edwin Herbert Hall (1855–1938), American physicist; discovered in 1879 the effect bearing his name.

where $R_k = h/e^2$ is the von Klitzing constant[5]. From 1st January 1990 the value of the Klitzing constant was fixed at exactly [5]

$$R_{H-90} = 25\,812.807 \text{ ohm} \qquad (2.5)$$

with a reproducibility better than $1 \cdot 10^{-8}$. Hall quantum effect devices can be constructed with very small dimensions (chip size measures a few mm only).

One could think about a system of units completely based on fundamental constants. For instance, the electron mass, the electron charge, Planck's constant, the speed of light, or others. However, this will result in very small or large measures [9]. For instance, the unit of length would measure $5.3 \cdot 10^{-11}$ meters, the unit of time about $2.4 \cdot 10^{-17}$ s, and hence the unit for velocity $2.2 \cdot 10^{+6}$ m/s, obviously very impractical values for daily use.

## 2.3. Quantities and properties

A quantity is a property ascribed to phenomena, processes or objects, to which a value or a class can be assigned. Quantities can be classified in many ways into groups. We give some of these classifications in brief.

Quantities that have magnitude only are called *scalar* quantities (for instance temperature); a quantity with magnitude and direction is called a *vector* quantity (for instance velocity).

We distinguish energy related quantities (associated with energetic phenomena, for instance force, electric current) and quantities that are not associated with energy (static quantities or simply properties, for instance resistivity, length).

Properties that are independent of the dimensions or the amount of matter are pure material properties; others have a value that is determined not only by the substance but also by the size or the construction layout. For instance, the resistivity $\rho(\Omega\text{m})$ is a pure material property, whereas the resistance $R(\Omega)$ depends on the material as well as the dimensions of the resistor body. In general, the value of a material property is orientation dependent. This dependency is expressed by subscripts added to the symbols. For instance, $s_x$ or $s_1$ is the elasticity in the $x$-direction of a specified coordinate system; $s_{xy}$ or $s_6$ denotes the shear elasticity around the $z$-axis (in the $xy$-plane).

Variables characterizing the state of a system are related by physical laws. Variables acting upon the system are called *independent* variables. Variables describing the system's state are related to these input variables, hence they are called *dependent* variables. Obviously, a particular variable can be dependent or independent,

---

[5] Klaus von Klitzing (1943), German physicist (born in Poland); 1985 Nobel Prize for physics for the discovery of the quantized Hall effect.

## 22   Measurement Science for Engineers

**Table 2.2** *Geometric quantities*

| Quantity | Symbol | SI unit | Abbreviation |
|---|---|---|---|
| Length | $l$ | Meter | m |
| Surface area | $A$ | | $m^2$ |
| Projected area | $A_p$ | | $m^2$ |
| Volume | $V$ | | $m^3$ |
| Plane angle | $\alpha \ldots \varphi$ | Radian | rad |
| Solid angle | $\Omega$ | Steradian | sr |

according to its purport in the system. This classification will further be detailed in Section 2.3.6 where we present a general system for the description of quantities in the electrical, thermal and mechanical domains, based on thermodynamics. In that section we also discuss another classification of quantities, based on an energetic consideration: *through-variables* and *across-variables*. First, we offer a brief review of quantities in the geometric, electrical, magnetic, thermal, mechanical, and optical domains.

### 2.3.1. The geometric domain

Table 2.2 lists the quantities for the geometric domain. Its main purpose is to fix notations for these parameters and to give some definitions.

The radian and steradian are supplementary units (see also Appendix B1). The *radian* is defined as the plane angle between two radii of a circle which cuts off on the circumference an arc equal in length to the radius. The unit is m/m, so the radian is dimensionless. For clarity we will use the unit radian.

The *steradian* is the solid angle which, having its vertex in the centre of a sphere, cuts off an area of the surface of the sphere equal to that of a square with sides of length equal to the radius of the sphere. The unit is $m^2/m^2$, hence the steradian too is dimensionless. Again, we will use the steradian for the purpose of clarity.

Projected area is defined as the rectilinear projection of a surface of any shape onto a plane normal to the line of sight along which the surface is being viewed. The projected or apparent area is used to facilitate equations for radiance in optical systems (Section 2.3.5). If $\varphi$ is the angle between the local surface normal and the line of sight, the apparent area is the integral over its perceptible surface $S$, hence:

$$A_p = \iint_S \cos \varphi \, dS \tag{2.6}$$

**Example 2.1**
The apparent area of a flat surface $S$ viewed under an angle $\varphi$ is $S \cdot \cos \varphi$. The surface area of a sphere with radius $r$ is $4\pi r^2$. Looking from a large distance we observe, from any angle, a circular projection with radius $r$, so the apparent area equals $\pi r^2$.

**Table 2.3** *Electrical and magnetic quantities*

| Quantity | Symbol | Unit |
| --- | --- | --- |
| Electric current | $I$ | A (ampere) |
| Current density | $J$ | $A \cdot m^{-2}$ |
| Electric charge | $Q$ | $C$ (coulomb) $= A \cdot s$ |
| Dielectric displacement | $D$ | $C \cdot m^{-2}$ |
| Electric polarization | $P$ | $C \cdot m^{-2}$ |
| Electric field strength | $E$ | $V \cdot m^{-1}$ |
| Potential difference | $V$ | $W \cdot A^{-1}$ |
| Magnetic flux | $\Phi$ | Wb (weber) $= J \cdot A^{-1}$ |
| Magnetic induction | $B$ | T (tesla) $= Wb \cdot m^{-2}$ |
| Magnetic field strength | $H$ | $A \cdot m^{-1}$ |
| Magnetization | $M$ | $A \cdot m^{-1}$ |

## 2.3.2. The electrical and magnetic domain

Table 2.3 lists the most important quantities for the electrical and magnetic domain, together with the symbols and units.

Moving electric charges generate a magnetic field $H$. Its strength, the *magnetic field strength*, is given by the relation

$$I = \oint_C H \cdot dl \qquad (2.7)$$

where $I$ is the current through a closed contour $C$. For each configuration with current flow, the field strength can be calculated by solving the integral equation (2.7). Only for simple, highly symmetric structures can an analytical solution be found; in general, the field strength must be calculated by numerical methods (for instance finite element methods). Induction phenomena are described by the variables *magnetic induction B* and *magnetic flux* $\Phi$. By definition, the flux is

$$\Phi = \iint B \cdot dA \qquad (2.8)$$

For a homogeneous field $B$, the flux through a surface with area $A$ equals $\Phi = B \cdot A \cdot \cos\alpha$, with $\alpha$ the angle between the surface's normal and the direction of $B$. Hence, the flux through a surface in parallel with the induction is zero.

Table 2.4 shows the major properties for the electrical and magnetic domain.

*Conductivity* is the inverse of *resistivity*, and *conductance* the inverse of *resistance*. The electric *permittivity* is the product of the permittivity of free space $\varepsilon_0$ and the relative permittivity (or dielectric constant) $\varepsilon_r$. Similarly, the magnetic *permeability* is the product of the magnetic permeability of free space $\mu_0$ and the relative

**Table 2.4** *Electrical and magnetic properties*

| Property | Symbol | Unit | Definition |
|---|---|---|---|
| Resistivity | $\rho$ | $\Omega \cdot m$ | $E = \rho \cdot J$ |
| Resistance | $R$ | $\Omega$ (ohm) | $V = R \cdot I$ |
| Conductivity | $\sigma$ | $S \cdot m^{-1} = \Omega^{-1} \cdot m^{-1}$ | $\sigma = 1/\rho$ |
| Conductance | $G$ | S (siemens) | $G = 1/R$ |
| Capacitance | $C$ | F (farad) | $Q = C \cdot V$ |
| Permittivity | $\varepsilon$ | $F \cdot m^{-1}$ | $D = \varepsilon \cdot E$ |
| Self-inductance | $L$ | H (henry) | $\Phi = L \cdot I$ |
| Mutual inductance | $M$ | H | $\Phi_{ab} = M \cdot I$ |
| Reluctance | $R_m$ | $H^{-1}$ | $n \cdot I = R_m \cdot \Phi$ |
| Permeability | $\mu$ | $H \cdot m^{-1}$ | $B = \mu \cdot H$ |

**Table 2.5** *Analogies between electrical and magnetic expressions*

| Electrical domain | Magnetic domain |
|---|---|
| $E = \dfrac{1}{\sigma} \cdot J$ | $H = \dfrac{1}{\mu} \cdot B$ |
| $V = \int E \cdot dl$ | $n \cdot I = \int H \cdot dl$ |
| $I = \iint J \cdot dA$ | $\Phi = \iint B \cdot dA$ |
| $V = R_e \cdot I$ | $n \cdot I = R_m \cdot \Phi$ |
| $R_e = \dfrac{1}{\sigma} \cdot \dfrac{l}{A}$ | $R_m = \dfrac{1}{\mu} \cdot \dfrac{l}{A}$ |
| $C = \varepsilon \dfrac{A}{d}$ | $L = n^2 \mu \dfrac{A}{l}$ |

permeability $\mu_r$. Numerical values of $\varepsilon_0$ and $\mu_0$ are:

$$\varepsilon_0 = (8.85416 \pm 0.00003) \cdot 10^{-12} \text{ (F/m)}$$

$$\mu_0 = 4\pi \cdot 10^{-7} \text{ (Vs/Am)}$$

The relative permittivity and relative permeability account for the dielectric and magnetic properties of the material; this will be explained further in the sections on capacitive and inductive sensors (Chapter 8).

Table 2.5 shows the analogy between expressions for various electrical and magnetic quantities. In a magnetic circuit the magnetic resistance (also called *reluctance*) is equivalent to the electrical resistance in an electrical circuit. The parameters $A$ and $l$ in the expression for $R_e$ represent the cross section area and the length of a cylindrical conductor; the parameters $A$ and $l$ in the expression for $R_m$ are defined likewise, but now for a magnetic material. The term $n \cdot I$ in

**Table 2.6** *Time signal quantities*

| Signal quantity | Symbol | Unit |
|---|---|---|
| Time | $t$ | s |
| Frequency | $f$ | Hz (hertz) = s$^{-1}$ |
| Period | $T$ | s |
| Phase difference | $\varphi$ | rad |
| Duty cycle | $\delta$ | |
| Pulse width | $\tau$ | s |

**Table 2.7** *Thermal quantities*

| Quantity | Symbol | Unit |
|---|---|---|
| (Thermodynamic) temperature | $\Theta, T$ | K (Kelvin) |
| Celsius temperature | $T, t$ | °C (degree) |
| Quantity of heat | $Q_{th}$ | J (joule) |
| Heat flow rate | $\Phi_{th}$ | W (watt) |

the equation for the reluctance $R_m$ is the product of the number of turns of a coil carrying a current $I$, and is therefore expressed with the "unit" ampere-turns. In the expression for the selfinductance $L$, these parameters refer to the cross section area and the length of a wire-wound coil, with $n$ the number of turns.

In the electrical domain, some particular quantities associated with time apply; they are listed in Table 2.6. The duty cycle is defined as the high-low ratio of one period in a periodic pulse signal:

$$\delta = \frac{\tau}{T - \tau} \qquad (2.9)$$

It varies from 0 (whole period low) to 100% (whole period high). A duty cycle of 50% refers to a symmetric square wave signal.

### 2.3.3. The thermal domain

Table 2.7 and Table 2.8 review the major thermal quantities and properties. The symbol for temperature is $T$, but in the literature covering various domains, the symbol $\Theta$ is used, to draw a distinction between the symbols for temperature and for stress ($T$). The symbol for Celsius temperature is $t$, but to avoid confusion with the symbol $t$ for time, $T(°C)$ is also used.

The unit of Celsius temperature is the degree Celsius, symbol °C. The numerical value of the Celsius temperature is given by $T(°C) = T(K) - 273.15$. So, the degree Celsius is equal in magnitude to the Kelvin, hence temperature differences may be expressed in either °C or in K.

26  Measurement Science for Engineers

**Table 2.8** *Thermal properties*

| Property | Symbol | Unit | Definition |
|---|---|---|---|
| Heat capacitance | $C_{th}$ | $\text{J} \cdot \text{K}^{-1}$ | $Q_{th} = C_{th} \cdot \Delta\Theta$ |
| Specific heat capacity | $c_p$ | $\text{J} \cdot \text{kg}^{-1}\text{K}^{-1}$ | $(1/m) \cdot dQ_{th}/dT$ |
| Thermal resistance | $R_{th}$ | $\text{K} \cdot \text{W}^{-1}$ | $\Delta\Theta = R_{th} \cdot \Phi_{th}$ |
| Thermal conductivity | $g_{th}$ | $\text{W} \cdot \text{m}^{-1}\text{K}^{-1}$ | |
| Coefficient of heat transfer | $h$ | $\text{W} \cdot \text{m}^{-2}\text{K}^{-1}$ | |

The thermal conductivity $g_{th}$ and the thermal resistance $R_{th}$ are analogous to their electrical equivalent. A temperature difference $\Theta$ across a cylindrical object with cross section $A$ and length $l$ causes a heat flow $Q_{th}$ through the object equal to $\Theta/R_{th}$, with $R_{th} = (1/g_{th})(l/A)$, similar to the expression for the electrical resistance.

The interfacial heat transfer coefficient $h$ accounts for the heat flow across the boundary of an object whose temperature is different from that of its (fluidic) environment. Its value strongly depends on the flow type along the object and the surface properties.

**Example 2.2**
The cooling capacity of a flat thermoelectric cooler (Peltier cooler)[6] is limited by the heat flowing towards the cold plate (temperature $T_c$) from both the hot plate (temperature $T_h$) and the ambient (temperature $T_a$). Consider a cooler with plate area $A = 4 \text{ cm}^2$. The material between the plates has a thermal conductance of $G_{th} = 15 \text{ mW/K}$; the interfacial heat transfer coefficient between the cold plate and the ambient is $h = 25 \text{ Wm}^{-2}\text{K}^{-1}$. Calculate the heat flow towards the cold plate, when $T_a = 20°\text{C}$, $T_h = 25°\text{C}$ and $T_c = 0°\text{C}$.

Heat from hot to cold plate: $\Phi_{hc} = G_{th}(T_h - T_c) = 425 \text{ mW}$; heat from ambient to cold plate: $\Phi_{ac} = h \cdot A \cdot (T_a - T_c) = 200 \text{ mW}$.

### 2.3.4. The mechanical domain

Quantities in the mechanical domain describe state properties related to distance, force and motion. A possible categorization of these quantities is a division into:

- displacement quantities;
- force quantities;
- flow quantities.

---

[6] Jean Charles Athanase Peltier (1785–1845), French physicist; demonstrated in 1834 the converse Seebeck effect, called the Peltier effect.

**Table 2.9** *Displacement quantities*

| Quantity | Symbol | Unit | Relation |
|---|---|---|---|
| Length | $l$ | m | |
| Strain | $S$ | | |
| Velocity | $v$ | m·s$^{-1}$ | $v = dx/dt$ |
| Acceleration | $a$ | m·s$^{-2}$ | $a = dv/dt$ |
| Angular displacement | $\varphi$ | rad | |
| Angular velocity | $\omega$ | rad·s$^{-1}$ | $\omega = d\varphi/dt$ |

**Table 2.10** *Force quantities*

| Quantity | Symbol | Unit |
|---|---|---|
| Force | $F$ | N |
| Tension, stress | $T$ | Pa = N·m$^{-2}$ |
| Shear stress | $\tau$ | Pa |
| Pressure | $p$ | Pa |
| Moment of force | $M$ | Nm |

**Table 2.11** *Mechanical properties*

| Property | Symbol | Unit | Definition |
|---|---|---|---|
| Mass density | $\rho$ | kg·m$^{-3}$ | $M = \rho \cdot V$ |
| Mass | $m$ | kg | $F = m \cdot a$ |
| Weight | $G$ | kg | $F = G \cdot g$ |
| Elasticity | $c$ | N·m$^{-2}$ | $T = c \cdot S$ |
| Compliance | $s$ | m$^2$·N$^{-1}$ | $S = 1/c$ |

We will briefly summarise these groups. The basic *displacement quantities* and their units and symbols are listed in Table 2.9. Table 2.10 shows the most common *force quantities*, symbols and units. Some commonly used material properties in the mechanical domain are listed in Table 2.11.

The relation between tension $T$ and strain $S$ is given by Hooke's law[7]:

$$T = c \cdot S \qquad (2.10)$$

or:

$$S = s \cdot T \qquad (2.11)$$

---

[7] Robert Hooke (1635–1703), British physicist and architect.

Note that compliance $s$ is the reciprocal of the elasticity $c$. A force (tension) results in a deformation (strain): longitudinal forces cause translational deformation; shear forces cause rotational deformation. Some crystalline materials also show shear deformation due to translational forces and vice versa. Actually, these material parameters are orientation dependent, and should be provided with suffixes denoting the orientation.

Some flow quantities are listed in Table 2.12. They are not further discussed in this book.

### 2.3.5. The optical domain

In the optical domain two groups of quantities are distinguished: radiometric and photometric quantities. The radiometric quantities are valid for the whole electromagnetic spectrum. Photometric quantities are only valid within the visible part of the spectrum ($0.35 < \lambda < 0.77$ μm). They are related to the standardized mean sensitivity of the human eye and are applied in photometry (see also Section 10.1.1 on radiometry). Table 2.13 shows the major radiometric and corresponding photometric quantities. We use the subscripts $s$ and $d$ to indicate equal quantities but with

**Table 2.12** *Fluidic quantities*

| Quantity | Symbol | Unit |
| --- | --- | --- |
| Flow velocity | $v$ | $m \cdot s^{-1}$ |
| Mass flow | $\Phi_m$ | $kg \cdot s^{-1}$ |
| Volume flow | $\Phi_v$ | $m^3 \cdot s^{-1}$ |
| Dynamic viscosity | $\eta$ | $Pa \cdot s$ |
| Kinematic viscosity | $\nu$ | $m^2 \cdot s^{-1}$ |

**Table 2.13** *Radiometric and photometric quantities*

| | Radiometric quantities | | | Photometric quantities | |
| --- | --- | --- | --- | --- | --- |
| | Quantity | Symbol | Unit | Quantity | Unit |
| Power | Radiant flux | $P = dU/dt$ | W | Luminous flux | lumen (lm) |
| Energy | Radiant energy | $U$ | J | Luminous energy | lm·s |
| Emitting power per unit area | Radiant emittance | $E_s = dP_s/dS$ | W/m² | Luminous emittance | lm/m² = lux |
| Incident power per unit area | Irradiance | $E_d = dP_d/dS$ | W/m² | Illuminance | lm/m² = lux |
| Radiant power per solid angle | Radiant intensity | $I = dP_s/d\Omega$ | W/sr | Luminous intensity | lm/sr = candela (cd) |
| Power per solid angle per unit of projected surface area | Radiance | $L = dI/dS_p$ | W/m²sr | Luminance, brightness | lm/m²sr = cd/m² |

respect to a light *source* and a light *detector*, respectively. Some of these quantities will be explained in more detail now.

The *radiant intensity* $I_s$ (W/sr) is the emitted radiant power of a point source per unit of solid angle. For a source consisting of a flat radiant surface, the emitted power is expressed in terms of W/m², called the *radiant emittance* or *radiant exitance* $E_s$. It includes the total emitted energy in all directions. The quantity *radiance* accounts also for the direction of the emitted light: it is the emitted power per unit area, per unit of solid angle (W/m²sr).

A flat surface can receive radiant energy from its environment, from all directions. The amount of incident power $E_d$ is expressed in terms of W/m², and is called the *irradiance*.

In radiometry, the unit of (emitting or receiving) surface area is often expressed in terms of projected area $A_p$ (see equation (2.6)). A surface with (real) area $A$ that is irradiated from a direction perpendicular to its surface receives power equal to $E_d \cdot A$ (W). When that surface is rotated over an angle $\varphi$, relative to the direction of radiation, it receives only an amount of radiant power equal to $E_d \cdot A_p = E_d \cdot A \cos \varphi$. To make these radiometric quantities independent of the viewing direction, they are expressed in terms of projected area instead of real area, thus avoiding the $\cos \varphi$ in the expressions.

In Table 2.13, the *radiance L* of a surface is defined following this idea: it is the emitted radiant energy in a particular direction, per solid angle (to account for the directivity) and per unit of projected area, that is, per unit area when the emitting surface is projected into that direction. Figure 2.3 illustrates the relation between several of the optical quantities.

### Example 2.3 Irradiance of a lamp above a table-top

A light bulb has a radiant flux of $P = 5$ W. The construction of the lamp is such that it emits light isotropically (i.e. independent of orientation). A plane (e.g. a table-top) is located at a distance $h$ of 1 m from the bulb. What is the irradiance at the surface of the plane?

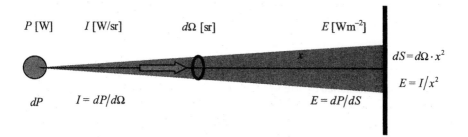

**Figure 2.3** *Illustration of the relation between various radiometric quantities*

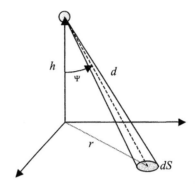

**Figure 2.4** *Irradiance of a point source on a planar surface*

The situation is as shown in Figure 2.4. Since the lamp is isotropic, the radiant intensity is: $I = P/(4\pi) = 5/(4\pi) = 0.4$ W/sr. The normal of a small area $dS$ on the plane at a distance $d$ from the lamp makes an angle $\psi$ with respect to the direction of the lamp. Hence, the solid angle corresponding to this area is: $d\omega = dS \cos \psi/d^2$. The radiant flux through this angle is $I d\omega$. Therefore, the irradiance on the surface at a distance $r$ from the centre of the plane is: $E_d = dP/dS = I d\omega/dS = I \cos \psi/d^2$. Expressing $\cos \psi$ and $d$ in $h$ and $r$ we finally obtain: $E_d = Ih/(h^2 + r^2)^{3/2} = I \cos^3 \psi/h^2$ Wm$^{-2}$. For instance, if $h = 1$ m, at the centre of the table ($r = 0$) we have $E = 0.4$ W $\cdot$ m$^{-2}$.

All radiometric units have their counterpart in the photometric domain (right-hand side of Table 2.13). However, we will not use them further in this book. The radiometric quantities in the table do not account for the frequency (or wavelength) dependence. All of them can be expressed also in spectral units, for example the spectral irradiance $E'_d(\lambda)$, which is the irradiance per unit of wavelength interval (W/m$^2$ µm). The total irradiance over a spectral range from $\lambda_1$ to $\lambda_2$ then becomes:

$$E_d = \int_{\lambda_1}^{\lambda_2} E'_d(\lambda)\, d\lambda \qquad (2.12)$$

Note that we use primed symbols for the spectral quantities, to make distinction between the different dimensions: W $\cdot$ m$^{-2}$ and W $\cdot$ m$^{-3}$, respectively.

Light falling on the surface of an object scatters back, and part of the scattered light reaches the sensor or an observer. Actually, the object acts as a secondary light source, in the same way as we see the moon by the sunlight scattered from its surface. If the energy is retransmitted isotropically, i.e. if the surface is a perfect diffuser and its radiance or brightness is not a function of angle, then $I = I_0 \cdot \cos \varphi$ (see Figure 2.5). This is known as *Lambert's cosine law*[8]. Many surfaces scatter

---

[8] Johann Heinrich Lambert (1728–1772), German mathematician, physician and philosopher.

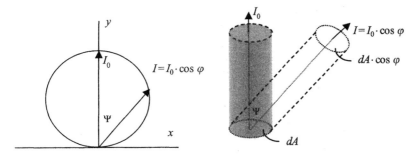

**Figure 2.5** *Directivity diagram of a Lambertian surface, and illustration of Lambert's law*

**Table 2.14** *Characteristic parameters of optical sensors*

| Parameter | Symbol | Unit |
|---|---|---|
| Output signal (voltage, current) | $S(V, I)$ | V, A |
| Signal output power | $P_o$ | W |
| Signal input power | $P_i$ | W |
| Output noise power | $N$ | W |
| Responsivity | $R = S/P_i$ | $V \cdot W^{-1}; A \cdot W^{-1}$ |
| Output signal-to-noise ratio | $P_o/N$ | – |
| Noise equivalent power | $NEP = S/R$ | W |
| Detectivity | $D = 1/NEP$ | $W^{-1}$ |

the incident light in all directions equally. A surface which satisfies Lambert's law is called *Lambertian*.

The apparent brightness of a Lambertian surface is the same when viewed from any angle. The relation between the radiant emittance $E$ and the radiance $L$ for a Lambertian surface is simply given by.

$$E = \pi \cdot L \tag{2.13}$$

Hence, the radiant emittance $E$ (power per unit area emitted by a Lambertian surface) equals $\pi$ times the radiance $L$ (power per solid angle per unit projected area).

To characterize optical sensors, special signal parameters are used, related to the noise behaviour of these components (Table 2.14). Note that signals are expressed in volts or amperes, and power in watts. Irrespective of this choice, the signal-to-noise ratio is a dimensionless quantity. All quantities are wavelength dependent: $S = S(\lambda)$ etc. If the wavelength dependency should be taken into account, spectral parameters are considered, with units V/m, A/m and W/m. All signals are also frequency dependent: $S = S(f)$; they can be normalized with respect to bandwidth, in which case the units are $V/\sqrt{Hz}$, $A/\sqrt{Hz}$ and W/Hz.

The parameters *NEP* and *D* represent the lowest perceivable optical signal. The *detectivity D* is just the reciprocal of *NEP*.

The *noise equivalent power* or *NEP* is equal to the radiation that produces a signal power at the output of an ideal (noise free) sensor equal to the noise power of the real sensor. When specifying *NEP*, also the wavelength $\lambda$, the sensor bandwidth $\Delta f$, the temperature and the sensitive surface area $A$ should be specified, as all these parameters are included in the noise specification. In many systems the radiation comes from a modulated source. In that case the *NEP* is defined as the power of sinusoidally modulated monochromatic radiation that produces an RMS signal at the output of an ideal sensor, equal to the RMS noise signal of the real sensor.

When the output signal is a current or a voltage, the NEP can be written as output signal over responsivity: $NEP = S/R$ (W). Some manufacturers specify the signal parameters per unit of bandwidth; in that case *NEP* is in $W/\sqrt{Hz}$, since $NEP = S(f)/R(f)$. This is rather confusing, so when considering specifications of a manufacturer of optical sensors, dimensions should be checked carefully.

The *NEP* of many optical sensors is proportional to the sensitive area of the sensor, which means that the noise current and noise voltage are proportional to $\sqrt{A}$. A parameter that accounts for the area dependence as well as the frequency dependence is the *specific detectivity* $D^*$, defined as

$$D^* = \frac{\sqrt{A \cdot \Delta f}}{NEP} \tag{2.14}$$

where *NEP* is in W. So, when $D^*(m\sqrt{Hz}/W)$ and the active surface area $A$ of an optical detector are given, the signal-to-noise ratio at a specified irradiance $E_d$ (W/m$^2$) is found to be $E_d \cdot D^* \cdot \sqrt{(A/\Delta f)}$.

**Example 2.4**
A particular optical detector with surface area $A = 10$ mm$^2$ and specific detectivity $D^* = 10^8$ cm$\sqrt{Hz}/W$ receives light with irradiance $E = 50$ mW/cm$^2$. Calculate the *NEP* and the signal-to-noise ratio at the detector output for a signal bandwidth of 1 Hz.

$NEP = \sqrt{(A \cdot \Delta f)}/D^* = 3.16 \cdot 10^{-9}$ (W); the signal-to-noise ratio is $E \cdot A/NEP = 5 \cdot 10^3$. In the case of a much larger signal bandwidth, for instance a pulsed signal with 1 MHz bandwidth, NEP is $10^3$ larger, so the signal-to-noise ratio is only 5.

### 2.3.6. Relations between quantities in different domains

A sensor performs the exchange of information (thus energy) from one domain to another and therefore it operates on the interface between different physical domains. In this section we derive relations between quantities from various domains, to obtain parameters describing the transfer properties of sensors. The relation between variables is governed by physical effects, by material properties or by a particular system lay-out. It either acts within one physical domain or crosses domain boundaries. Such relations describe the fundamental operation of sensors.

Several attempts have been made to set up a consistent framework for quantities and properties. Most of these descriptions are based on energy considerations, or more precisely, on the first and second law of thermodynamics. For each domain two variables can be defined in such a way that their product equals the energy (or energy density) of the domain type, the *conjugate quantities*. For instance:

- mechanical: tension $T$ [N/m$^2$] and deformation $S$ [–];
- electrical: field strength $E$ [V/m] and dielectric displacement $D$ [C/m$^2$];
- magnetic: magnetic induction $B$ [Wb/m$^2$] and magnetic field $H$ [A/m];
- thermal: temperature $\Theta$ [K] and entropy $\sigma$ [J/Km$^3$].

The variables $E$, $D$, $B$, $H$, $T$ and $S$ are vector variables, whereas $\sigma$ and $\Theta$ are scalars (so often denoted as $\Delta\sigma$ and $\Delta\Theta$). Note that the dimension of the product of each domain pair is $J \cdot m^{-3}$ (energy per unit volume) in all cases.

The variables in this list belong to either of two classes: through-variables or across-variables. To explain this classification, we first introduce the term lumped element. A *lumped element* symbolizes a particular property of a physical component; that property is thought to be concentrated in that element between its two end points or nodes. Exchange of energy or information only occurs through these terminals.

A *through-variable* is a physical quantity which is the same for both terminals of the lumped element. An *across-variable* describes the difference with respect to a physical quantity between the terminals. They are also called *extensive variables* and *intensive variables*, respectively. In an electronic system, current is a through-variable, voltage (or potential difference) is an across-variable. Therefore, through-variables are called *generalized I-variables* and across-variables are called *generalized V-variables*. However, this is just a matter of viewpoint. It is perfectly justified to call them *generalized forces* and *displacements*.

In the above groups of variables, $T$, $E$ and $\Theta$ are *across-variables*. On the other hand, $S$, $D$ and $\sigma$ are *through-variables*. They are related to each other through physical material properties or system lay-out. Based on the laws of energy conservation, several frameworks can be built up for a systematic description of variables. The energy content of an infinitely small volume of an elastic dielectric material changes by adding or extracting thermal energy and the work exerted upon it by electrical and mechanical forces. If only through-variables affect the system, its internal energy $U$ changes as

$$dU = TdS + EdD + \Theta d\sigma \qquad (2.15)$$

Likewise, if only across-variables affect the energy content, the energy change is written as

$$dG = -SdT - DdE - \sigma d\Theta \tag{2.16}$$

$G$ is the Gibbs potential, which can be found from the free energy $U$ by a Legendre transformation. We continue the discussion with the last expression because the resulting parameters are more in agreement with experimental conditions. After all, it is easier to have the across-variables as inputs or independent quantities (temperature, electric field or voltage, force) and to measure the resulting through-variables or dependent quantities (strain, dielectric displacement or current).

The through-variables $S$, $D$ and $\sigma$ in this equation can be written as partial derivatives of the Gibbs potential:

$$S(T, E, \Theta) = \left(\frac{\partial G}{\partial T}\right)_{\Theta, E}$$
$$D(T, E, \Theta) = \left(\frac{\partial G}{\partial E}\right)_{T, \Theta} \tag{2.17}$$
$$\sigma(T, E, \Theta) = \left(\frac{\partial G}{\partial \Theta}\right)_{T, E}$$

From these equations we can derive the various material and sensor parameters. To that end, the variables $S$, $D$ and $\Delta\sigma$ are approximated by linear functions, that is we take only the first term of the Taylor series expansion at the points $T = 0$, $E = 0$ and $\Theta = \Theta_0$ of the functions $S(T, E, \Theta)$, $D(T, E, \Theta)$, and $\sigma(T, E, \Theta)$:

$$dS = \left.\frac{\partial S}{\partial T}\right|_{E,\Theta} dT + \left.\frac{\partial S}{\partial E}\right|_{\Theta} dE + \left.\frac{\partial S}{\partial \Theta}\right|_{E} \Delta\Theta$$
$$dD = \left.\frac{\partial D}{\partial T}\right|_{E,\Theta} dT + \left.\frac{\partial D}{\partial E}\right|_{\Theta} dE + \left.\frac{\partial D}{\partial \Theta}\right|_{E} \Delta\Theta \tag{2.18}$$
$$d\sigma = \left.\frac{\partial \sigma}{\partial T}\right|_{E,\Theta} dT + \left.\frac{\partial \sigma}{\partial E}\right|_{\Theta} dE + \left.\frac{\partial \sigma}{\partial \Theta}\right|_{E} \Delta\Theta$$

Combination of equations (2.17) and (2.18) results in:

$$S = \left(\frac{\partial^2 G}{\partial T^2}\right)_{\Theta, E} T + \left(\frac{\partial^2 G}{\partial T \partial E}\right)_{\Theta} E + \left(\frac{\partial^2 G}{\partial T \partial \Theta}\right)_{E} \Delta\Theta$$
$$D = \left(\frac{\partial^2 G}{\partial E \partial T}\right)_{\Theta} T + \left(\frac{\partial^2 G}{\partial E^2}\right)_{\Theta, T} E + \left(\frac{\partial^2 G}{\partial E \partial \Theta}\right)_{T} \Delta\Theta \tag{2.19}$$
$$\Delta\sigma = \left(\frac{\partial^2 G}{\partial \Theta \partial T}\right)_{E} T + \left(\frac{\partial^2 G}{\partial \Theta \partial E}\right)_{T} E + \left(\frac{\partial^2 G}{\partial \Theta^2}\right)_{E, T} \Delta\Theta$$

## Basics of Measurement

Now we have a set of equations connecting the (dependent) through-variables $S$, $D$ and $\sigma$ with the (independent) across-variables $T$, $E$ and $\Theta$. The second-order derivatives represent material properties; they have been given special symbols. The constant variables are put as superscripts, to make place for the subscripts denoting orientation, as will be discussed later.

$$S = s^{E,\Theta}T + d^{\Theta}E + \alpha^E \Delta\Theta$$
$$D = d^{\Theta}T + \varepsilon^{\Theta,T}E + p^T \Delta\Theta \tag{2.20}$$
$$\Delta\sigma = \alpha^E T + p^T E + \frac{\rho}{T}c^{E,T}\Delta\Theta$$

These superscripts denote constancy with respect to the indicated parameters, for instance: $s^{E,\Theta}$ is the compliance at zero electric field $E$ and constant temperature $\Theta$. The nine associated effects are displayed in Table 2.15.

Table 2.16 shows the corresponding properties. The parameters for just a single domain ($\varepsilon$, $c_p$ and $s$) correspond to those in Tables 2.4 and 2.8. The other parameters ($p$, $\alpha$ and $d$) denote "cross effects". The parameters $p$ and $d$ will be discussed in more detail in the section on piezoelectric sensors (Section 8.9).

Note that direct piezoelectricity and converse piezoelectricity have the same symbol ($d$), because the dimensions are equal (m/V and C/N). The same holds for the pair pyroelectricity and converse pyroelectricity, as well as for thermal expansion and piezocaloric effect. Further, the symbol $c_p$ from Table 2.8 has been replaced by $c^{E,T}$ for consistency of notation.

**Table 2.15** *Nine physical effects corresponding to the parameters in equation (2.20)*

| | | |
|---|---|---|
| Elasticity | Converse piezoelectricity | Thermal expansion |
| Direct piezoelectricity | Permittivity | Pyroelectricity |
| Piezocaloric effect | Converse pyroelectricity | Heat capacity |

**Table 2.16** *Symbols, property names and units of the effects in Table 2.15*

| Symbol | Property | Unit |
|---|---|---|
| $s$ | Compliance, elasticity | $m^2 N^{-1}$ |
| $d$ | Piezoelectric constant | $mV^{-1} = CN^{-1}$ |
| $\alpha$ | Thermal expansion coefficient | $K^{-1}$ |
| $p$ | Pyroelectric constant | $Cm^{-2}K^{-1}$ |
| $\varepsilon$ | Permittivity; dielectric constant | $Fm^{-1}$ |
| $c^{E\cdot T}$ | (Specific) heat capacity | $Jkg^{-1}K^{-1}$ |

Equations (2.15) and (2.16) can be extended just by adding other couples of conjugate quantities, for instance from the chemical or the magnetic domain. Obviously, this introduces many other material parameters. With three couples we have 9 parameters, as listed in Table 2.15. With four couples of intensive and extensive quantities we get 16 parameters, so 7 more (for instance the magneto-caloric effect, expressed as the partial derivative of entropy to magnetic field strength).

There are other ways to systematically describe variables and parameters. One of these is the Bond-graph notation [10]. This notation claims to be even more general: the variables are grouped into effort and flow variables. The notation is applicable for all kind of technical systems.

A systematic representation of sensor effects is described in [11]. It is based on the six energy domains (radiant, mechanical, thermal, magnetic, chemical and electrical), arranged in a "sensor cube" with three axes: input energy domain, output energy domain and interrogating input energy. The latter applies for so-called *modulating sensors*. For instance the Hall effect: the input is magnetic induction $B$, the output a voltage and on the interrogating axis is the current through the Hall plate, that carries the energy that is modulated by the input. Direct and modulating sensors are discussed further in the next section.

## 2.4. Transducers

A sensor or transducer[9] performs the conversion of information from the physical domain of the measurand to the electrical domain. Signal conditioning may be added or included to protect the sensor from being loaded, to fit the sensor output to the input range of the ADC, or to enhance the S/N (signal-to-noise ratio) prior to further signal processing.

Many authors have tried to build up a consistent categorization of sensors [12, 13], with more or less success. It seems not easy to create a closed systematic description encompassing all sensor principles. There is at least consensus on a division into two groups of sensors: direct and modulating sensor types. The distinguishing property is the need for auxiliary energy (Figure 2.6). Direct sensors do not require additional energy for the conversion. As a consequence, the sensor withdraws energy from the measurement object, which may cause loss of information about the original state of the object. Examples of direct sensors are the thermocouple and the piezoelectric force and acceleration sensor. Indirect or modulating sensors use an additional energy source that is modulated by the measurand; the sensor output energy mainly comes from this auxiliary source, and just a fraction of energy is withdrawn from the measurement object. In this respect, modulating sensors do not significantly load the measurement object and hence are more accurate than

---

[9] In many text books on transducers a distinction is made between a sensor and a transducer. However, literature is not consistent in this respect. Therefore, we will consider these words as synonyms in this book, to avoid further confusion.

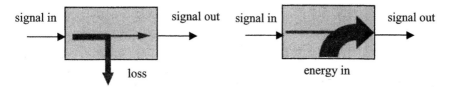

**Figure 2.6** *Energy and signal flows in direct (left) and modulating (right) sensors. Thick arrows symbolize energy flow*

**Figure 2.7** *Two-port model for a sensor*

direct sensors. Most sensors are of the modulating type, for instance all resistive and capacitive sensors and many inductive sensors.

A direct sensor can be described by a two port model, according to Figure 2.7. The input port is connected to the measurand, the output port corresponds to the electrical connections of the sensor. Likewise, a modulating sensor can be conceived as a system with three ports: an input port, an output port and a port through which the auxiliary energy is supplied. However, since in most cases the auxiliary signal has a fixed value, this third port will not be considered further here.

At each of the two signal ports we can define two variables: an across-variable and a through-variable (compare 2.3.6). In this section we call them effort variable $\mathcal{E}$ and flow variable $\mathcal{F}$, respectively. For each domain the two variables $\mathcal{E}$ and $\mathcal{F}$ are defined in such a way that their product equals the energy density of the domain type.

In systems without internal energy sources the relation between the four quantities can be described in various ways. For instance, we can express both flow variables as a function of the effort variables and vice versa, or the two output variables as a function of the two input variables or vice versa. We will use the first possibility, mainly for practical reasons.

$$\mathcal{F}_i = f_1(\mathcal{E}_i, \mathcal{E}_o)$$
$$\mathcal{F}_o = f_2(\mathcal{E}_i, \mathcal{E}_o) \tag{2.21}$$

Assuming a linear system this is approximated by:

$$\mathcal{F}_{ii} = \alpha \cdot \mathcal{E}_i + \gamma \cdot \mathcal{E}_o$$
$$\mathcal{F}_{oi} = \gamma \cdot \mathcal{E}_i + \beta \cdot \mathcal{E}_o \tag{2.22}$$

or, in short notation:

$$\begin{pmatrix} \mathcal{F}_{ii} \\ \mathcal{F}_{oi} \end{pmatrix} = \begin{pmatrix} \alpha & \gamma \\ \gamma & \beta \end{pmatrix} \begin{pmatrix} \mathcal{E}_i \\ \mathcal{E}_o \end{pmatrix} \quad (2.23)$$

Apparently, the properties of the system are described by three parameters: $\alpha$, $\beta$ and $\gamma$. Moreover, the system is reversible: the output can act as input or vice versa: the system behaves either as a sensor (transduction from the measurand to an electrical – sensor – signal) or as an actuator (transduction from an electrical – control – signal to an actuation). In this description the matrix elements have a clear physical meaning: $\alpha$ and $\beta$ represent admittances in the domain concerned, and $\gamma$ represents the transfer; they are directly related to the material properties as introduced in Section 2.3 (Table 2.15).

Together with the source and load impedances, $\alpha$ and $\beta$ determine the effect of loading errors. From (2.23) expressions for the input admittance (or impedance), the output admittance and the transfer of the system can be derived.

The input impedance is the ratio between the input effort variable and the input flow variable; likewise, the output impedance is the ratio between the output effort variable and the output flow variable. Note that the input admittance depends on the load at the output, and vice versa. We distinguish two extreme cases: open and short-circuited terminals (Figure 2.7). For short-circuited output ($\mathcal{E}_o = 0$) the input impedance is easily found from equation (2.23):

$$Y_{is} = \frac{1}{Z_{is}} = \alpha \quad (2.24)$$

Likewise, the output impedance at short-circuited input ($\mathcal{E}_i = 0$) is simply:

$$Y_{os} = \frac{1}{Z_{os}} = \beta \quad (2.25)$$

The situation is different for open terminals. The input impedance at open output ($\mathcal{F}_o = 0$) is found from:

$$Y_{io} = \frac{1}{Z_{io}} = \alpha \left(1 - \frac{\gamma^2}{\alpha\beta}\right) \quad (2.26)$$

and the output impedance at open input from

$$Y_{oo} = \frac{1}{Z_{oo}} = \beta \left(1 - \frac{\gamma^2}{\alpha\beta}\right) \quad (2.27)$$

Apparently, input properties depend on what is connected at the output and vice versa. The degree of this mutual influence is expressed by the *coupling factor* $\kappa$, defined as:

$$\kappa^2 = \frac{\gamma^2}{\alpha\beta} \tag{2.28}$$

The coupling factor of a stable two-port is between 0 and 1; $\kappa = 0$ means the output does not depend on the input; $\kappa = 1$ means total dependence. The coupling factor is also related to the energy transfer of the system: a large coupling factor means a high conversion efficiency.

**Example 2.5 Piezoelectric sensor**
The mechanical input quantities of a piezoelectric sensor are $\mathcal{E}_i = T$ and $\mathcal{F}_i = S$, and the electrical output quantities are $\mathcal{E}_o = E$ and $\mathcal{F}_o = D$. When we consider the extensive quantities as the independent variables and the intensive quantities as the dependent variables we can write the next set of equations:

$$\begin{pmatrix} S \\ D \end{pmatrix} = \begin{pmatrix} \alpha & \gamma \\ \gamma & \beta \end{pmatrix} \begin{pmatrix} T \\ E \end{pmatrix} \tag{2.29}$$

This is just a subset of equation (2.20). Comparing these equations we find for the three system parameters:

$\alpha = s^E$ (compliance at $E = 0$, that is short-circuited output terminals),

$\beta = \varepsilon^T$ (permittivity at $T = 0$, that is unclamped[10] mechanical terminals),

$\gamma = d$ (the piezoelectric constant),

and the coupling constant is found to be

$$\kappa = \frac{\gamma^2}{\alpha\beta} = \frac{d^2}{s^E \varepsilon^T} \tag{2.30}$$

Now, when we leave the electrical output port open ($D = 0$), the compliance becomes

$$s^D = s^E \cdot (1 - \kappa^2) \tag{2.31}$$

which means that the stiffness of a piezoelectric material depends on how the electrical terminals are connected. Likewise, the dielectric constant at clamped mechanical input ($S = 0$), becomes

$$\varepsilon^S = \varepsilon^T \cdot (1 - \kappa^2) \tag{2.32}$$

which means that the permittivity of a piezoelectric material depends on the state at the mechanical input (clamped or unclamped). We will use these results in the discussion on the interfacing of piezoelectric sensors.

---

[10] In the mechanical domain "unclamped" is preferred to "shortcircuited" and "clamped" to "open"; this is a consequence of the choice for $S$ (deformation) as a through-variable and $T$ (tension) as an across-variable.

## 2.5. Further reading

In this chapter various topics have been brought up only casually, in particular some that are related to thermodynamics. A discussion in more depth is far beyond the scope of this book, and the reader is referred to the many text books on these subjects. We just list here some useful sources for further information on units and standards.

B.S. Massey, *Units, Dimensional Analysis and Physical Similarity*, Van Nostrand Reinold, 1971, ISBN 0.442.05178.6.
The first half of this book is on units, their physical background and some historical facts about the international system of units. The rest of the book discusses physical algebra, a theory based on dimensional equations (as an extension of numerical equations). It also contains a useful list of many dimensionless parameters, arranged in alphabetical order, from Ab (absorption number) to Ws (Weissenberg number).

B.W. Petley, *The Fundamental Physical Constants and the Frontier of Measurement*, Bristol [etc.]: Hilger, 1985; ISBN 0-85274-427-7.
A fundamental discussion about physical constants, how they are measured and why they can or cannot be used as the basis for a system of units. Reading this book requires some knowledge of physics and quantum mechanics.

R.A. Nelson, *The International System of Units – Its History and Use in Science and Industry*, Via Satellite, February 2000.
A short but clear overview of the history of the International System of Units SI, available on the Internet.

Bureau International des Poids et Mesures (publ.), *The International System of Units (SI)*; 7th edition, 1998, Organisation Intergouvernementale de la Convention du Mètre.
An overview of the definitions of units and aspects of the practical realization of the most important standards. It also contains a list of resolutions and recommendations by the Comité International des Poids et Mesures (CIPM). An excellent guide for the proper use of symbols and prefixes. Also published by the National Institute of Standards and Technology, as the NIST special publication 330; U.S. Government Printing Office, Washington (2001). Both documents are also available on the Internet.

## 2.6. Exercises

1. For which of the following pairs of quantities the product of their dimensions equals energy density $(J/m^3)$?
   a. electric field strength $E$ and dielectric displacement $D$
   b. voltage $V$ and current $I$
   c. stress $T$ and strain $S$
   d. magnetic flux $\Phi$ and magnetic field strength $H$
   e. pressure $p$ and volume $V$.

2. For all sensors listed below indicate the class they belong to: direct or modulating.
   a. photodiode
   b. photoresistor
   c. Hall sensor
   d. piezoelectric sensor
   e. thermocouple
   f. strain gauge

3. Calculate the apparent area of a cube with edge length $a$, viewed exactly along the direction of a body diagonal.

4. Derive the nine material parameters from equation (2.15), in terms of second order derivatives (as in equation (2.19)).

5. The heat withdrawn from the cold plate of a Peltier cooler satisfies the equation $\Phi = \alpha \cdot T_c \cdot I_p$, with $\alpha$ the Seebeck coefficient of the junction materials that make up the cooler, $T_c$ the temperature of the cold plate and $I_p$ the electric current through the junction. Further parameters of the Peltier cooler are the electrical resistance $R_e$ between the plates and the thermal conductance $K_a$ between the plates. Calculate the maximal temperature difference between the hot and cold plate, using the heat balance with respect to the cold plate.

# Chapter 3

# Measurement Errors and Uncertainty

No matter what precautions are taken, there will always be a difference between the result of a measurement and the true (but unknown) value of a quantity. The difference is called the *measurement error*. A measurement is useless without a quantitative indication of the magnitude of that error. Such an indication is called the *uncertainty*. Without knowing the uncertainty, the comparison of a measurement result with a reference value or with results of other measurements cannot be made.

This chapter addresses the problem of how to define measurement errors, uncertainty and related terms. It raises the question of how to determine the uncertainty in practice. For that purpose, a short summary of the related statistical issues is first given. This chapter discusses some of the various sources of errors and the various types of errors. It also pays attention to the question of how to specify these errors. The chapter concludes with some techniques used to reduce the effects of errors.

## 3.1. True values, errors and uncertainty

The objective of this section is twofold. The first goal is to arrive at *accurate* definitions of the terms. The reason is that for appropriate communication on errors and uncertainty we need accurate and internationally accepted definitions. The second goal is to arrive at *operational* definitions. In other words, the definitions should not end up as theoretical concepts, but they must be such that they can be implemented and used in practice. The definitions and terminology used in this chapter, especially those in Sections 3.1, 3.4 and 3.5, are based on a publication of the International Organization for Standardization [1].

### 3.1.1. *Context*

For the moment, the context will be that of a measurement of a *single* quantity of a physical object in a *static* situation according to a *cardinal* scale. The discussion can be extended to multiple quantities. For instance, the determination of the 3D position of an object involves the simultaneous measurement of three quantities instead of one. However, such an extension is beyond the scope of this book.

The restriction to static situations implies that we are in the luxurious position that we can repeat the measurements in order to assess the uncertainty. The dynamic situation is much more involved. The assessment of uncertainty would require the availability of models describing the dynamic behaviour of the process. This too is beyond the scope of the chapter.

Different scales require different error analyses. For instance, the uncertainty of a burglar alarm (nominal scale) is expressed in terms of "false alarm rate" and "missed event rate". However, in this chapter, measurements other than with a cardinal scale are not considered.

### 3.1.2. Basic definitions

The particular quantity to be measured is called a *measurand*. Its (*true*) *value*, $x$, is the result that would be obtained by a perfect measurement. Since perfect measurements are only imaginary, a true value is always indeterminate and unknown. The value obtained by the measurement is called the *result of the measurement* and is denoted here by the symbol $z$. The (*measurement*) *error*, $e$, is the difference between measurement result and the true value:

$$e = z - x \tag{3.1}$$

Because the notion of error is defined in terms of the unknown "true value", the error is unknown too. Nevertheless, when evaluating a measurement, the availability of the error can be so urgent that in the definition of "measurement error" the phrase "true value" is sometimes replaced by the so-called *conventional true value*. This is a value attributed to the quantity, and that is accepted as having an uncertainty small enough for the given purpose. Such a value could, for instance, be obtained from a reference standard.

The *relative error* $e_{rel}$ is the error divided by the true value, i.e. $e_{rel} = e/x$. In order to distinguish the relative error from the ordinary error, the latter is sometimes called *absolute error*. The term is ambiguous since it can also mean $|e|$, but in this book the former denotation will be used.

The term *accuracy* of a measurement refers to a qualification of the expected closeness of the result of a measurement and the true value. The generally accepted, quantitative measure related to accuracy is uncertainty. The *uncertainty* of a measurement is a parameter that characterizes the range of the values within which the true value of a measurement lies. It expresses the lack of knowledge that we have on the measurand. Before proceeding to a precise definition of uncertainty some concepts from probability theory must be introduced.

## 3.2. Measurement error and probability theory

The most common definition of uncertainty is based on a probabilistic model in which the actual value of the measurement error is regarded as the outcome of a so-called "*stochastic experiment*". The mathematical definition of a stochastic experiment is complicated and as such not within the scope of this book. See [2] for a comprehensive treatment of this subject. However, a simple view is to regard a stochastic experiment as a mathematical concept that models the generation of random numbers. The generation of a single random number is called a *trial*. The particular value of that number is called the *outcome* or *realization* of that trial. After each trial, a so-called *random variable* takes the outcome as its new value. Random variables are often denoted by an underscore. Hence, $\underline{z}$ is the notation for the result of a measurement, and $\underline{e}$ is the notation of the measurement error, both regarded as random variables. Each time that we repeat our measurement, $\underline{z}$ and $\underline{e}$ take different values.

A random number occurs with a given probability. For instance, in throwing a dice, the six possible numbers come with probability $\frac{1}{6}$. However, many random variables can have any value within a given range of real numbers. For instance, a round-off error can take any (real) value in the interval $(-0.5, +0.5]$. Since the set of real numbers of such an interval is infinite, the probability that a particular value will be generated is always zero. There is no sense in defining the probability for these types of random variables. But we can still define the probability that such a random variable is smaller than or equal to a value. Let $\underline{a}$ be a real random variable, and let $a$ be some value. The probability that $\underline{a}$ is smaller than or equal to $a$ is denoted by $\Pr(\underline{a} \leq a)$. Obviously, this probability depends on the particular choice of $a$. Regarded as a function of $a$, $\Pr(\underline{a} \leq a)$ is called the (*cumulative*) *distribution function*, and denoted by $F_{\underline{a}}(a)$.

Figure 3.1(a) shows the distribution function of the experiment "throwing a dice". We see that the probability of throwing less or equal than, for instance, 3 is 0.5, as expected. Another example is given in Figure 3.1(b). Here, the experiment is "rounding off an arbitrary real number to its nearest integer". $\underline{a}$ is the round-off

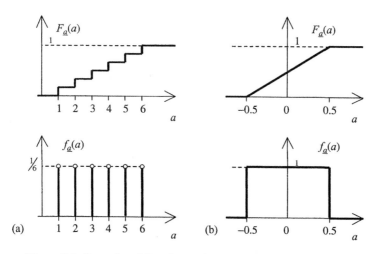

**Figure 3.1** *Examples of distribution functions and density functions: (a) throwing a dice; (b) round-off error*

error. (These types of errors occur in AD-conversion; see Chapter 6.) Obviously, the main difference between the distribution functions of Figures 3.1(a) and (b) is that the former is *discrete* (stepwise) and the latter is *continuous*. This reflects the fact that the number of possible outcomes in throwing a dice is finite, whereas the number of outcomes of a round-off error is infinite (i.e. any real number within $(-0.5, +0.5]$).

Figure 3.1 illustrates the fact that $F_{\underline{a}}(a)$ is a non-decreasing function with $F_{\underline{a}}(-\infty) = 0$, and $F_{\underline{a}}(\infty) = 1$. Simply stated, $F_{\underline{a}}(-\infty) = 0$ means that $a$ is never smaller than $-\infty$. $F_{\underline{a}}(\infty) = 1$ means that $a$ is always smaller than $+\infty$.

The distribution function facilitates the calculation of the probability that a random variable takes values within a given range:

$$\Pr(a_1 < \underline{a} \leq a_2) = F_{\underline{a}}(a_2) - F_{\underline{a}}(a_1) \quad (3.2)$$

If $F_{\underline{a}}(a)$ is continuous, then, in the limiting case as $a_1$ approaches $a_2$, equation (3.2) turns into:

$$\lim_{a_1 \to a_2} \Pr(a_1 < \underline{a} \leq a_2) = (a_2 - a_1) \left. \frac{dF_{\underline{a}}(a)}{da} \right|_{a=a_1} \quad (3.3)$$

The first derivative of $F_{\underline{a}}(a)$ is called the (*probability*) *density function*[1] $f_{\underline{a}}(a)$:

$$f_{\underline{a}}(a) = \frac{dF_{\underline{a}}(a)}{da} \quad (3.4)$$

Figure 3.1(b) shows the density function for the "round-off error" cases. The density function shows that the error is uniformly distributed in the interval $(-0.5, +0.5]$. Hence, each error within that interval is evenly likely.

The first derivative of a step function is called a Dirac-function $\delta(\cdot)$. With that, even if the random variable is discrete, the definition of the density function holds. Each step-like transition in the distribution function is seen as a Dirac function in the density function, the weight of which equals the step height. Figure 3.1(a) shows the density function belonging to the "dice" experiment. This density is

---

[1] For brevity, the shorter notation $f(a)$ is used instead of $f_{\underline{a}}(a)$, but this is only allowed in a context without any chance of confusion with other random variables. Sometimes, one uses $p(a)$ instead of $f(a)$, for instance, if the symbol $f(\cdot)$ has already been used for another purpose.

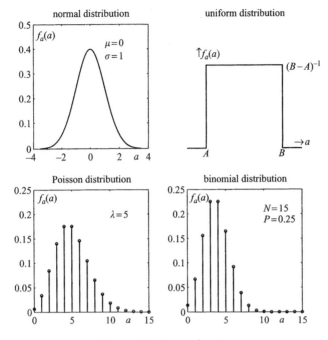

**Figure 3.2** *Density functions*

expressed as $\sum_{n=1}^{6} \frac{1}{6}\delta(a - n)$. The figure shows the six Dirac functions with weights $\frac{1}{6}$.

### 3.2.1. *Some important density functions*

Some density functions frequently met in measurement systems are (see Figure 3.2):

The normal distribution with parameters $\mu$ and $\sigma$:

$$f_{\underline{a}}(a) = \frac{1}{\sigma\sqrt{2\pi}} \exp\left(-\frac{(a-\mu)^2}{2\sigma^2}\right) \quad (3.5)$$

The uniform distribution with parameters $A$ and $B$:

$$f_{\underline{a}}(a) = \begin{cases} \frac{1}{B-A} & \text{if } A < a \leq B \\ 0 & \text{elsewhere} \end{cases} \quad (3.6)$$

The Poisson distribution with parameter $\lambda$:

$$f_{\underline{a}}(a) = \sum_{n=0}^{\infty} \frac{\lambda^n \exp(-\lambda)}{n!} \delta(a-n) \qquad (3.7)$$

The binomial distribution with parameter $N$ and $P$

$$f_{\underline{a}}(a) = \sum_{n=0}^{\infty} \frac{N!}{n!(N-n)!} P^n (1-P)^{N-n} \delta(a-n) \qquad (3.8)$$

Normal distributions (also called Gaussian distributions) occur frequently, both in physical models and in mathematical analysis. Any physical quantity that is an additive accumulation of many independent random phenomena is normally distributed. An example is the noisy current of a resistor. Such a current is induced by the thermal energy that causes random movements of the individual electrons. Altogether, these individual movements give rise to a random current that obeys a normal distribution. An advantage of the normal distribution is that some of its properties (to be discussed later) facilitate mathematical analysis whereas in the case of other distributions such an analysis is difficult.

As said before, the uniform distribution is used in AD-conversion and other round-off processes in order to model quantization.

The Poisson distribution and the binomial distribution are both discrete. The Poisson distribution models the process of counting events during a period of time. An example is a photo detector. The output of such a device is proportional to the number of photons that are intercepted on a surface during a period of time (similar to how rain intensity is measured by accumulating raindrops in a rain gauge). The parameter $\lambda$ is the mean number of counts.

The binomial distribution occurs whenever the experiment consists of $N$ independent trials of a Boolean random variable. $P$ is the probability that such a variable takes the value "true", and hence, $1 - P$ is the probability of "false". The random number returned by the binomial experiment is the number of times that the Boolean variable gets the value "true".

Since both the Poisson distribution and the binomial distribution find their origin in a counting process of individual events (and thus phenomena that are caused by a number of independent random sources) we might expect that these distributions both approximate the normal distribution. This is indeed the case. The Poisson distribution can be approximated by a normal distribution with $\mu = \lambda$ and $\sigma = \sqrt{\lambda}$ if $\lambda$ is large enough. The binomial distribution corresponds to a normal distribution with $\mu = NP$ and $\sigma = \sqrt{[NP(1-P)]}$ provided that $N$ is sufficiently large. However, we have to take into account that both the Poisson and the binomial distribution are discrete, whereas the normal distribution is continuous. In fact, it is the envelope of the discrete distributions that is well approximated by the normal distribution.

### 3.2.2. Expectation, variance and standard deviation

A random variable is – in a probabilistic sense – completely defined if its density function is specified. Therefore, complete probabilistic knowledge of the measurement error requires knowledge of $f_{\underline{a}}(a)$. Usually, this requirement is too strong. We may confine ourselves to a few parameters that roughly characterize the density function. Two parameters of particular interest are the *expectation* $\mu_a$ (often called *mean*) and the *standard deviation* $\sigma_a$, defined as follows:

$$E\{\underline{a}\} = \mu_a = \int_{-\infty}^{\infty} a\, f_{\underline{a}}(a)\, da \qquad (3.9)$$

and:

$$\sigma_a = \sqrt{\mathrm{Var}\{\underline{a}\}} \quad \text{with: } \mathrm{Var}\{\underline{a}\} = \int_{-\infty}^{\infty} (a - \mu_a)^2 f_{\underline{a}}(a)\, da \qquad (3.10)$$

where $E\{\underline{a}\}$ is an alternative notation for the expectation, and $\sigma_a^2 = \mathrm{Var}\{\underline{a}\}$ is the *variance*. The expectation is roughly the value $a$ around which the density function is situated. The standard deviation is a rough measure of the width of the function. Table 3.1 shows the expectation and standard deviation of the density functions of the previous section.

The standard deviation is useful in order to establish an interval around the mean such that the probability that an outcome falls inside that interval has a prescribed value $\alpha$, i.e.:

$$\Pr(\mu - k\sigma < \underline{a} \leq \mu + k\sigma) = \alpha \qquad (3.11)$$

The parameter $k$ is the *coverage factor*. $\alpha$ is the *confidence level*. For a normal distribution the following relations hold:

| $\alpha = 0.9$ | $\alpha = 0.95$ | $\alpha = 0.955$ | $\alpha = 0.99$ | $\alpha = 0.9972$ | $\alpha = 0.999$ |
|---|---|---|---|---|---|
| $k = 1.645$ | $k = 1.960$ | $k = 2$ | $k = 2.576$ | $k = 3$ | $k = 3.291$ |

Table 3.1 *Expectation and standard deviation of some distributions*

| | Expectation | Standard deviation |
|---|---|---|
| Normal distribution | $\mu$ | $\sigma$ |
| Uniform distribution | $\frac{1}{2}(A + B)$ | $\frac{1}{\sqrt{12}}(B - A)$ |
| Binomial distribution | $NP$ | $\sqrt{NP(1 - P)}$ |
| Poisson distribution | $\lambda$ | $\sqrt{\lambda}$ |

**Example 3.1 Coverage factor and confidence level**
A sugar factory sells its products in quantities of 1 kg. For that purpose, the factory uses a packaging machine that weighs the sugar with a standard deviation of 5 mg. According to legislation, the factory must assure that at most 0.5% of its packs contain less than 1 kg. What should be the set point of the packaging machine?

If $z$ is the measurement result, and the true weight of a pack is $\underline{x}$, then $\underline{x} = z - \underline{e}$ where $\underline{e}$ is the measurement error. The legislation requires that $\Pr(\underline{x} < 1\,\text{kg}) = 0.005$. Thus, $\Pr(z - \underline{e} < 1\,\text{kg}) = 0.005$, or equivalently $\Pr(\underline{e} > z - 1\,\text{kg}) = 0.005$. We assume a zero mean, normal distribution for $\underline{e}$ with $\sigma = 5\,\text{mg}$. The normal distribution is symmetric. Therefore, the requirement for $z$ is equivalent to: $\Pr(|\underline{e}| > z - 1\,\text{kg}) = 0.01$. In turn, this is equivalent to $\Pr(|\underline{e}| \leq z - 1\,\text{kg}) = 0.99$. Equation (3.11) shows that this requirement is fulfilled if $k\sigma = z - 1$, where $k$ is the coverage factor associated with a confidence level of $\alpha = 0.99$. According to the table we have $k = 2.576$. Consequently, the requirement leads to $z = 1.0129\,\text{kg}$ which should be used as the set point.

## 3.3. Statistical inference

The natural question that follows is how to determine the probability density function of the measurement error. Two methods are available. The first is to identify all possible sources of errors by means of appropriate physical and mathematical modelling, and then to analyze how these errors propagate through the measurement chain and how they affect the final measurement error. Typical error sources are thermal noise in a resistor, Poisson noise (or shot noise) in a photo detector, quantization noise of an AD converter, and so on. All these sources affect the probability density function of the measurement error in their own way. Sections 3.4.2 and 3.4.3 deal with the propagation of errors.

The current section deals with the second method, i.e. the statistical approach. It is applicable only if the measurement can be repeated under the same repeatability conditions. These conditions can include, for instance:

- the same temperature;
- the same measurement procedure;
- the same measuring instrument;
- the same location.

*Histogramming* is a technique that uses repeated measurements in order to estimate the corresponding density. However, instead of estimating the probability density of the error, it often suffices to estimate only some of its parameters, e.g. expectation and standard deviation. In this section is first discussed the issue of histogramming. It then proceeds with statistical methods for the estimation of expectation and standard deviation.

Suppose that in a series of $N$ measurements $z_n$ is the $n$-th result with $n$ ranging from 1 to $N$. In statistical literature [3] each measurement $z_n$ is called a *sample*. These samples can be used to calculate a *normalized histogram*, being an estimate of the probability density of $\underline{z}$. Often, we are interested in the probability density of $\underline{e}$, not in $\underline{z}$. If this is the case, we need to have a "conventional true value" $x$ (see Section 3.1.2) so that we are able to calculate the error associated with each measurement: $e_n = z_n - x$; see equation (3.1). These errors can be used to estimate the probability density of $\underline{e}$.

### 3.3.1. *Estimation of density functions: histograms*

Lets assume that we want to estimate the probability density of a real random variable $\underline{a}$. Here, $\underline{a}$ stands for the measurement result $\underline{z}$, or for the measurement error $\underline{e}$, or for whatever is applicable. We have a set of $N$ samples $a_n$ that are all realizations of the random variable $\underline{a}$. Furthermore, we assume that these samples are statistically independent. Let $a_{\min}$ and $a_{\max}$ be the minimum and maximum value of the series $a_n$. We divide the interval $[a_{\min}, a_{\max}]$ into $K$ equally sized sub-intervals $(a_k, a_k + \Delta a]$ where $\Delta a = (a_{\max} - a_{\min})/K$. Such a sub-interval is called a *bin*. We count the number of outcomes that fall within the $k$-th bin, and assign that count to the variable $h_k$. The probability that a sample falls in the $k$-th bin is:

$$P_k = \int_{a_k}^{a_k + \Delta a} f_{\underline{a}}(a) da \qquad (3.12)$$

Since we have $N$ samples, $h_k$ has a binomial distribution with parameters $(N, P_k)$ (see Section 3.2.1). Thus, the expectation of $h_k$ is $E\{h_k\} = NP_k$ and its variance is $NP_k(1 - P_k)$; see Table 3.1.

The normalized histogram is:

$$H_k = \frac{h_k}{N \Delta a} \qquad (3.13)$$

The expectation of $H_k$ is:

$$E\{H_k\} = \frac{E\{h_k\}}{N \Delta a} = \frac{P_k}{\Delta a} = \frac{1}{\Delta a} \int_{a_k}^{a_k + \Delta a} f_{\underline{a}}(a) da \approx f_{\underline{a}}\left(a_k + \frac{1}{2}\Delta a\right) \qquad (3.14)$$

Therefore, $H_k$ is an estimate for $f_{\underline{a}}(a_k + \frac{1}{2}\Delta a)$.

Figure 3.3(a) shows 100 outcomes of a normally distributed, random variable. Two (normalized) histograms have been derived from that data set and plotted in the figure together with the true probability density. In the real world, the true density is always unknown. Here, we have used data whose density function is known, so that we can evaluate the accuracy of the estimated density. One histogram is calculated with $K = 4$, the other with $K = 25$. The normalized histogram provides estimates for $f_{\underline{a}}(\cdot)$ only at a *finite* number of values of $a$, i.e. at $a_k + \frac{1}{2}\Delta a$. At all other values of $a$ we need to interpolate. In Figure 3.3 this is done by so-called *zero order interpolation* (also called *nearest neighbour interpolation*): if $a$ falls in the $k$-th bin, then $f_{\underline{a}}(a)$ is estimated by $H_k$. This kind of interpolation gives rise to the blockish appearance of the histograms, as shown in the figure.

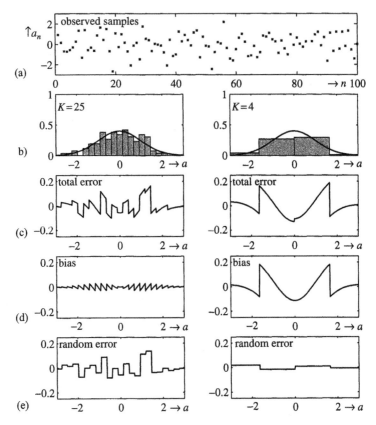

**Figure 3.3** *Estimation of probability density functions (pdfs) by means of histogramming: (a) set with samples (N = 100); (b) histogram with K = 25 (left) and K = 4 (right) bins together with the true density function; (c) the estimation error; (d) the non-random part of the error, i.e. the bias; (e) the random part of the error*

Figure 3.3 illustrates the fact that an estimate always comes with an *estimation error*, i.e. a difference between the true value of the probability density and the estimated one. The estimation error consists of two parts. The first one is the *random error*. The random error is the difference between $H_k$ and its expectation. Since $h_k$ has a binomial distribution, the statistical fluctuations of $H_k$ around its expectation has a standard deviation of $\sqrt{[P_k(1-P_k)/N]}/\Delta a$. Therefore, the relative deviation $\sigma_{H_k}/\text{E}\{H_k\}$ of the random error is $\sqrt{[(1-P_k)/(NP_k)]}$. In order to keep the random error small, the denominator $NP_k$ must be large enough. This is achieved by having a large $N$, and by not allowing $P_k$ to be too small. 'Having a large $N$' means: the data set must be large enough to have the statistical significance. $P_k$ depends on the selection of $\Delta a$, or equivalently by the selection of $K$. If $K$ is too large, then $P_k$ will become too small. Since it is the product $NP_k$ that matters, $K$ is allowed to grow as $N$ increases.

Measurement Errors and Uncertainty 53

The second type of error is the *bias*. The bias is the estimation error that persists with a fixed value of $K$, even if $N$ becomes infinitely large, and thus even if the random error is negligible. In other words, the bias is an error that is inherent to the estimation method, and that is not caused by the randomness of the data set.

The first reason for the bias is the approximation that is used in equation (3.14). The approximation is only accurate if $\Delta a$ is small. In other words, $K$ must be large to have a close approximation. If $\Delta a$ is not sufficiently small, then the approximation causes a large bias. The second reason for the bias is the interpolation. The nearest neighbour interpolation gives rise to an interpolation error. Especially, the interpolation can become large for values of $a$ that are far away from a bin centre. Therefore, here too, $K$ must be large to have a small error.

Figure 3.3 illustrates the influence of $K$ on the two types of errors. If $K$ is small, and the bins are wide, the random error is relatively small, but then the bias is large. If $K$ is large, and the bins are narrow, the random error is large, and the bias is small. A trade-off is necessary. A reasonable rule of thumb is to choose $K = \sqrt{N}$.

**Example 3.2 Bias and random errors in density estimation**
The table below shows the maximum values of the various errors for the example of Figure 3.3. The maximum is calculated over all values of $a$. The maximum errors which occur when $K = \sqrt{N} = 10$ is selected are also shown:

|  | $K = 25$ | $K = 10$ | $K = 4$ |
|---|---|---|---|
| Max total error | 0.16 | 0.12 | 0.19 |
| Max bias | 0.03 | 0.07 | 0.18 |
| Max random error | 0.14 | 0.05 | 0.02 |

Clearly, when $K = 25$, the random error prevails. But when $K = 4$, the bias prevails. However, when $K = 10$, the bias and random error are about of the same size. Such a balance between the two error types minimizes the total error.

### 3.3.2. Estimation of mean and standard deviation

The obvious way to estimate the mean of the error $\underline{a}$ from a set of samples $a_n$ is to use the average $\bar{a}$ of the set[2]:

$$\hat{\mu}_a = \bar{a} = \frac{1}{N}\sum_{n=1}^{N} a_n \qquad (3.15)$$

---

[2] The notation $\bar{a}$ is often used to indicate the result of an averaging operator applied to a set $a_n$.

The symbol "^" on top of a parameter is always used to indicate that it is an estimate of that parameter. Thus $\hat{\mu}_a$ is an estimate of $\mu_a$.

The set $a_n$ can be associated with a set of random variables $\underline{a}_n$ (all with the same distribution as $\underline{a}$). Therefore, the average (being a weighted sum of $a_n$) can also be associated with a random variable: $\hat{\underline{\mu}}_a$. Consequently, the average itself has a mean, variance and standard deviation:

$$E\{\hat{\underline{\mu}}_a\} = \mu_a \tag{3.16}$$

$$\text{Var}\{\hat{\underline{\mu}}_a\} = \frac{1}{N}\text{Var}\{\underline{a}_n\} \tag{3.17}$$

$$\sigma_{\hat{\mu}_a} = \frac{1}{\sqrt{N}}\sigma_a \tag{3.18}$$

The proof can be found in Appendix C. Equation (3.17) is valid under the condition that the cross products $E\{(\underline{a}_n - \mu_a)(\underline{a}_m - \mu_a)\}$ vanish for all $n \neq m$. If this is the case, then the set $a_n$ is said to be *uncorrelated*.

The *bias*, being the non-random part of the estimation error, is the difference between the true parameter and the mean of the estimate. Equation (3.16) shows that the bias of the average is zero. Therefore, the average is said to be an *unbiased* estimator. The standard deviation of the estimator is proportional to $1/\sqrt{N}$. In the limiting case, as $N \to \infty$, the estimation error vanishes. An estimator with such a property is said to be *consistent*.

The unbiased estimation of the variance $\text{Var}\{a\}$ is more involved. We have to distinguish two cases. The first case is when the mean $\mu_a$ is fully known. An efficient estimator is:

$$\hat{\sigma}_a^2 = \frac{1}{N}\sum_{n=1}^{N}(a_n - \mu_a)^2 = \overline{(a - \mu_a)^2} \tag{3.19}$$

In the second case, both the mean and the variance are unknown, so that they must be estimated simultaneously. The mean is estimated by the average $\hat{\mu}_a = \bar{a}$ as before. Then, the variance can be estimated without bias by the so-called *sample variance* $S_a^2$:

$$S_a^2 = \frac{1}{N-1}\sum_{n=1}^{N}(a_n - \bar{a})^2 = \frac{N}{N-1}\overline{(a - \bar{a})^2} \tag{3.20}$$

Apparently, we need a factor $N/(N-1)$ to correct the bias in $\overline{(a - \bar{a})^2}$. Without this correction factor, $S_a^2$ would be too optimistic, i.e. too small. The background of this is that, in the uncorrected estimate, i.e. in $\overline{(a - \bar{a})^2}$, the term $\bar{a}$ absorbs a part of the variability of the set $a_n$. In fact, $\bar{a}$ is the value of the parameter $\alpha$ that minimizes the expression $\overline{(a - \alpha)^2}$. For instance, in the extreme case $N = 1$, we have $\bar{a} = a_1$ and $\overline{(a - \bar{a})^2} = 0$. The latter would be a super-optimistic estimate for the variance. However, if $N = 1$, then $S_a^2$ is indeterminate. This exactly represents our state of knowledge about the variance if we have only one sample.

As said before, these estimators are unbiased. The estimation error that remains is quantified by their variances (see Appendix C):

$$\text{Var}\{\hat{\sigma}_a^2\} = \frac{2\sigma_a^4}{N} \quad \text{Var}\{S_a^2\} = \frac{2\sigma_a^4}{N-1} \tag{3.21}$$

Strictly speaking, these last expressions are valid only if $a_n$ is normally distributed.

Equations (3.18) and (3.21) impose the problem that in order to calculate $\sigma_{\hat{\mu}_a}$, $\text{Var}\{\hat{\sigma}_a^2\}$ or $\text{Var}\{S_a^2\}$ we have to know $\sigma_a$. Usually, this problem is solved by substituting $\hat{\sigma}_a$ or $S_a$ for $\sigma_a$.

**Example 3.3 Estimation of the mean and variance from 100 samples**
We apply the estimators for the mean and the variance to the 100 samples given in Figure 3.3. These samples are drawn from a random generator whose true mean value and variance are $\mu_a = 0$ and $\sigma_a^2 = 1$.

Application of equations (3.15) to (3.21) yields the following estimates including the estimated standard deviations of the estimates:

|  | True value | Estimate | True std deviation of the estimate | Estimated std deviations |
|---|---|---|---|---|
| Mean | $\mu_a = 0$ | $\bar{a} = -0.05$ | $\sigma_{\hat{\mu}_a} = 0.1$ | $\frac{1}{\sqrt{N}} S_a \approx 0.106$ |
| Variance (mean known) | $\sigma_a^2 = 1$ | $\hat{\sigma}_a^2 = 1.056$ | $\sqrt{\frac{2\sigma_a^4}{N}} = 0.14$ | $\sqrt{\frac{2\hat{\sigma}_a^4}{N}} \approx 0.15$ |
| Variance (mean unknown) | $\sigma_a^2 = 1$ | $S_a^2 = 1.064$ | $\sqrt{\frac{2\sigma_a^4}{N-1}} = 0.14$ | $\sqrt{\frac{2S_a^4}{N-1}} \approx 0.15$ |

The column with the heading "true std deviation of the estimate" is added to allow the comparison between the true standard deviations and the corresponding estimated one. The latter are derived from the sample data only, i.e. without prior knowledge.

## 3.4. Uncertainty analysis

The uncertainty in a measurement is often expressed as the standard deviation of the measurement error (the so-called *standard uncertainty*). Alternatively, one can use the term *expanded uncertainty* which defines an interval around the measurement result. This interval is expected to encompass a given fraction of the probability density associated with the measurement error; see equation (3.11). Assuming a normal distribution of the measurement error, the standard uncertainty can be transformed to an expanded uncertainty by multiplication of a coverage factor.

In practice, there are two methods to determine the standard uncertainty. The first method is the statistical approach, based on repeated measurements (sometimes called *type A evaluation of uncertainty*). The second approach is based on a priori distributions (the *type B evaluation*).

### 3.4.1. Type A evaluation

The goal of type A evaluation is to assess the components that make up the uncertainty of a measurement. It uses statistical inference techniques applied to data acquired from the measurement system. Examples of these techniques are histogramming (Section 3.3.1), and estimation of mean and standard deviation (Section 3.3.2). Many other techniques exist, such as curve fitting and analysis of variance, but these techniques are not within the scope of this book. Type A evaluation is applicable whenever a series of repeated observations of the system are available. The results of type A evaluation should be taken with care because statistical methods easily overlook systematic errors (see Section 3.5.1).

**Example 3.4 Calibration of a level sensor**
The measurement problem is to calibrate a level sensor against a stable reference level. The goal is to obtain the best estimate of the difference between the measurement result and the reference level. For that purpose $N = 50$ observations $z_n$, $n = 1, \ldots, N$ are made. The reference level is considered to be the true value $x$. Each observation is associated with an error $e_n = z_n - x$. The errors are plotted in Figure 3.4.

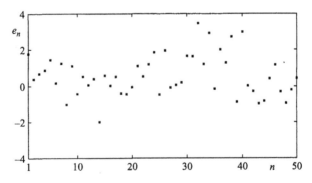

**Figure 3.4** *Errors obtained from repeated measurements for the calibration of a level sensor*

If we assume that the observations are independent, the best estimate for the difference is the average of $e_n$. Application of equation (3.15) yields: $\bar{e} = 0.58$ mm. The estimated standard deviation of the set $e_n$ follows from equation (3.20): $S_e = 1.14$ mm. According to equation (3.18) the standard deviation of the mean is $\hat{\sigma}_{\bar{e}} = \hat{\sigma}_e/\sqrt{N} \approx S_e/\sqrt{N} = 0.16$ mm.

The conclusion is that, at the reference level, the sensor should be corrected by a term $\bar{e} = 0.58$ mm. However, this correction introduces an uncertainty of 0.16 mm.

### 3.4.2. Type B evaluation

The determination of the uncertainty based on techniques other than statistical inference is called *type B evaluation*. Here, the uncertainty is evaluated by means of a pool of information that is available on the variability of the measurement result:

- The specifications of a device as reported by the manufacturer.
- Information obtained from a certification procedure, for instance, on behalf of calibration.
- Information obtained from reference handbooks.
- General knowledge about the conditional environment and the related physical processes.

The principles will be clarified by means of an example:

**Example 3.5 Uncertainty due to temperature effects on a pressure sensor**
The MPX2010 is a pressure sensor. In a given application, the sensor will be used at 20°C with a temperature deviation of 2°C. The measurement system is calibrated at 20°C. What will be the uncertainty due to temperature effects?

The output voltage $V$ is linearly related to the (differential) input pressure $P$, that is $V = \text{offset} + \text{gain} \times P + \text{errors}$. Its specifications mention the following issues related to temperature variations:

*Operating characteristics MPX2010*

| Characteristic | Symbol | Min | Typ | Max | Unit |
|---|---|---|---|---|---|
| Pressure range | P | 0 | — | 10 | kPa |
| Full scale span | $V_{FSS}$ | 24 | 25 | 26 | mV |
| Temperature effect on offset | $TC_{off}$ | −1.0 | — | 1.0 | mV |

The *full scale span* is the voltage range that corresponds to the pressure range. Hence, $V_{FSS}$ corresponds to 10 kPa. The parameter $TC_{off}$ is the output deviation with minimum rated pressure applied over the temperature range of 0° to 85°C, relative to 25°C. Converted to the pressure domain this parameter ranges from −0.4 kPa to 0.4 kPa.

In order to determine the uncertainty some assumptions must be made. The first assumption is that a change of temperature manifests itself as an offset, i.e. an additive term, that depends linearly on the temperature. Formulated mathematically:

$$z = x + a(T - T_{ref}) = x + a\Delta T \tag{3.22}$$

$x$ is the true pressure. $z$ is the output of the system with modelled temperature influence. $a$ is the temperature coefficient of the offset. $T_{ref}$ is the temperature at which the device is calibrated (in our case $T_{ref} = 20°C$). The second assumption is that the standard deviation of $a$ can be derived from $TC_{off}$. In the specifications, the minimum and maximum are mentioned. In such cases, often a uniform distribution between these limits is understood. Therefore, the standard deviation of $TC_{off}$ is $2/\sqrt{12} \approx 0.6\,mV \hat{=} 0.2\,kPa$. The output deviation is measured over the ranges from $0°C$ to $25°C$, and from $25°C$ to $85°C$. The worst case scenario is to use the former range to determine the standard deviation of $a$, i.e. $\sigma_a \approx 0.2/25 = 0.01\,kPa/°C$.

The uncertainty in $z$ occurs due to the term $a\Delta T$. Both factors in this term, $a$ and $\Delta T$, are uncertain with standard deviations of $\sigma_a \approx 0.01\,kPa/°C$ and $\sigma_{\Delta T} = 2°C$. Since $a$ and $\Delta T$ are independent (in a probabilistic sense), the standard deviation of $z$ due to temperature variations is $\sigma_z = \sigma_a \sigma_{\Delta T} = 0.02\,kPa$.

### 3.4.3. Combined uncertainty and error propagation

A measurement result $z$ often depends on a number of parameters such as the parameters of the sensor (size, material parameters), or the physical parameters of the environment (temperature, pressure, humidity, etc.). These parameters will be denoted by $y_1, \ldots, y_N$. The function $g(\cdot)$ describes the relationship between $z$ and the $y_n$'s:

$$z = g(y_1, \ldots, y_N) \tag{3.23}$$

However, the determination of these parameters is prone to errors. We denote the true parameters by $\phi_1, \ldots, \phi_N$ so that the true value of the measurand is given by:

$$x = g(\phi_1, \ldots, \phi_N) \tag{3.24}$$

Equation (3.24) expresses the fact that if the $\phi_n$'s were known, the measurand could be determined without any error. Unfortunately, we have only inexact knowledge about the $\phi_n$. Errors in the $y_n$'s propagate through the measurement chain and finally introduce errors in the output $z$. The uncertainty of $z$ is the so-called *combined uncertainty*. This section addresses the problem of how to relate the uncertainties in the $y_n$'s to the final, combined uncertainty in $z$.

**Linear relationship**
The measurement system is linear if

$$x = a_1 \phi_1 + a_2 \phi_2 + \cdots \tag{3.25}$$

and (consequently):
$$z = a_1 y_1 + a_2 y_2 + \cdots \quad (3.26)$$

$a_n$ are assumed to be known constants whose uncertainties are negligible.

The error associated with $y_n$ is $e_n = y_n - \phi_n$. Let $\sigma_n$ be the uncertainties (standard deviations) associated with $y_n$. Then, the final error $e = z - x$ is found to be:

$$e = a_1 e_1 + a_2 e_2 + \cdots \quad (3.27)$$

Mean values of $e_n$, denoted by $\mu_n$, result in a mean value $\mu$ of $e$ according to:

$$\mu = a_1 \mu_1 + a_2 \mu_2 + \cdots \quad (3.28)$$

The variance of $e$, and thus the squared uncertainty of the final measurement, follows from:

$$\sigma_e^2 = \sum_{n=1}^{N} a_n^2 \sigma_n^2 + 2 \sum_{n=1}^{N} \sum_{m=n+1}^{N} a_n a_m E\{(e_m - \mu_m)(e_n - \mu_n)\} \quad (3.29)$$

where $E\{(e_m - \mu_m)(e_n - \mu_n)\}$ is the so-called *covariance* between $e_n$ and $e_m$ (see Appendix C). The covariance is a statistical parameter that describes how accurately one variable can be estimated by using the other variable. If the covariance between $e_n$ and $e_m$ is zero, then these variables are uncorrelated, and knowledge of one variable does not amount to knowledge of the other variable. Such a situation arises when the error sources originate from phenomena that are physically unrelated. For instance, the thermal noise voltages in two resistors are independent, and therefore are uncorrelated. On the other hand, if one of the two phenomena influences the other, or if the two phenomena share a common factor, then the two variables are highly correlated, and the covariance deviates from zero. For instance, interference at a shared voltage supply line of two operational amplifiers causes errors in both outputs. These errors are highly correlated because they are caused by one phenomenon. The errors will have a non-zero covariance.

If all $y_n$ and $y_m$ are uncorrelated, i.e. if $E\{(e_m - \mu_m)(e_n - \mu_n)\} = 0$ for all $n$ and $m$, equation (3.29) simplifies to:

$$\sigma_e^2 = \sum_{n=1}^{N} a_n^2 \sigma_n^2 \quad (3.30)$$

From now on, we assume uncorrelated measurement errors in $y_n$. Thus, their covariances are zero (unless explicitly stated otherwise).

## Nonlinear relationship

In the nonlinear case equation (3.23) applies. Assuming that $g(\cdot)$ is continuous, a Taylor series expansion yields:

$$e = z - x = g(y_1, \ldots, y_N) - g(\phi_1, \ldots, \phi_N)$$

$$= \sum_{n=1}^{N} e_n \frac{\partial g}{\partial y_n} + \frac{1}{2} \sum_{n=1}^{N} \sum_{m=1}^{N} e_n e_m \frac{\partial^2 g}{\partial y_n \partial y_m} + \cdots \quad (3.31)$$

where the partial derivatives are evaluated at $y_1, \ldots, y_N$.

The mean value $\mu$ of the final measurement error is approximately found by application of the expectation operator to equation (3.1), and by truncating the Taylor series after the second order term:

$$\mu = E\{e\} \approx E \left\{ \sum_{n=1}^{N} e_n \frac{\partial g}{\partial y_n} + \frac{1}{2} \sum_{n=1}^{N} \sum_{m=1}^{N} e_n e_m \frac{\partial^2 g}{\partial y_n \partial y_m} \right\}$$

$$= \sum_{n=1}^{N} \mu_n \frac{\partial g}{\partial y_n} + \frac{1}{2} \sum_{n=1}^{N} \sigma_n^2 \frac{\partial^2 g}{\partial y_n^2} \quad (3.32)$$

Thus, even if the measurements $y_n$ have zero mean, the final measurement result may have a nonzero mean due to the nonlinearity. Figure 3.5(a) illustrates this phenomenon.

For the calculation of the variance $\sigma_e^2$ of $e$ one often confines oneself to the first order term of the Taylor series. In that case, $g(\cdot)$ is locally linearized, and thus,

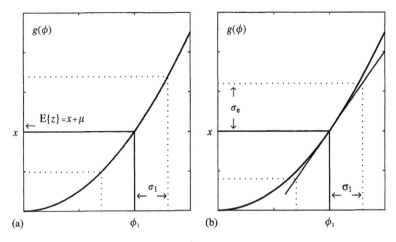

**Figure 3.5** *Propagation of errors in a nonlinear system: (a) noise in a nonlinear system causes an offset; (b) propagation of the noise*

equation (3.30) applies:

$$\sigma_e^2 \approx \sum_{n=1}^{N} \sigma_n^2 \left( \frac{\partial g}{\partial y_n} \right)^2 \quad (3.33)$$

See Figure 3.5(b).

A special case is when $g(\cdot)$ has the form:

$$g(y_1, y_2, \ldots) = c y_1^{p_1} y_2^{p_2} \cdots \quad (3.34)$$

with $c, p_1, p_2, \ldots$ known constants with zero uncertainty. In that case, equation (3.33) becomes:

$$\frac{\sigma_e^2}{z^2} \approx \sum_{n=1}^{N} \frac{p_n^2 \sigma_n^2}{y_n^2} \quad (3.35)$$

In this particular case, equation (3.35) gives an approximation of the *relative combined uncertainty* $\sigma_e/z$ in terms of *relative uncertainties* $\sigma_n/y_n$. Specifically, if all $p_n = 1$, it gives the uncertainty associated with *multiplicative errors*.

**Example 3.6 Uncertainty of a power dissipation measurement**
The power dissipated by a resistor can be found by measuring its voltage $V$ and its current $I$, and then multiplying these results: $P = VI$. If the uncertainties of $V$ and $I$ are given by $\sigma_V$ and $\sigma_I$, then the relative uncertainty in $P$ is:

$$\frac{\sigma_P}{P} = \sqrt{\frac{\sigma_V^2}{V^2} + \frac{\sigma_I^2}{I^2}} \quad (3.36)$$

**The error budget**
In order to assess the combined uncertainty of a measurement one first has to identify all possible error sources by means of the techniques mentioned in Sections 3.4.1 and 3.4.2. The next step is to check to see how these errors propagate to the output with the techniques mentioned above. This is systematically done by means of the *error budget*, i.e. a table which catalogues all contributions to the final error.

As an example, we discuss a measurement based on a linear sensor and a two-point calibration. The model of the measurement is as follows:

$$y = ax + b + n$$
$$z = \frac{y - B}{A} \quad (3.37)$$

$x$ is the measurand. $y$ is the output of the sensor. $a$ and $b$ are the two parameters of the sensor. $n$ is the sensor noise with standard deviation $\sigma_n$. A linear operation on $y$ yields the final measurement $z$. The parameters $A$ and $B$ are obtained by means of a two-point calibration. In fact, they are estimates of the sensor parameters

**Table 3.2** *Error budget of a system with a linear sensor*

| Error source | Uncertainty | Sensitivity | Contribution to variance |
|---|---|---|---|
| Sensor noise | $\sigma_n = 1$ | 1.667 | $1.667^2 \times 1^2$ |
| Uncertainty reference 1 | $\sigma_1 = 2$ | $-0.01x + 1$ | $(-0.01x + 1)^2 \cdot 2^2$ |
| Uncertainty reference 2 | $\sigma_2 = 2$ | $0.01x$ | $(0.01x)^2 \cdot 2^2$ |
| Combined uncertainty | | | $\sqrt{4.8 + 0.0008(x - 50)^2}$ |

$a$ and $b$. They are calculated by means of two reference values $x_1$ and $x_2$, whose uncertainties are $\sigma_1$ and $\sigma_2$, and of the corresponding observed values $y_1$ and $y_2$ (which are also uncertain). After having assured ourselves that no other error source exists with a significant contribution to the final error, we conclude that there are five independent error sources that must be included in the error budget. These errors propagate through the system to the output via the estimated parameters $A$ and $B$.

### Example 3.7 The error budget of a linear sensor

Suppose that we have the following reference values: $x_1 = 0$ and $x_2 = 100$ with uncertainties $\sigma_1 = \sigma_2 = 2$. The uncertainty in $y$ due to the noise is $\sigma_n = 1$. In addition, suppose that $y_1 = 20$ and $y_2 = 80$. For simplicity, we assume that the uncertainties of $y_1$ and $y_2$ are negligible. This reduces the number of error sources to three. These sources and their uncertainties are listed in the first two columns of Table 3.2.

The parameters $A$ and $B$ are obtained from the reference values and their observations by solving the following equations:

$$\left. \begin{array}{l} y_1 = Ax_1 + B \\ y_2 = Ax_2 + B \end{array} \right\} \Rightarrow \begin{array}{l} A = \dfrac{y_1 - y_2}{x_1 - x_2} = 0.6 \\ B = \dfrac{y_2 x_1 - x_2 y_1}{x_1 - x_2} = 20 \end{array}$$

The errors of $x_1$ and $x_2$ propagate to $A$ and $B$. Application of equation (3.33) yields $\sigma_A = 0.017$ and $\sigma_B = 2$. It is tempting to use these results to calculate the uncertainty of the final error. However, $A$ and $B$ share common error sources. Thus, we suspect that $A$ and $B$ are correlated. A better approach is to substitute the expressions that we have found for $A$ and $B$ into equation (3.37):

$$z = \frac{ax + b + n - (y_2 x_1 - x_2 y_1)/(x_1 - x_2)}{(y_1 - y_2)/(x_1 - x_2)}$$

This expression really shows how the three error sources propagate to the output. The third column in Table 3.2 shows the sensitivity of $z$ with respect to these errors, i.e. $\partial z/\partial n$, $\partial z/\partial x_1$ and $\partial z/\partial x_2$. Application of equation (3.33), performed in the fourth column, gives the combined uncertainty.

## 3.5. Characterization of errors

The purpose of this section is to distinguish between the different types of errors that can occur in a measurement. A first dichotomy arises from the distinction between *systematic* and *random* errors (Section 3.5.1). A second dichotomy is according to whether the measurement is static or dynamic. Section 3.5.2 discusses some aspects of the dynamic case. Section 3.5.3 is an inventory of all kinds of physical phenomena that causes different types of errors. Section 3.5.4 is about the specifications of errors.

### 3.5.1. *Systematic errors and random errors*

Usually, the error of a measurement is composed of two components, which together are responsible for the combined uncertainty of the measurement. One component is the *systematic error*. The other is the *random error*. A random error is an error that can be reduced by averaging the results of repeated measurements. The systematic error is the error that persists even after averaging an infinite number of repeated measurements. Thus, we have:

$$\text{measurement error} = \text{systematic error} + \text{random error}$$

Once a systematic error is known, it can be quantified, and a correction can be applied. An example of that is recalibration. However, the determination of a correction factor always goes with some uncertainty. In other words, the remainder of a corrected systematic error is a *constant random error*. Such an error is *constant* because if the measurement is repeated, the error does not change. At the same time the error is also *random* because if the calibration is redone, it takes a new value. Hence, the remainder of a corrected systematic error is still a systematic error. Its uncertainty should be accounted for in the combined uncertainty of the measurement.

**Example 3.8 Error types of a linear sensor**
In order to illustrate the concepts, we consider the errors of the linear sensor discussed in example 3.7. The output of the system is (see equation (3.37)):

$$z = \frac{ax + b + n - B}{A} \tag{3.38}$$

$a$ and $b$ are two sensor parameters. $A$ and $B$ are two parameters obtained from a two-point calibration procedure. Deviations of $A$ and $B$ from $a$ and $b$ give rise to two systematic errors.

$$z \approx x - \underbrace{\frac{\Delta B}{a}}_{\text{I}} - \underbrace{\frac{\Delta A}{a}x}_{\text{II}} + \underbrace{\frac{1}{a} \cdot noise}_{\text{III}} \tag{3.39}$$

The error budget of $z$ consists of three factors:

I   a constant, additive random error due to the uncertainty in $B$
II  a constant, multiplicative random error due to the uncertainty in $A$
III an additive, fluctuating random error due to the noise term.

According to Table 3.2, the non-constant random error brings an uncertainty of 1.667. The two constant random errors are not independent because they both stem from the calibration error. The uncertainty of the total constant random error is $\sqrt{[2 + 0.0008(x-50)^2]}$.

A systematic effect that is not recognized as such leads to unknown systematic errors. Proper modelling of the physical process and the sensor should identify this kind of effect. The error budget and the sensitivity analysis, as discussed in Section 3.4.3, must reveal whether potential defects should be taken seriously or not.

Another way to check for systematic errors is to compare some measurement results with the results obtained from alternative measurement systems (that is, systems that measure the same measurand, but that are based on other principles). An example is the determination of the position of a moving vehicle based on an accelerometer. Such a sensor essentially measures acceleration. Thus, a double integration of the output of the sensor is needed to get the position. Unfortunately, the integrations easily introduce a very slowly varying error. By measuring the position now and then by means of another principle, for instance GPS, the error can be identified and corrected.

In order to quantify a systematic error it is important to define the repeatability conditions carefully. Otherwise, the term "systematic error" is ambiguous. For instance, one can treat temperature variations as a random effect, such that its influence (thermal drift) leads to random errors. One could equally well define the repeatability conditions such that the temperature is constant, thus letting temperature effects lead to a systematic error.

### 3.5.2. Signal characterization in the time domain

In this section, the context of the problem is extended to a *dynamic* situation in which measurement is regarded as a process that evolves in time. The measurand is a signal instead of a scalar value. Consequently, the measurement result and the measurement error now also become signals. They are functions of the time $t$:

$$e(t) = z(t) - x(t) \qquad (3.40)$$

The concept of a random variable must be extended to what are called *stochastic processes* (also called *random signals*). The stochastic experiment alluded to in Section 3.2 now involves complete signals over time, where the time $t$ is defined in some interval. Often this interval is infinite, as in $t \in (-\infty, \infty)$, or in $t \in [0, \infty)$.

The outcome of a trial is now a signal instead of a number. Such a generated signal is called a *realization*. After each trial, a so-called stochastic process $\underline{e}(t)$ takes the produced realization as its new signal. The set of all possible realizations is the *ensemble*.

If the ensemble is finite (that is, there is only a finite number of realizations to choose from), we could define the probability that a particular realization is generated. However, the ensemble is nearly always infinite. The probability of a particular realization is then zero.

We can bypass this problem by freezing the time to some particular moment $t$, and by restricting our analysis to that very moment. With frozen time, the stochastic process $\underline{e}(t)$ becomes a random variable. By that, the concept of a probability density function $f_{\underline{e}(t)}(e,t)$ of $\underline{e}(t)$ is valid again. Also, parameters derived from $f_{\underline{e}(t)}(e,t)$, such as expectation (=mean) $\mu(t)$ and variance $\sigma_e^2(t)$ can then be used.

In principle, $f_{\underline{e}(t)}(e,t)$ is a function not only of $e$, but also of $t$. Consequently, $\mu(t)$ and $\sigma_e^2(t)$ depend on time. In the special case, where $f_{\underline{e}(t)}(e,t)$ and all other probabilistic parameters[3] of $\underline{e}(t)$ do not depend on $t$, the stochastic process $\underline{e}(t)$ is called *stationary*.

Often, a non-stationary situation arises when the *integration* of a noisy signal is involved. A simple example is the measurement of a speed $v(t)$ by means of an accelerometer. Such a sensor actually measures the acceleration $a(t) = \dot{v}(t)$. Its output is $y(t) = a(t) + \underline{e}(t)$ where $\underline{e}(t)$ models the errors produced by the sensor, e.g. noise. In order to measure the speed, the output of the accelerometer has to be integrated: $z(t) = \int y(t)\,dt = v(t) + C + \int \underline{e}(t)\,dt$. The measurement error is $C + \int \underline{e}(t)\,dt$, where $C$ is an integration constant that can be neutralized easily. The second term $\int \underline{e}(t)\,dt$ is integrated noise. It can be shown [2] that the variance of integrated noise is proportional to time. Therefore, if $\underline{e}(t)$ is stationary noise with standard deviation $\sigma_e$, the standard deviation of the error in $z(t)$ is $k\sigma_e\sqrt{t}$. The standard deviation grows continuously. This proceeds until the voltage clips against the power supply voltage of the instrumentation.

As said before, stationary random signals have a constant expectation $\mu$ and a constant standard deviation $\sigma$. Loosely phrased, the expectation is the non-fluctuating part of the signal. The standard deviation quantifies the fluctuating part. The power of an electrical random signal $\underline{s}(t)$ that is dissipated in a resistor is proportional to $E\{\underline{s}^2(t)\} = \mu^2 + \sigma^2$. Hence, the power of the fluctuating part and the non-fluctuating part add up. The quantity $E\{\underline{s}^2(t)\}$ is called the *mean square* of the signal. The *RMS* (root mean square) of a random signal is the square root of the

---

[3] The complete characterization of a stochastic process (in probabilistic sense) does involve much more than that what is described by $f_{\underline{e}(t)}(e,t)$ alone [2]. However, such a characterization is beyond the scope of this book.

mean square:

$$RMS = \sqrt{E\{\underline{s}^2(t)\}} \qquad (3.41)$$

A true *RMS*-voltmeter is a device that estimates the *RMS* by means of averaging over time. Suppose that $s(t)$ is a realization of a stationary random signal, then the output of the *RMS*-voltmeter is:

$$RMS(t) = \sqrt{\frac{1}{T} \int_{\tau=t-T}^{t} s^2(\tau) \, d\tau} \qquad (3.42)$$

If $u(t)$ is a periodic signal with period $T$, that is $u(t) = u(t-T)$, then we can form a stationary random signal by randomizing the phase of $u(t)$. Let $\underline{\tau}$ be a random variable that is uniformly distributed between 0 and $T$. Then $\underline{s}(t) = u(t - \underline{\tau})$ is a stationary random signal whose *RMS* is given by equation (3.42).

**Example 3.9 *RMS* of a sinusoidal waveform and a block wave**
Suppose that $\underline{s}(t)$ is a sinusoidal waveform with amplitude $A$ and period $T$ and random phase, e.g. $\underline{s}(t) = A \sin(2\pi(t - \underline{\tau})/T)$. Then, the *RMS* of $\underline{s}(t)$ is:

$$RMS = \sqrt{\frac{1}{T} \int_{t=0}^{T} \left( A \sin\left( 2\pi \frac{t}{T} \right) \right)^2 dt} = \frac{A}{\sqrt{2}} \qquad (3.43)$$

If $\underline{s}(t)$ is a periodic block wave whose value is either $+A$ or $-A$, the *RMS* is simply $A$. Note that the true *RMS*-voltmeter from equation (3.42) only yields the correct value if the averaging time equals a multiple of the period of the waveform.

The *RMS* is a parameter that roughly characterizes the magnitude of a random signal, but it does not characterize the time dependent behaviour. Figure 3.6 shows

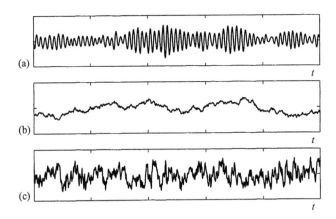

**Figure 3.6** *Different types of noise: (a) narrow band noise; (b) low frequency noise; (c) broad band noise*

three different types of noise. In Figure 3.6(a) the noise has a preference frequency. In fact, this type of noise can be regarded as the sum of a number of harmonics (sinusoidal waveforms) with random amplitudes and phases, but with frequencies that are close to the preference frequency. This type of noise is called *narrow band noise*, because all frequencies are located in a narrow band around the preference frequency. The noise in Figure 3.6(b) is slowly varying noise. It can be regarded as the sum of a number of harmonics whose frequencies range from zero up to some limit. The amplitudes of these harmonics are still random, but on an average, the lower frequencies have larger amplitudes. It is so-called *low frequency noise*. The noise realization shown in Figure 3.6(c) is also built from harmonics with frequencies ranging from zero up to a limit. But here, on an average, the amplitudes are equally distributed over the frequencies.

These examples show that the time dependent behaviour can be characterized by assuming that the signal is composed of a number of harmonics with random amplitudes and random phases, and by indicating how, on average, these amplitudes depend on the frequency. The function that establishes that dependence is the so-called *power spectrum* $P(f)$ where $f$ is the frequency. The power spectrum is defined such that the power (= mean square; see page 66) of the signal measured in a small band around $f$ and with width $\Delta f$ is $P(f)\Delta f$.

The term *white noise* refers to noise which is not bandlimited, and where, on an average, all frequencies are represented equally. In other words, the power spectrum of white noise is flat: $P(f) = Constant$. Such a noise is a mathematical abstraction, and cannot exist in the real world. In practice, if the power spectrum is flat within a wide band, such as in Figure 3.6(c), it is already called "white".

### Example 3.10 Noise characterization of an amplifier

Figure 3.7 shows the model that is used to characterize the noise behaviour of an amplifier. Such a device is used to amplify the signal generated by a sensor, here modelled by a voltage $V_i$ and a resistor $R$. The output of the amplifier is $V_o$. The noise produced by the amplifier is modelled with two noise sources, a voltage source $V_n$ and a current source $I_n$. These sources are at the input side of the amplifier.

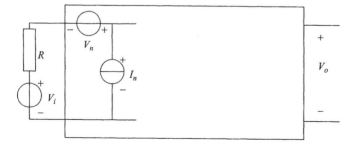

**Figure 3.7** *The noise characterization of an amplifier*

The manufacturer specifies the square root of the power density at a given frequency. For instance, the LM6165 is a high speed, low noise amplifier. The noise specifications are: $V_n = 5\,\text{nV}/\sqrt{\text{Hz}}$ and $I_n = 1.5\,\text{pA}/\sqrt{\text{Hz}}$ measured at $f = 10\,\text{kHz}$. In fact, these figures are the *RMS*s of the noise voltage and current if the bandwidth of the amplifier was 1 Hz centred around 10 kHz. In other words, the power spectra at $f = 10\,\text{kHz}$ is $P_{V_n}(10\,\text{kHz}) = 25\,(\text{nV})^2/\text{Hz}$ and $P_{I_n}(10\,\text{kHz}) = 2.25\,(\text{pA})^2/\text{Hz}$.

If the noise is white, then the noise spectra are flat. Thus, $P_{V_n}(f) = C_{V_n}$ with $C_{V_n} = 25\,(\text{nV})^2/\text{Hz}$ and $P_{I_n}(f) = C_{I_n}$ with $C_{I_n} = 2.25\,(\text{pA})^2/\text{Hz}$. Suppose that the bandwidth of the amplifier is $B$, then the noise power is $BC_{V_n}$ and $BC_{V_n}$. For instance, if $B = 10\,\text{MHz}$, then the power is $2.5 \times 10^{-10}\,\text{V}^2$ and $2.25 \times 10^{-17}\,\text{A}^2$, respectively. The *RMS*s are the square root of that: $16\,\mu\text{V}$ and $5\,\text{nA}$.

Most amplifiers show an increase of noise power at lower frequencies. Therefore, if the bandwidth of the amplifier encompasses these lower frequencies, the calculations based on the "white noise assumption" are optimistic.

### 3.5.3. Error sources

The uncertainties in measurement outcomes are caused by physical and technical constraints. In order to give some idea of what effects might occur, this section mentions some types of error sources. The list is far from complete.

**Systematic effects**

*Saturation, dead zone, hysteresis*
These phenomena are characterized by a static, nonlinear mapping of the measurand $x$ and the output of a sensor: $y = g(x)$. The effects are illustrated in Figure 3.8. *Saturation* occurs when the sensitivity $\partial g/\partial x$ decreases as $|x|$ increases (see Figure 3.8(a)). If $|x|$ becomes too large, $y$ clips against the power voltage or ground. The *dead zone* of a sensor is an interval of $x$ for which the output of the sensor does not change. In a mechanical system, static friction (called *stiction*) could be the cause. Stiction forms a barrier that a force must overcome before the relative motion of two bodies in contact can begin. *Hysteresis* is characterized by the phenomenon that $g(x)$ depends on whether $x$ is increasing or decreasing.

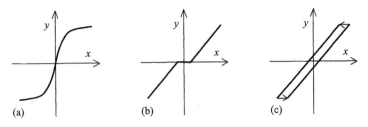

**Figure 3.8** *Non-linear systematic errors*

It occurs in mechanical systems when two parts have a limited space for free movement.

*Loading errors*
A *loading error* occurs when the interaction between the physical process and the sensor is not compensated. A well-known example is the measurement of the potential difference over a resistor by means of a voltmeter. The internal impedance of the voltmeter influences this potential difference, and thus, without compensation, an error occurs.

Another example is the measurement of the temperature in a furnace. The temperature sensor is located either in the wall of the furnace, or outside the furnace connected to a thermal conductor. Due to thermal leakage, there will be a temperature difference between the centre of the furnace and the sensor.

### Example 3.11 Loading error of a voltmeter
Figure 3.9 shows a model of a sensor with output voltage $V_y$ and internal resistor $R$ connected to the input of a voltmeter. The input voltage $V_i$ of the voltmeter is affected by its input impedance $R_i$:

$$V_i = \frac{R_i}{R_i + R} V_y \approx \left(1 - \frac{R}{R_i}\right) V_y \qquad (3.44)$$

The approximation is accurate if $R_i \gg R$. Clearly, the load of the sensor can cause a systematic error that must be compensated (see Section 7.1).

*Calibration errors*
Calibration errors are induced by:

- The uncertainty of the reference values (the conventional true values) (see Example 3.7).
- Random errors that occur during calibration.
- Erroneous calibration curves.

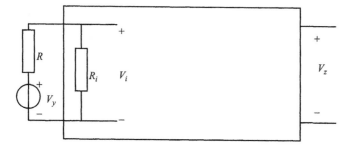

**Figure 3.9** *The loading error of a voltmeter*

Errors in a two-point calibration procedure lead to a multiplicative and an additive error. See Example 3.8. A wrong calibration curve occurs when a two-point calibration procedure is used based on a linear model of the sensor, whereas in reality the sensor is nonlinear. A solution might be to use some polynomial as a calibration curve, and to fit that curve using multiple-point calibration.

*Dynamic errors*
*Dynamic errors* occur when the measurement instrument is too slow to be able to keep track of a (fast changing) measurand. The change of the output of an amplifier is limited. This is one cause of a dynamic error.

Another dynamic error occurs when the bandwidth of the instrument is not sufficient. The measuring device should have an amplitude transfer that is constant over the full bandwidth of the measurand. In that region, the phase characteristic should hold a linear relationship with the frequency.

In time-discrete systems, a too low sampling frequency gives rise to aliasing errors (see Chapter 5).

**Random effects**

*Environmental errors*
External influences may cause errors that are often random. An example is electromagnetic interference (EMI). The interfering noise with power concentrated at multiples of 50 (or 60) Hz often originates from the mains of the AC power supply. Pulse-like interferences occur when high voltages or currents are switched on or off. These types of interferences are caused by electrical discharges in a gas (e.g. as in the ignition system of a petrol engine) or by the collector of an AC electromotor. At much higher frequencies, EMI occurs, for instance, from radio stations.

*Resolution error*
The *resolution error* is the generic name for all kinds of quantization effects that might occur somewhere in the measurement chain. Examples are:

- A wire wound potentiometer used for position measurement or for attenuation.
- A counter used for measuring distances (i.e. an odometer).
- AD converters.
- Finite precision arithmetic.

Essentially, the resolution error is a systematic error. However, since the resolution error changes rapidly as the true value changes smoothly, it is often modelled as a random error with a uniform distribution function.

*Drift*
The generic name for slowly varying, random errors is *drift*. Possible causes for drift are: changes of temperature and humidity, unstable power supplies, $1/f$ noise (see below), and so on.

*Noise*
The generic name for more or less rapidly changing, random errors is *noise*. The *thermal noise* (also called *Johnson noise*) of a resistor is caused by the random, thermal motion of the free electrons in conductors. In most designs, this noise can be considered as white noise, because the power spectrum is virtually flat up to hundreds of MHz. The mechanical equivalent of Johnson noise is the so-called *Brownian motion* of free particles in a liquid.

Another physical phenomenon that is well described by white noise is the so-called *Poisson noise* or *shot noise*. It arises from the discrete nature of physical quantities, such as electrical charge and light. For instance, in a photodetector, light emitted on a surface is essentially a bombardment of that surface by light photons. The number of photons received by that surface within a limited time interval has a Poisson distribution.

Some noise sources have a power spectrum that depends on the frequency. A well-known example is the so-called $1/f$ noise. It can be found in semiconductors. Its power spectrum is in a wide frequency range inversely proportional to some power of the frequency, i.e. proportional to $1/f^p$ where $p$ is in the range from 1 to 2.

### 3.5.4. Specifications

The goal of this section is to explain the generally accepted terms that are used to specify sensor systems and measurement instruments.

**Sensitivity, dynamic range and resolution**
The *sensitivity* $S$ of a sensor or a measurement system is the ratio between output and input. If $z = f(x)$, then the sensitivity is a differential coefficient $S = \partial f/\partial x$ evaluated at a given set point. If the system is linear, $S$ is a constant.

The *range* of a system is the interval $[x_{\min}, x_{\max}]$ in which a device can measure the measurand within a given uncertainty $\sigma_e$. The *dynamic range* $D$ is the ratio $D = (x_{\max} - x_{\min})/\sigma_e$. Often, the dynamic range is given on a logarithmic scale in dB. The expression for that is $D = 20\,^{10}\log[(x_{\max} - x_{\min})/\sigma_e]$.

The *resolution* $R$ expresses the ability of the system to detect small increments of the measurand. This ability is limited due to the resolution errors, dead zones and hysteresis of the system. The resolution is defined as $R = x/\Delta x$ where $\Delta x$ is the smallest increment that can be detected. Often, the specification is given in terms of maximum resolution: $R_{max} = x_{max}/\Delta x$.

### Accuracy, repeatability and reproducibility
The *accuracy* is the closeness of agreement between the measurement result and the true value. The quantification of the accuracy is the uncertainty, where uncertainty comprises both the random effects and the systematic effects.

The *repeatability* is the closeness of agreement between results of measurement of the same value of the same measurand, and which are obtained under the same conditions (same instrument, same observer, same location, etc.). Hence, the repeatability relates to random errors, and not to systematic errors. The quantification is the uncertainty, where uncertainty comprises only the random effects.

The *reproducibility* is the closeness of agreement between results of measurements of the same value of the same measurand, but that are obtained under different conditions (different instruments, different locations, possibly different measurement principles, etc.).

### Offset, gain error and nonlinearity
If $z = g(x)$, then the *offset error* (or *zero error*) is $\Delta A = g(0)$. The offset is an additive error. The *gain error* $\Delta B$ in a linear measurement system, $z = Bx$, is the deviation of the slope from the unit value: $\Delta B = B - 1$. The gain error gives rise to a multiplicative error $x \Delta B$.

The *nonlinearity error* is $g(x) - A - Bx$ where $A$ and $B$ are two constants that can be defined in various ways. The simplest definition is the *end point* method. Here, $A$ and $B$ are selected such that $g(x_{min}) = A + Bx_{min}$ and $g(x_{max}) = A + Bx_{max}$. Another definition is the *best fit* method. $A$ and $B$ are selected such that $A + Bx$ fits $g(x)$ according to some closeness criterion (e.g. a least squares error fitting). The *nonlinearity* is the maximum nonlinearity error observed over the specified range expressed as a percentage of the range $L = \max(|g(x) - A - Bx|)/(x_{max} - x_{min})$. These concepts are illustrated in Figure 3.10.

### Bandwidth, indication time and slew rate
These are the three most important parameters used to characterize the dynamic behaviour. The *bandwidth* is the frequency range in which a device can measure. Usually, it is defined as the frequency range for which the transfer of power is not less than half of the nominal transfer. If $H(f)$ is the (signal) transfer function, and $H(f) = 1$ is the nominal transfer, then half of the power is transferred at frequencies for which $|H(f)|^2 = \frac{1}{2}$. The cut-off frequency $f_c$ follows from $|H(f_c)| = 1/\sqrt{2}$ which corresponds to $-3$ dB with respect to the nominal response. The response for zero frequency (i.e. steady state) is for most systems nominal. For these systems, the bandwidth equals $f_c$. Obviously, for systems with a band-pass characteristic, two cut-off frequencies are needed.

Measurement Errors and Uncertainty   73

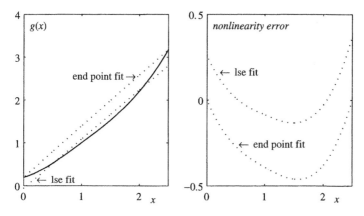

**Figure 3.10** *Nonlinearity errors*

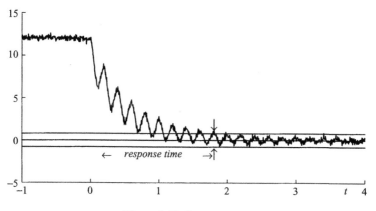

**Figure 3.11** *Response time*

The *indication time* or *response time* is the time that expires from a steplike change of the input up to the moment that the output permanently enters an interval $[z_0 - \Delta z, z_0 + \Delta z]$ around the new steady state $z_0$. For $\Delta z$, one chooses a multiple of the uncertainty (see Figure 3.11).

The *slew rate* is the system's maximum change rate of output $\max[|dz/dt|]$.

**Common mode rejection ratio**
It often occurs that the measurand is the difference between two physical quantities: $x_d = x_1 - x_2$. In the example of Figure 3.12(a), the measurand is the level $l$ of a liquid. The sensor measures the difference between pressures at the surface and at the bottom of the tank. The relation $l = (p_1 - p_2)/\rho g$ where $\rho$ is the mass density can be used to measure $l$. Other examples are differences between: floating voltages, forces, magnetic flux, etc. Such a measurement is called a *differential measurement*. The difference $x_d$ is the *differential mode input*. The average of the

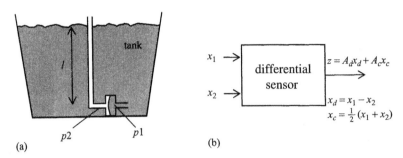

**Figure 3.12** *Differential measurements: (a) level measurement using a differential pressure sensor; (b) a differential sensor*

two quantities is called the *common mode input* $x_c = \frac{1}{2}(x_1 + x_2)$. A *differential sensor* is a sensor which produces the difference by means of its construction. Another option is to have two separate sensors, one for each quantity. In that case, the subtraction must be done afterwards by means of a differential amplifier (see Section 4.1.4).

The output of a linear differential sensor or amplifier is given by the following expression:

$$z = A_d x_d + A_c x_c \qquad (3.45)$$

where $A_d$ and $A_c$ are the differential and common mode amplification factors. The purpose of the device is to amplify the differential mode and to suppress the common mode. The ability of the device to do so is expressed in the so-called *common mode rejection ratio*: $CCMR = A_d/A_c$.

### 3.6. Error reduction techniques

This section reviews various methods that can be used to reduce the effects due to the shortcomings of sensors and the influences from an interfering environment. For that purpose, we model the measurement process including the influence of noise and other error sources as:

$$z = g(x, e_1, e_2, \ldots) \qquad (3.46)$$

As before, $x$ is the measurand, and $z$ is the measurement result. $e_1, e_2, \ldots$ are all kind of error sources, e.g. noise, thermal drift, environmental influences, and so on. In the linearized case, the equation becomes:

$$z = \Delta b + (1 + \Delta a)x + S_1 e_1 + S_2 e_2 + \ldots \qquad (3.47)$$

where $\Delta b$ is the offset error, $\Delta a$ is the gain error, and $S_i$ are the sensitivities for the error sources $e_i$, i.e. $S_i = \partial z/\partial e_i$. With that, the final error becomes:

$$\Delta b + \Delta a\, x + S_1 e_1 + S_2 e_2 + \cdots \qquad (3.48)$$

This expression shows that the error is composed of two parts. The first part is the calibration error $\Delta b + \Delta a\, x$. It can be minimized by a proper calibration procedure, or by application of feedback (see Section 3.6). Due to ageing and wear, the transformation characteristic, $a$ and $b$, may change over time. Therefore, now and then, the system needs recalibration. Smart sensor systems have built-in facilities that enable automatic recalibration even during operation. If the transformation characteristic is nonlinear, a *correction curve* should be applied. The parameters of this curve should also be refreshed by recalibration.

The second part, $S_1 e_1 + S_2 e_2 + \cdots$, is the result of the errors that are induced somewhere in the measurement chain, and that propagate to the output. Each term of this part consists of two factors: the error sensitivity $S_i$ and the error source $e_i$. Essentially, there are three basic methods to reduce or to avoid this type of error [4]:

- Elimination of the error source $e_i$ itself.
- Elimination of the error sensitivity $S_i$.
- Compensation of an error term $S_i e_i$ by a second term $-S_i e_i$.

These techniques can be applied at the different stages of the measurement process. *Preconditioning* is a collection of measures and precautions that are taken *before* the actual measurement takes place. Its goal is either to eliminate possible error sources, or to eliminate their sensitivity coefficients. *Postconditioning* or *postprocessing* are techniques that are applied *afterwards*. The compensation techniques take place *during* the actual measurement.

The following sections present overviews of these techniques.

### 3.6.1. *Elimination of error sources*

Obviously, an effective way to reduce the errors in a measurement would be to take away the error source itself. Often, the complete removal is not possible. However, a careful design of the system can already help to diminish the magnitude of many of the error sources. An obvious example is the selection of the electronic components, such as resistors and amplifiers. Low-noise components must be used for those system parts that are most sensitive to noise (usually the part around the sensor).

Another example of error elimination is the conditioning of the room in which the measurements take place. Thermal drift can be eliminated by stabilizing the temperature of the room. If the measurement is sensitive to moisture, then the regularization of the humidity of the room might be useful. The influence of air and air pressure is undone by placing the measurement set up in vacuum. Mechanical vibrations and shocks are eliminated for a greater part by placing the measurement set up in the basement of a building.

### 3.6.2. *Elimination of the error sensitivities*

The second method to reduce errors is blocking the path from error source to measurement result. Such can be done in advance, or afterwards. Examples of the former are:

- Heat insulation of the measurement set up in order to reduce the sensitivity to variations of temperature.
- Mounting the measurement set up on a mechanically damped construction in order to reduce the sensitivity to mechanical vibrations, i.e. a granite table top resting on shock absorbing rubbers.
- Electronic filtering before any other signal processing takes place. For instance, prefiltering before AD-conversion (Chapter 6) can reduce the noise sensitivity due to aliasing. Electronic filters are discussed in Section 4.2.

**Postfiltering**
An example of reducing the sensitivity afterwards is postfiltering. In the static situation, repeated measurements can be averaged so as to reduce the random error. If $N$ is the number of measurements, then the random error of the average reduces by a fraction $1/\sqrt{N}$ (see equation (3.18)).

In the dynamic situation $z(t)$ is a function of time. The application of a filter is of use if a frequency range exists with much noise and not much signal. Therefore, a filter for noise suppression is often a low-pass filter or a band-pass filter. Section 4.2 discusses the design of analogue filters. Here, we introduce the usage of digital filters.

Suppose that the measurements are available at discrete points of time $t_n = n\Delta T$, as in digital measurement systems. This gives a time sequence of measurement samples $z(n\Delta T)$. We can apply averaging to the last $N$ available samples. Hence, if $n\Delta T$ denotes the current time, the average over the last $N$ samples is:

$$\bar{z}(n\Delta T) = \frac{1}{N} \sum_{m=0}^{N-1} z((n-m)\Delta T) \qquad (3.49)$$

Such a filter is called a *moving average filter* (MA-filter). The reduction of random errors is still proportional to $1/\sqrt{N}$. Often, however, the random errors in successive samples are not independent. Therefore, usually, moving averaging is less effective than averaging in the static case.

For the reduction of random errors it is advantageous to choose $N$ very large. Unfortunately, this is at the cost of the bandwidth of the measurement system. In the case of MA-filtering the cut-off frequency of the corresponding low-pass filter is inversely proportional to the parameter $N$. If $N$ is too large, not only the noise is suppressed, but also the signal.

There are formalisms to design filters with optimal transfer function. Such a filter yields, for instance, maximum signal-to-noise ratio at its output. Examples of these filters are the Wiener filter and the Kalman filter [2].

**Guarding and grounding**

As mentioned in Section 3.5.3, many external error sources exist that cause an electric or magnetic pollution of the environment of the measurement system. Without countermeasures, these environmental error sources are likely to enter the system by means of stray currents, capacitive coupling of electric fields, or inductive coupling of magnetic fields. The *ground* of an electric circuit is a conducting body, such as a wire, whose electric potential is – by convention – zero. However, without precautions, the ground is not unambiguous. The environmental error sources may induce stray currents in the conductors of the electric circuits, in the metallic housing of the instrumentation, or in the instrumental racks. *Guarding* is the technique of grounding a system, such that these stray currents do not occur, or do not enter the system.

The *earth* of the mains power usually consists of a connection to a conducting rod which is drilled into the ground such that the impedance between rod and earth is smaller than 0.2 Ω. Unfortunately, the potential difference between two earths from two main plugs can be substantial because of the environmental error sources. Therefore, it is not appropriate to consider both earths as one ground. One can define one of the earths of these plugs as the ground of the system, but it is not necessary (and sometimes undesirable) to do so. In fact, the main reason for having an earth is the safety that it offers when connected to the housings of the instruments[4].

As an example, Figure 3.13(a) shows an instrumentation system with two electrical connections to earth points. Such a configuration induces disturbances that cannot be fully compensated by a differential amplifier. The disturbances arise from stray currents that always exist between different earth points. The impedance $R_A$ between the two earth points is on the order of 0.1 Ω. Together with the stray currents, it forms a potential difference between the earth points. The potential difference is largely a common mode disturbance. Therefore, it will be compensated

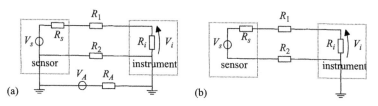

**Figure 3.13** *Grounding: (a) creation of a ground loop by two earth points; (b) star point grounding*

---

[4] Many instruments have a terminal marked with "earth" or ⊥. Such a terminal is connected to the housing of the instruments and can be connected to the earth.

for a large part if the CMRR of the instrumentation amplifier is large enough. However, a small part of the potential difference will be seen as a differential mode voltage due to the asymmetries that arise from the internal sensor impedance, the cable impedance, and the input impedance of the amplifier. Therefore, a part of the disturbances propagates to the output.

The errors that arise from stray currents can be eliminated simply by avoiding two-point grounding. The rule is: *never* create ground loops. Instead, use *star point grounding*. Figure 3.13(b) shows the correct way to ground the system.

**Shielding**
Environmental interferences enter the system by means of either capacitive coupling or inductive coupling. *Shielding* is the technique of reducing these coupling factors by placing metal cases around the electronics.

The model for capacitive coupling of environmental inferences is an AC voltage source that is coupled to the various parts of the electric circuit via parasitic capacitors. Figure 3.14(a) shows an example consisting of a sensor wired to an instrument, e.g. a voltmeter. The ungrounded wire is coupled to the AC voltage $V_{AC}$ by means of the capacitor $C_p$. A fraction of $V_{AC}$ appears at the input of the instrument. The error sensitivity equals the transfer function from $V_{AC}$ to $V_i$:

$$\frac{R_s \| R_i}{\left|(R_s \| R) + [(1/(j\omega C_p))]\right|} \tag{3.50}$$

Figure 3.14(b) shows the same circuit, but now with the wires replaced by a rotationally symmetric coaxial cable. The inner wire of the cable connects $R_s$ to $R_i$. The outer wire (the coating) is connected to the ground. The parasitic capacitor between the inner wire and $V_{AC}$ is very small, because the ground keeps the potential of the outer wire fixed at zero. Consequently, the error sensitivity is much lower.

A disadvantage of using a coaxial cable is that the capacity between the two wires is increased. Consequently, the (capacitive) load of the sensor is enlarged.

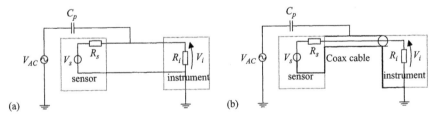

**Figure 3.14** *Shielding against capacitive coupling: (a) an unshielded connection between sensor and instrument; (b) a coaxial cable reduces the effect of the parasitic capacitance*

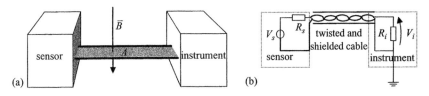

**Figure 3.15** *Shielding against inductive coupling: (a) a changing magnetic field induces a voltage in the loop area created by the two connecting wires; (b) a twisted cable reduces the inductive coupling*

**Example 3.12 The sensitivity of an unshielded input lead**
Suppose that the parasitic capacity per unit length of a wire is 5 pF/m. The length of the wire is 1 m. The *RMS* of the error source is $V_{AC} = 220$ V at a frequency of 50 Hz. $R_s = 100\ \Omega$. The input impedance of the instrument is large compared with $R_s$. The *RMS* voltage of the error at the input of the instrument is 35 μV. If with coaxing the remaining capacitance can be reduced to 0.05 pF, the error will be 0.35 μV.

Induction is another mechanism for coupling to environmental interferences. Magnetic fields are caused by large currents in the neighbourhood of the measurement set-up. A magnetic field induces a voltage in a loop area of a wire according to:

$$V_{mag} = -\frac{d}{dt}\int_{area} \vec{B} \cdot \vec{dA} \qquad (3.51)$$

Figure 3.15(a) illustrates an example. Here, the connecting wires between a sensor and the instrument forms a loop with area $A$.

The coupling can be reduced in the following ways:

- Shielding the wires by metal cases.
- Arranging the geometry of the wiring such that $\vec{B} \perp \vec{dA}$.
- Minimizing the loop area.

Shielding the wires is not always very effective because for most metals the penetration depth of a magnetic field is large (except for mumetal). The second solution is also difficult because the direction of the magnetic field is often unknown. The third solution is often applied. The wires in a cable are often twisted in order to alternate the sign of the magnetic field across the loop area (see Figure 3.15(b)). Another possibility is to use coaxial cable. Here, the loop area is zero because of the rotationally symmetry.

### 3.6.3. *Compensation*

Here, we accept that an error enters the system and propagates to the output, yielding $S_i e_i$. The strategy is to add another term $-S_i e_i$ that compensates $S_i e_i$. This principle

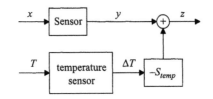

**Figure 3.16** *Compensation for thermal drift*

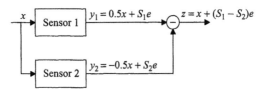

**Figure 3.17** *Balanced compensation*

is often applied directly in order to correct for thermal drift. See Figure 3.16. The output of the sensor is modelled with $y = x + S_{temp}\Delta T$ where $\Delta T$ is the deviation of the temperature from its nominal value. $S_{temp}$ is the sensitivity to the temperature. The thermal drift $S_{temp}\Delta T$ is compensated by measuring $\Delta T$ with a temperature sensor and applying a correction: $z = y - S_{term}\Delta T = x$.

Of course, this technique is applicable only if the sensitivity to the error is known. One possible method to find the sensitivity is by measuring the output $z$ at zero input ($x = 0$), and then trimming the proportionality factor of the correction term such that $z$ becomes zero. Some integrated sensors have built-in facilities for compensation. The trimming of the proportionality factor is sometimes done by the manufacturer by adjustment of some built-in resistor (so-called laser trimmed temperature compensation).

**Balanced compensation**
An advanced method of compensation is achieved if two sensors are used. Both measure the same measurand, but with opposite sign: $y_1 = 0.5x + S_1 e$ and $y_2 = -0.5x + S_2 e$. See Figure 3.17. The *differential measurement* is $z = y_1 - y_2 = x + (S_1 - S_2)e$. If the arrangement is such that the errors enter the sensors with equal sensitivities, then the two error terms will cancel.

Such a differential measurement also improves the linearity of the system. Suppose that the input-output relation of a sensor is given by a Taylor series expansion

$$y = a_0 + a_1 x + a_2 x^2 + a_3 x^3 + \cdots \tag{3.52}$$

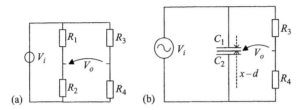

**Figure 3.18** Wheatstone bridge: (a) general layout; (b) application to capacitive displacement measurement

Then, the differential measurement based on two sensors configured with opposite signs with respect to the measurand is:

$$z = y_1 - y_2$$
$$= 2a_1 x + 2a_3 x^3 + \cdots \qquad (3.53)$$

All even terms are cancelled, thus improving the nonlinearity.

An example is the *Wheatstone bridge* with two active elements. See Figure 3.18(a). $R_3$ and $R_4$ are two fixed resistors with impedance $R$. $R_1$ and $R_2$ are two resistive sensors. Their impedances depend on the measurand $x$, but also on some error $e$. If the sensors are considered to be linear (possibly by means of a truncated Taylor series), then the impedances are modelled by:

$$R_1 = R(1 + S_{x1}x + S_1 e)$$
$$R_2 = R(1 + S_{x2}x + S_2 e) \qquad (3.54)$$

The principle of balanced compensation is achieved if the sensitivities to the measurand have opposite sign, $S_{x1} = -S_{x2}$, and those of the error are equal $S_1 = S_2$. That can be achieved by a symmetric sensor design. Under these conditions, the output voltage of the bridge is:

$$V_o = \left( \frac{1 + S_{x2}x + S_1 e}{2 + 2 S_1 e} - \frac{1}{2} \right) V_i$$
$$\approx \tfrac{1}{2} V_i S_{x2} x (1 - S_1 e) \qquad (3.55)$$

Thus, if $x$ is small, then $V_o = V_i S_{x2} x / 2$.

### Example 3.13 Capacitive measurement of displacements

Figure 3.18(b) shows a set-up for a balanced measurement of displacements. The sensor consists of three parallel plates. The plate in the middle can be displaced relative to the other plates. The goal of the sensor is to measure this displacement. For that purpose, the capacitive impedances between the outer plates and the inner plate are used in a bridge.

The capacity of two parallel plates is approximately inversely proportional to the distance between the plates. Suppose that in the neutral position the distance between the plates is $d$, then in an arbitrary position the distances are $d - x$ and $d + x$, respectively. Consequently, the capacities are $C_1 = K/(d - x)$ and $C_2 = K/(x + d)$. If the input voltage is a sine wave with an angular frequency of $\omega$, then the impedance of the two capacitors are $1/j\omega C_1$ and $1/j\omega C_2$. The output voltage is found to be:

$$V_o = \frac{x}{2d} V_i \tag{3.56}$$

A single capacitor bears a strong nonlinear relationship with $x$, but the balanced configuration of two sensors appears to be linear.

### Compensation method and feedback

These two techniques are applicable for measurements of physical quantities for which accurate and stable actuators exist. The advantage of the techniques is that the sensor is not required to be very stable. The only requirement is that the sensor is highly sensitive in the proximity of zero. Figure 3.19 illustrates the principles.

The compensation method is accomplished manually. See Figure 3.19(a). The output of the actuator is subtracted from the measurand. The input signal of the actuator is manipulated such that the difference becomes zero. Figure 3.20 shows an example. The goal is to measure the voltage $x$ of a source whose internal impedance $R_i$ is unknown. The measurement is done with a reference voltage $V_{ref}$ that is attenuated by a calibrated attenuator $y$. The value of $y$ is adjusted such that the difference $x - yV_{ref}$ is zero. Such a measurement is insensitive to the internal impedance of the source.

A feedback-based measurement system can be regarded as an implementation of the compensation method in which the compensation is done automatically. The ideal case is when the output of the sensor is directly fed back to the measurand. See Figure 3.19(b). In practical situations, the measurand itself cannot be fed

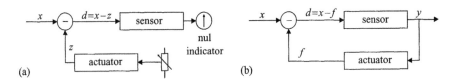

**Figure 3.19** *The principle of the compensation method and feedback: (a) compensation method; (b) feedback*

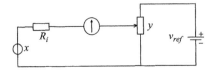

**Figure 3.20** *Voltage measurement with the compensation method*

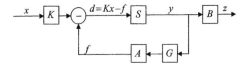

**Figure 3.21** *General lay-out of a feedback measurement system*

back. Instead, some derived variable $Kx$ is used (see Figure 3.21). In this figure, the actuator itself is represented by $A$. In addition, we have two design parameters $G$ and $B$. The factor $G$ is used to influence the feedback signal. The factor $B$ transforms the sensor signal $y$ to the final measurement result $z$. The analysis of this system is as follows:

$$\left. \begin{array}{l} d = Kx - f \\ y = Sd \\ f = GAy \\ z = By \end{array} \right\} \quad \begin{array}{l} y = \dfrac{SK}{1 + SGA} x \\ z = By \end{array} \quad (3.57)$$

If $G$ is selected such that $SGA$ is large, then $z = BK(GA)^{-1}x$. Thus, a suitable choice of $B$ is $K^{-1}GA$. Then $z = x$, and the result does not depend on the actual value of $S$.

It must be noticed that the subsystems $K$, $S$ and $A$ depend on the frequency of the measurand $x$. A stability analysis must be done to find the choices of $G$ that assure a stable system that covers the desired frequency band.

An error source that enters the system at the output of the actuator, i.e. $f = GAy + e_1$ induces an error at the output of $-e_1/K$. Thus, the sensitivity of this error is $S_1 = -1/K$. An error source that enters the system at the output of the sensor, i.e. $y = Sd + e_2$ induces an error at the output of $e_2/KS$. The sensitivity is $1/KS$.

Chapters 7, 8 and 9 present various examples of sensor system in which feedback plays an important role.

## 3.7. Further reading

[1] A.J. Wheeler, A.R. Ganji, *Introduction to Engineering Experimentation*, Second Edition, Pearson Prentice-Hall, Upper Saddle River, NJ, 2004.
[2] C.F. Dietrich, *Uncertainty, Calibration and Probability – The Statistics of Scientific and Industrial Measurement*, Second Edition, Adam Hilger, Bristol, 1991.
[3] K.S. Shanmugan, A.M. Breipohl, *Random Signals, Detection, Estimation and Data Analysis*, J. Wiley & Sons, 1988.
[4] A. Gelb (editor): *Applied Optimal Estimation*, M.I.T. Press, Cambridge (Mass.), 1974.

[5] F. van der Heijden, D. de Ridder, D.M.J. Tax, R.P.W. Duin, *Classification, Parameter Estimation and State Estimation – An engineering approach using MATLAB*, J. Wiley & Sons, Chichester, 2004.
[6] A.F.P. van Putten, *Electronic Measurement Systems: Theory and practice*, Second Edition, IOP Publ., 1996.

**Some bibliographical notes**

The concept of uncertainty is defined formally in a report of the International Organization for Standardization (ISO) [1]. Wheeler and Ganji's book contains a chapter with detailed coverage of uncertainty analysis. Dietrich's book also pays great attention to uncertainty analysis, but unfortunately this book is less accessible.

There is an overwhelming amount of literature on probability theory, but the book by Papoulis [2] is the classic text. We also mention the book by Shanmugan et al. because of its accessibility for first- or second-year undergraduate students. The topic of statistical inference is also covered by many textbooks. Kreyszig's book [3] has been selected here because of its accessibility.

The topic of Kalman filtering, i.e. the problem of simultaneously estimating dynamic variables using a multi-sensor system, is covered by Gelb's book, which in this area is considered the "bible". However, the latest developments, such as particle filtering, are not treated. For a more recent introduction of the topic see the book by van der Heijden et al.

The book by van Putten contains chapters about noise sources and about error reduction techniques. It goes into more detail than is possible here.

### 3.8. Exercises

1. A measurement instrument consists of a sensor/amplifier combination and an AD-converter. The noise at the output of the amplifier has a standard deviation of about 2.0 mV. The quantization step of the AD-converter is 3.00 mV. The AD-converter rounds off to the nearest quantization level. What is the standard uncertainty of the measurement result?

2. A measuring instrument consists of a sensor/amplifier combination and an AD-converter. The sensor/amplifier shows a systematic error with a standard uncertainty of 1 mV. The standard deviation of the noise at the output of the amplifier is 1 mV. Both the systematic error and the noise are normally distributed. The quantization step of the AD-converter is 1 mV. The AD converter rounds off to the nearest quantization level. What is the approximate distribution of the resulting error, and what is the standard deviation?

3. A measuring instrument consists of a sensor/amplifier combination and an AD-converter. The sensor/amplifier shows a systematic error with a standard uncertainty of 0.01 mV. The standard deviation of the noise at the output of the amplifier

is 0.01 mV. The quantization step of the AD-converter is 1 mV. The AD converter rounds off to the nearest quantization level. The measurand is static. In order to improve the accuracy, the measurement is repeated 100 times, and the results are averaged. What is the resulting uncertainty?

4. The postal service wants to know the volume of a block shaped parcel. For that purpose the three edges of the parcel are measured with a relative uncertainty of 2%. What is the relative uncertainty of the estimated volume?

5. A cylindrical rain gauge measures the amount of rain by collecting the rain drops. The top area of the gauge is 5 cm². After twenty-four hours, the level in the gauge is 2.0 cm. Assume that the mean volume of a rain drop is 0.10 cm³. What is the standard uncertainty of the level?

6. The length of an object is repeatedly measured with the following results:

$$8, 8, 9, 7, 6, 8, 9, 8, 6, 7 \text{(units: mm)}$$

The results are averaged in order to reduce the random error.
a. What is the estimated mean?
b. What is the sample variance of the sequence?
c. What is the estimated uncertainty of the mean?
d. What is the estimated uncertainty of the sample variance?

# Chapter 4

# Analogue Signal Conditioning

The output of any information conveying system must satisfy conditions with respect to range, dynamic properties, signal-to-noise ratio, immunity to load, robustness of transmission, etc. Conditioning of measurement signals prior to further processing can be accomplished in various ways, both in the analogue and digital domain. Major conditioning functions are:

- Signal range matching: comprising signal amplification (or attenuation) and level shift. For example, the output signal of a sensor ranges from $-10\,\text{mV}$ to $+10\,\text{mV}$; matching this range to an ADC with input range from 0 to 5V requires a gain of 250 and a voltage shift of $+10\,\text{mV}$ prior to amplification. Both functions can easily be realized using operational amplifiers.
- Impedance matching: for optimal voltage transfer the sensor needs to be connected to a circuit with infinite input impedance; optimal current transfer requires zero impedance and optimal power transfer an impedance equal to that of the sensor. Impedance matching can easily be combined with gain or level shift using operational amplifiers.
- Improving SNR (signal-to-noise ratio): whenever possible, noise and interference on a sensor signal should be suppressed as much as possible in the analogue domain, prior to AD conversion. Filtering, too, can be combined with amplification, by means of operational amplifiers.
- Modulation and demodulation: to convert the signal to another, more appropriate part of the spectrum.

In this chapter we consider the following conditioning functions: amplification (Section 4.1); analogue filtering (Section 4.2); modulation and demodulation (also called detection) (Section 4.3).

## 4.1. Analogue amplifiers

### 4.1.1. *Operational amplifiers*

Most analogue amplifier circuits consist of operational amplifiers, combined with resistance networks. An operational amplifier is in essence a differential amplifier with very high voltage gain, a high common mode rejection ratio and a very low input current and offset voltage. The amplifier is composed of a large number of transistors and resistors and possibly some capacitors, but no inductors. All components are integrated on a single chip, mounted in a metal or plastic

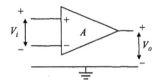

**Figure 4.1** *Symbol of an operational amplifier; voltage supply connections are omitted*

encapsulation. The inputs are called *inverting* and *non-inverting* inputs (indicated by a minus and plus sign, respectively). An *ideal* operational amplifier (Figure 4.1) has the following properties:

- infinite gain, common mode rejection ratio and input impedance;
- zero offset voltage, bias currents and output impedance.

Otherwise stated: everything that should be high is infinite, and what should be small is zero.

An actual operational amplifier has a number of deviations from the ideal behaviour. These deviations limit the performance of circuits designed with operational amplifiers. It is important to be aware of the limitations of operational amplifiers and to know how to estimate their influence on the configuration built up with these amplifiers.

The specifications of the various available types of operational amplifiers show strong divergence. Table 4.1 gives an overview of the main properties of five different types. The specifications are typical values and are valid for 25°C. Often, also minimum and maximum values are specified. Type I in Table 4.1 is characterized by a very low price; type II is designed for excellent input characteristics (low offset voltage, bias current and noise); type III is suitable for low supply voltage operation (down to 1.8 V, with "rail-to-rail" input and output swing, and less than 250 $\mu$A supply current); type IV is designed for high frequency applications; and type V is a chopper-stabilized amplifier.

The input components of type I are bipolar transistors, of type II JFETs and of types III, IV and V MOSFETs. The favourable high-frequency characteristics of type IV are reflected in the high values for the unity gain bandwidth and the slew-rate. The chopper amplifier has an ultra-low offset; the principle of chopper stabilization is explained in Section 4.3.3.

Note that the specifications given in Table 4.1 are typical values for each class. There exists a variety of operational amplifiers with widely divergent specifications. The designer has to consult the data sheets of manufacturers to find a proper type for the intended application.

The most important imperfections of an operational amplifier are the *offset voltage* and the input *bias currents*: they limit the accuracy of measurements where small

**Table 4.1** Selection of typical specifications for five types of operational amplifiers

| Parameter | Type I | Type II | Type III | Type IV | Type V | Unit |
|---|---|---|---|---|---|---|
| $V_{off}$ | 1 | 0.5 | 1 | 0.5 | adjustable | mV |
| t.c. $V_{off}$ | 20 | 2–7 | 5 | 10 | 0.003 | $\mu$V/K |
| $I_{bias}$ | 80 n | 10–50 p | 250 n | 2 p | 2 p | A |
| $I_{off}$ | 20 n | 10–20 p | 150 n | 2 p | 4 p | A |
| Input noise | 4 | 25 | 25 | 20 | 20 | nV/$\sqrt{\text{Hz}}$ |
|  | – | 0.01 | 2 | – | 0.005 | pA/$\sqrt{\text{Hz}}$ |
| $A_0$ | $2 \cdot 10^5$ | $2 \cdot 10^5$ | $2 \cdot 10^5$ | 1500 | $10^7$ | – |
| $R_i$ | $2 \cdot 10^6$ | $10^{12}$ | $10^{12}$ | $10^{12}$ | – | $\Omega$ |
| CMRR | 90 | 80–100 | 70 | 100 | 140 | dB |
| SVRR | 96 | 80–100 | 90 | 70 | 120 | dB |
| $f_t$ | 1 | 1–4 | 4.3 | 300 | 2 | MHz |
| Slew-rate | 0.5 | 3–15 | 2.5 | 400 | 2 | V/$\mu$s |

$V_{off}$: input offset voltage
$I_{bias}$: bias current
$I_{off}$: current offset, difference between both bias currents
t.c.: temperature coefficient
$A_0$: DC gain
$R_i$: resistance between input terminals
SVRR: Supply Voltage Rejection Ratio
$f_t$: unity gain bandwidth
slew-rate: max. value of $dV_o/dt$.

sensor voltages or currents are involved. Therefore, their effect on the performance of various amplifier configurations will be analysed, by including them in the model of the operational amplifier. The offset voltage $V_{off}$ can be taken into account by putting a voltage source in series with either of the input terminals. Voltage offset originates from asymmetry in the input stage of the amplifier (due to mask tolerances and misalignment). Bias currents are the base or gate currents of the transistors located at the input; they are essentially a consequence of the biasing of the amplifier's transistors. The two bias currents $I_{b1}$ and $I_{b2}$ can be modelled by two current sources (Figure 4.2).

The operational amplifier is nearly always used in combination with a feedback network. An open amplifier will easily run into saturation because of the high voltage gain and the non-zero input offset voltage: its output is either maximal positive or maximal negative, limited by the power supply voltage. The open amplifier acts as a comparator: the output is either high or low, depending on the polarity of the input voltage difference. In all other applications, the operational amplifier is connected in feedback mode, resulting in a lower, but much more stable overall transfer.

Many operational amplifiers are designed in such a way that even with unity feedback stability is guaranteed. Such amplifiers have, however, a relatively small bandwidth and for this reason it is also important to analyse the circuit performance with respect to the finite bandwidth.

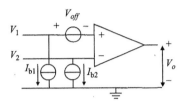

**Figure 4.2** *Model accounting for input offset voltage $V_{off}$ and bias currents $I_{b1}$ and $I_{b2}$*

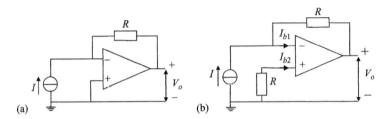

**Figure 4.3** *Current-to-voltage converter: (a) basic configuration; (b) with bias current compensation*

In feedback mode, the voltage difference between both input terminals is (almost) zero: this is a consequence of the infinite gain: at finite output the input must be zero. Further, an ideal operational amplifier has infinite input impedance; hence the input *signal* currents are zero as well. Since both input voltage and currents are zero (ideally), the operational amplifier behaves as a "nullor". This property will be used when analysing interface circuits built up with operational amplifiers.

### 4.1.2. Current-to-voltage converter

Figure 4.3(a) shows the circuit diagram of a current-to-voltage converter with operational amplifier. Assuming ideal properties, the voltage at the inverting input is zero (because the non-inverting input is grounded); hence the transfer of this configuration is simply:

$$V_o = -R \cdot I_i \qquad (4.1)$$

When the gain has a finite value $A$, the transfer is found to be

$$V_o = -\frac{A}{1+A} \cdot R \cdot I_i \qquad (4.2)$$

which, obviously, for $A$ infinity approaches the value in equation (4.1). Non-zero offset voltage and bias current contribute to an additional output voltage:

$$V_o(I_i = 0) = -R \cdot I_b + V_{off} \qquad (4.3)$$

where $I_b$ is the bias current through the inverting input of the operational amplifier. Compensation of the bias current makes use of the amplifier's symmetry: both bias currents have the same direction, and are about equal. Compensation is performed by inserting a resistance with the same value $R$ in series with the non-inverting terminal (Figure 4.3(b)). This resistance produces an output voltage $+R \cdot I_b$, cancelling the first term on the right-hand side of equation (4.3). However, since the two bias currents are not exactly equal, but differ by just the offset current $I_{off}$, the resulting error due to bias currents is $R \cdot |I_{b1} - I_{b2}| = R \cdot |I_{off}|$, which can be an order of magnitude lower than without this compensation.

For voltage noise and current noise similar expressions can be derived. Remember that in the case of uncorrelated noise sources (as is usually the case for the specified voltage and current noise sources of operational amplifiers) their contributions should be added according to RMS values, and so the root of the summed squares. When measuring small voltages, in particular a low voltage offset and low voltage noise is required; the measurement of small currents demand an operational amplifier with low bias and offset current and low current noise.

**Example 4.1 Current-to-voltage converter for a photo-diode**
In this example we design a current-to-voltage converter for the measurement of the photo-current from a photo-diode. The accuracy of the measurement should be better than 1 nA DC and 0.1 nA (RMS) noise over a 10 kHz bandwidth. Further, the maximum photo-current is 20 μA, and the output range of the interface circuit should be 0–10 V.

The range requirement is met with $R = 0.5\,\mathrm{M}\Omega$. Since the direction of the bias current and the offset voltage is not known a priori, the sum of $I_b$ and $V_{off}/R$ must be less than 10 nA. Taking a FET input amplifier (for instance type II in Table 4.1), the bias current can be neglected so the offset voltage may not exceed 0.5 mV (this is just the specified typical offset of that amplifier). Similarly, the noise specifications over the full bandwidth follow from $\sqrt{(V_n^2/R^2 + I_n^2)} < 10\,\mathrm{nA}$. The typical input voltage and current noise over the full bandwidth are 2.5 μV and 1 pA. The current noise is far below the specified maximum; the voltage noise is equivalent to $V_n/R = 0.05\,\mathrm{nA}$, so just small enough for this application.

### 4.1.3. Non-inverting amplifier

Figure 4.4(a) shows the most simple form of a non-inverting amplifier. Using the nullor properties of an operational amplifier in feedback mode, the (low frequency) voltage transfer of the configuration is found to be:

$$\frac{V_o}{V_i} = \frac{1}{\beta} = \frac{R_1 + R_2}{R_1} = 1 + \frac{R_2}{R_1} \tag{4.4}$$

Here, $\beta$ can be identified as the output fraction that is fed back to the input. The gain is determined by the ratio of two resistance values only. The input resistance is infinite, so the circuit is suitable for interfacing with sensors with voltage output.

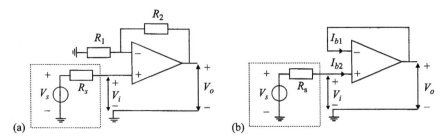

**Figure 4.4** *Non-inverting amplifier connected to a voltage source with resistance $R_s$ (a) arbitrary gain; (b) buffer, gain equals 1; bias currents are also indicated in this figure*

Equation (4.4) only holds for ideal properties of the operational amplifier, in particular for low frequencies only. The AC gain of a frequency compensated amplifier can be approximated by a first order transfer function:

$$A(\omega) = \frac{A_0}{1 + j\omega\tau_a} \tag{4.5}$$

where $A_0$ is the open loop DC gain and $\tau_a$ the first order time constant. Taking into account this finite gain, and applying the general transfer of a feedback system:

$$A_f(\omega) = \frac{A(\omega)}{1 + \beta A(\omega)} \tag{4.6}$$

the frequency dependent transfer of the non-inverting amplifier becomes:

$$\frac{V_o}{V_i} = \frac{A_0}{1 + A_0\beta} \cdot \frac{1}{1 + j\omega[\tau_a/(1 + A_0\beta)]} \tag{4.7}$$

Note that the DC gain is reduced by a factor $1 + A_0\beta$ with respect to the open loop value, and the bandwidth is increased by the same factor. Apparently, the product of gain and bandwidth has a fixed value, independent of $R_1$ and $R_2$. That is why the frequency dependence of the operational amplifier is often specified with its *unity gain bandwidth* (see Table 4.1), that is the bandwidth at unity feedback ($\beta = 1$, gain $= 1$). Note that only operational amplifiers with a first order frequency characteristic have a constant gain-bandwidth product.

For $R_2 = 0$ the transfer equals 1. The same holds for $R_1 = \infty$ (leaving out $R_1$) and for the combination of these cases (Figure 4.4(b)). This configuration is called a *buffer*; it has a voltage gain of 1, an infinite input impedance and a zero output impedance. When the finite gain of the operational amplifier is taken into account, the low-frequency voltage transfer is found to be $A_0/(1 + A_0) \approx 1 - (1/A_0)$, and so a fraction $1/A_0$ less than 1. Accurate buffering requires an operational amplifier with a large gain.

The offset voltage adds up to the input signal (it is in series with the non-inverting input terminal). The bias current through this terminal flows through the source resistance $R_s$, producing an additional input voltage equal to $I_b \cdot R_s$. The bias current through the *inverting* terminal produces an additional output voltage equal to $I_b \cdot R_2$. Accurate amplification of small input voltages requires a low offset voltage and low bias currents.

**Example 4.2 Amplification of a voltage signal**
A sensor produces a voltage between 0 to 10 mV, with a bandwidth from DC to 2 kHz. For a proper matching to an ADC, the voltage range should be enlarged to 0–10 V. The accuracy requirements are 0.1% full scale relative and 0.1% full scale absolute (offset), over a temperature range from 10 to 30°C. Derive the relevant specifications of the operational amplifier.

(a) Gain: the required gain is 10 V/10 mV = 1000, to be realized by a resistance ratio of 999; for a 0.1% accuracy each of these resistors should have 0.05% tolerance.
(b) Bandwidth: 2 kHz at a gain 1000 means 2 MHz unity gain bandwidth.
(c) Offset: 0.1% FS (10 mV) absolute means 10 μV maximum offset; when adjusted to zero at 20°C, the temperature change is ±10°C, so the maximum t.c. of the offset is 1 μV/K.

### 4.1.4. *Inverting amplifier*

Figure 4.5(a) shows an inverting amplifier configuration. Again, using the nullor properties of the operational amplifier, the transfer can easily be found:

$$\frac{V_o}{V_i} = -\frac{R_2}{R_1} \qquad (4.8)$$

Similar to the non-inverting amplifier, the gain of this amplifier is determined by the ratio of two resistance values, resulting in a stable transfer (under the condition of high loop gain). The input resistance equals $R_1$, since the voltage at the inverting input of the operational amplifier is zero. With the model of Figure 4.5(b), the

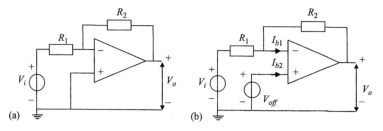

**Figure 4.5** *Inverting amplifier connected to an ideal voltage source ($R_s = 0$): (a) basic circuit; (b) offset voltage and bias currents*

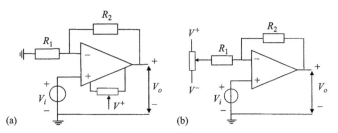

**Figure 4.6** *Offset compensation: (a) using internal compensation facility; (b) by adding an external voltage, derived from the supply voltages $V^+$ and $V^-$*

effects of offset voltage and bias currents can be analyzed:

$$V_o(V_i = 0) = \left(1 + \frac{R_2}{R_1}\right) \cdot V_{off} + I_{b1} \cdot R_2 \qquad (4.9)$$

which means that $V_{off}$ is amplified almost equally as the input signal.

It is possible to compensate for offset voltage and bias currents of the operational amplifier. Figure 4.6 gives two possibilities for offset compensation. Some operational amplifiers have additional terminals for this purpose. This is illustrated in Figure 4.6(a), where a potentiometer is connected to these terminals. The wiper is connected to the power supply voltage (either positive or negative, according the instructions of the manufacturer). In Figure 4.6(b) compensation is obtained by an additional, adjustable voltage derived from the supply voltages, and connected to the input terminal of the operational amplifier.

Compensation of the bias current is performed in a similar way to that in the current-voltage converter, now by inserting a resistance $R_3$ in series with the non-inverting terminal in Figure 4.5. Under the condition that $R_3$ has a value equal to that of $R_1$ and $R_2$ in parallel, the output offset due to bias currents is reduced to $(I_{b1} - I_{b2}) \cdot R_2 = I_{off} \cdot R_2$. All these compensation methods are only partially effective: when after correct compensation the temperature changes, the offset changes too. To estimate this effect the temperature coefficients of the offset voltage and bias currents as given by the manufacturer should be consulted.

### Example 4.3 Design of an inverting amplifier
Build up an amplifier with the following specifications: gain $-150$, minimum input resistance $2\,\text{k}\Omega$, maximum zero error (at the input) $0.5\,\text{mV}$.

A possible design is: $R_1 = 2.2\,\text{k}\Omega$; $R_2 = 150 \cdot 2.2 = 330\,\text{k}\Omega$. To meet the offset requirement, an operational amplifier should be chosen such that $V_o/150$ is less than $0.5\,\text{mV}$. Assume the bias current of the operational amplifier is $10\,\text{nA}$, this contributes to the input offset with only $22\,\mu\text{V}$. The offset voltage of the operational amplifier should therefore be less than $0.5\,\text{mV}$.

When the requirement on the input resistance is more severe, say 2 MΩ, then the contribution of the bias current to the offset is about 1000 times larger (22 mV). Moreover, $R_2$ should be 330 MΩ, which is an unusually large value. An alternative design is the extension of the previous design with a buffer amplifier connected to the input of the inverting amplifier circuit.

Note that the bias current and the input resistance are different properties of the operational amplifier. Even at infinite input resistance, there is a bias current. If an input terminal of an operational amplifier is left open (floating), the bias current can not flow anywhere, so the circuit will not function properly. Similarly, when an input is connected to a capacitor only, the bias current will charge this capacitor until saturation of the amplifier. Bias currents should always find a way to leave the amplifier, for instance via a (very) large resistance connected to ground.

### 4.1.5. *Differential amplifier*

Although an operational amplifier is essentially a differential amplifier, it is not useful as such, because of the very high gain. By proper feedback, a differential amplifier with stable gain can be created. The basic configuration is depicted in Figure 4.7.

The transfer can be calculated by using the superposition theorem: the output voltage is the sum of the outputs due to $V_1$ (at $V_2 = 0$) and $V_2$ (at $V_1 = 0$):

$$V_o = -\frac{R_2}{R_1} \cdot V_1 + \frac{R_4}{R_3 + R_4} \cdot \left(1 + \frac{R_2}{R_1}\right) \cdot V_2 \tag{4.10}$$

A special case occurs when $R_1 = R_3$ and $R_2 = R_4$. Under this condition the output voltage is

$$V_o = -\frac{R_2}{R_1} \cdot (V_1 - V_2) \tag{4.11}$$

which means a differential gain that equals $-R_2/R_1$ and an infinite CMRR (Chapter 3). Unfortunately, in practice the condition of equal resistances will never be fulfilled, resulting in a finite CMRR that can be calculated as follows.

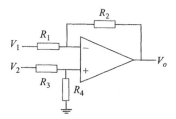

**Figure 4.7** *Differential amplifier; all voltages are relative to ground*

Suppose the relative uncertainty in the resistance value $R_i$ ($i = 1,...,4$) is $\delta_i$, so: $R_i = R_{ni}(1+\delta_i)$, with $R_{ni}$ the nominal value of $R_i$. Further, we assume $R_{n1} = R_{n3}$ and $R_{n2} = R_{n4}$, and $\delta_i \ll 1$.

The CMRR is the ratio between the common mode gain and the differential mode gain (Section 3.5.4). For a pure differential mode signal, $V_1 = -V_2 = \frac{1}{2}V_d$, the transfer is:

$$A_d = \frac{V_o}{V_d} \approx -\frac{R_{n2}}{R_{n1}} \qquad (4.12)$$

For a common mode input signal, $V_1 = V_2 = V_c$, the transfer is:

$$A_c = \frac{V_o}{V_c} = \frac{R_1 R_4 - R_2 R_3}{R_1(R_3 + R_4)} \approx \frac{R_{n2}}{R_{n1} + R_{n2}}(\delta_1 + \delta_4 - \delta_2 - \delta_3) \qquad (4.13)$$

Since the sign of $\delta_i$ is not known, we write $|\delta_i|$ which brings us to the rejection ratio:

$$CMRR = \frac{1 + R_2/R_1}{\sum_i |\delta_i|} \qquad (4.14)$$

### Example 4.4 CMRR of a differential amplifier
Find the CMRR of an amplifier according to Figure 4.7, with gain 100, and resistance tolerances ±1%.

Using equation (4.14), the CMRR becomes $101/(4 \cdot 0.01)$, that is about 2500. This means that a common mode voltage of 10 V produces the same output voltage as a 4.0 mV differential mode input.

The disadvantages of the differential amplifier in Figure 4.7 are a poor CMRR and low input impedances ($R_1$ and $R_2$). Connection of two buffers at both input terminals will increase the input impedances, but decreases the CMRR at the same time: the transfer of both buffers is never the same, and any difference reduces the CMRR. With a few additional components a much better configuration is obtained: the "instrumentation amplifier" (Figure 4.8). To analyze the transfer properties of

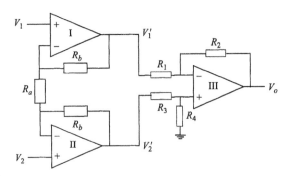

**Figure 4.8** *Instrumentation amplifier; all voltages are relative to ground*

this amplifier, we successively calculate the transfer for pure CM signals and for pure differential signals.

For common mode input voltages, $V_1 = V_2 = V_c$. Both inverting terminals of amplifiers I and II have this potential $V_c$, so no current will flow through $R_a$, and neither through $R_b$. In other words: $V_1'$ and $V_2'$ are equal to $V_c$. So this configuration behaves with respect to CM inputs similar to that of Figure 4.7.

The situation for differential mode (DM) input voltages is different: if $V_1 - V_2 = V_d$, the voltage across $R_a$ is $V_d$ too. The current through $R_a$ equals $V_d/R_a$ and flows through both resistances $R_b$. So $V_1' = V_1 + (R_b/R_a)V_d$ and $V_2' = V_2 - (R_b/R_a)V_d$. This makes the voltage between the inputs of the original differential amplifier (with operational amplifier III) to be $V_1' - V_2' = V_d(1 + 2R_b/R_a)$ which is just $1 + 2R_b/R_a$ times the input voltage.

Summarizing: the DM input is amplified by $1 + 2R_b/R_a$ whereas the CM input remains the same. The result is an increase in CMRR by a factor of $1 + 2R_b/R_a$.

An additional advantage of this circuit over that in Figure 4.7 is the adjustability of the gain by only a single resistance ($R_a$). In the simple configuration the gain must be varied by the simultaneous adjustment of two resistances to maintain a high CMRR.

The instrumentation amplifier has excellent properties with respect to input impedance and CMRR, and is low priced. The three operational amplifiers are integrated in a single chip; common terminals (for instance power supplies) are internally connected on the chip, reducing the number of external terminals. For these reasons the "instrumentation amplifier" is often used for the measurement of (small) differential voltages.

When the requirements with respect to offset voltage, bias currents, input impedance and noise cannot be met by a configuration with operational amplifiers, special amplification techniques should be considered, for instance chopper stabilized amplifiers. The basics of such amplifiers will be discussed in Section 4.3 on modulation and detection.

## 4.2. Analogue filters

An electronic filter enables the separation of signals based on the frequency of the constituent signal components. We distinguish between passive and active filters. Advantages of passive filters are:

- high linearity (passive components are highly linear);
- wide voltage and current range (not restricted by a supply voltage);
- no power supply required.

Disadvantages of passive filters are:

- the filter properties required cannot always be combined with other requirements, for instance with respect to the input and output impedances;
- inductors for applications at low frequencies are rather bulky; furthermore, an ideal (self)inductance is difficult to realize;
- not all kind of filter characteristics can be made with only resistors and capacitors.

There are four main filter types: lowpass, highpass, bandpass and bandreject or notch filters. Signal components with frequencies lying within the passband pass the filter unaltered, other frequency components (in the stopband) are attenuated as much as possible. Filters with an ideal frequency characteristic, that is a flat passband up to the cutoff frequency and zero transfer beyond it, do not exist. The selectivity of first order filters is poor: the slope of the frequency characteristic is only $\pm 20$ dB per decade ($\pm 6$ dB per octave). The selectivity can be improved by cascading a number of first order filters. The slope of a filter consisting of $n$ first order sections in series is $n \cdot 20$ dB per decade. However, the transfer within the passband is also affected, by mutual loading of the filter sections. A better approximation of the ideal filter characteristic is achieved using active circuits. According to the approximation criterion at a specified order, several filter types are available. The major filter types are:

- Butterworth filter: maximally flat in the passband;
- Chebysjev filter: steepest slope in the stopband from the cutoff frequency;
- Bessel filter: fastest step response.

A discussion on the mathematical background of these filters is beyond the scope of this book: the reader is referred to the literature on this topic.

An operational amplifier is a very useful device for the realization of analogue filters. As a starting point consider the configurations of Figure 4.4(a) and Figure 4.5(a) which are shown once more in Figure 4.9. The resistances $R_1$ and $R_2$ have been replaced by impedances $Z_1$ and $Z_2$. The complex transfer functions of these

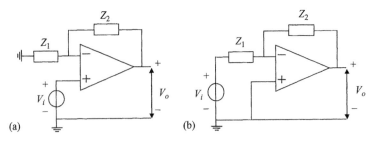

**Figure 4.9** *Basic filter configurations: (a) non-inverting transfer; (b) inverting transfer*

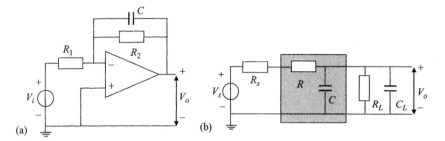

**Figure 4.10** *First order lowpass filter: (a) active; (b) passive (grey part), with source and load circuits*

circuits are:

$$\frac{V_o}{V_i} = 1 + \frac{Z_2}{Z_1} \quad (4.15)$$

and

$$\frac{V_o}{V_i} = -\frac{Z_2}{Z_1} \quad (4.16)$$

respectively. Various filter characteristics can be realized just by a proper choice of the impedances $Z_1$ and $Z_2$. This will be illustrated with the configuration of Figure 4.5(b).

### 4.2.1. *First order lowpass filter and integrator*

Let $Z_1$ in Figure 4.9(b) be a resistance $R_1$ and $Z_2$ a resistance $R_2$ in parallel to a capacitance $C$ (Figure 4.10(a)). The transfer function of this configuration appears to be

$$\frac{V_o}{V_i} = -\frac{Z_2}{Z_1} = -\frac{R_2}{R_1} \cdot \frac{1}{1 + j\omega R_2 C} \quad (4.17)$$

This filter has a low-frequency transfer of $-R_2/R_1$, a cutoff frequency at $\omega = 1/R_2C$, and a transfer in the stopband decreasing with 20 dB per decade (Figure 4.11(a)). Both the cutoff frequency and the low-frequency transfer can be given an arbitrary value (within limits) by the values of $R_1$, $R_2$ and $C$. An additional requirement of the minimum input impedance fixes these three component values.

Note that a possible source resistance will affect the cutoff frequency of the filter, since it is in series with the resistance of the filter. The filter will not change its transfer due to loading, since its output impedance is very low. The transfer of a passive lowpass filter on the contrary (Figure 4.11(b)) is sensitive to both source and load impedances (Figure 4.11(b)): a source resistance $R_s$ and a load capacitance

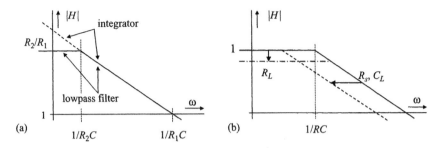

**Figure 4.11** Bode plots (amplitude characteristic only): (a) active lowpass filter and integrator; (b) passive lowpass filter and effects of source and load

$C_L$ will shift the stopband asymptote to the left (lowering the cutoff frequency); a load resistance $R_L$ will shift the passband asymptote downwards (increasing the cutoff frequency).

### Example 4.5 Design of a lowpass filter
A lowpass filter has to be designed with a LF transfer of 10, a minimum input resistance (low frequencies) of $2\,k\Omega$, and a cutoff frequency at 1600 Hz.

Make $R_1 = 2.2\,k\Omega$; so $R_2 = 22\,k\Omega$, and with $2\pi \cdot 1600 \cdot R_2 C = 1$ for $C$ the value 4.5 nF follows. If a much higher input impedance were required, for instance $200\,k\Omega$, then $R_2$ should be a 100 times larger and $C$ a 100 times smaller. Sometimes the requirements can result in very unpractical values. In such a case the functional demands should be spread over two or more circuits in series.

Without resistance $R_2$ the circuit behaves as an *integrator* with transfer function $V_o/V_i = -1/j\omega R_1 C$; in the time domain the transfer is: $v_o = -(1/R_1 C) \int v_i \, dt$. In practice this circuit will *not* function properly: the bias current and the offset voltage of the operational amplifier (which are never zero) will be integrated as well, resulting in a saturated output after a short time. A real integrator is always "tamed", by a resistance in parallel to $C$, as shown in Figure 4.11(a). So, a real integrator is nothing else than an active lowpass filter with very low cutoff frequency.

### Example 4.6 Design of an integrator
We want an integrator with transfer 1 at 10 kHz, and an integrating range from 10 Hz onwards. Find proper values for the capacitance and resistances.

According to Figure 4.11(a), $2\pi \cdot 10^4 \cdot R_1 C = 1$, to be realized by for instance $R_1 = 10\,k\Omega$ (reasonable input resistance) and $C = 1.6\,nF$. The transfer at 10 kHz is 1, so at 10 Hz (the cutoff frequency) the transfer is 1000. This is achieved by a resistance $R_2 = 10\,M\Omega$. Note that an offset voltage of 1 mV in this design results in an output offset of 1 V.

The frequency range over which signals are integrated is not only restricted to lower frequencies (by the taming resistance) but also to higher frequencies, due to the finite, frequency dependent transfer of the operational amplifier. An integrator for higher frequencies requires the application of a wide-band operational amplifier.

### 4.2.2. First order highpass filter and differentiator

Let $Z_1$ in Figure 4.9(b) be a resistance $R_1$ in series with a capacitance $C$, and $Z_2$ a resistance $R_2$ (Figure 4.12a). The transfer function of this circuit is:

$$\frac{V_o}{V_i} = -\frac{Z_2}{Z_1} = -\frac{R_2}{R_1} \cdot \frac{j\omega R_1 C}{1 + j\omega R_1 C} \tag{4.18}$$

In the frequency range $\omega \gg 1/R_1 C$ the transfer equals $-R_2/R_1$, whereas from DC to the cutoff frequency the amplitude transfer is proportional to the frequency (slope +20 dB per decade), as shown in Figure 4.13(a). Again, the two most important parameters of the filter characteristic (cutoff frequency and gain in the passband) are determined by the three component values.

Similar to the active lowpass filter, a load impedance will hardly affect the transfer characteristics of the active highpass filter; a source resistance will lower the cutoff frequency because it is in series with resistance $R_1$. The sensitivities of source and load impedances to the characteristics of a passive highpass filter (Figure 4.12(b)) are shown in Figure 4.13(b).

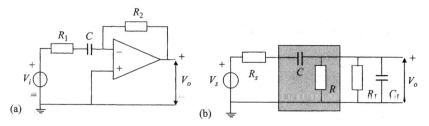

**Figure 4.12** First order highpass filter: (a) active; (b) passive (grey part) with source and load circuits

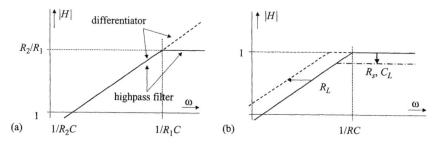

**Figure 4.13** Bode plots (amplitude characteristic only): (a) highpass filter and differentiator; (b) passive highpass filter and effects of source and load

For $R_1 = 0$, the circuit behaves as a *differentiator*, with complex transfer $V_o/V_i = -j\omega R_2 C$. The corresponding transfer in the time domain is $v_o = -(R_2 C) dv_i/dt$. Note that in the ideal case the transfer increases proportionally with frequency, which inevitably causes instability. So, as in the case of the integrator, the differentiator too must be "tamed": its gain should be limited from a specific frequency onwards. This is achieved by, for instance, a resistance in series with $C$, which brings us back to the highpass filter from Figure 4.12(a).

Even tamed in this way, instability may occur due to the second order behaviour of an operational amplifier (even when frequency-compensated, at sufficiently high frequencies the transfer is of a higher order). In such a case the gain above the cutoff frequency should further be diminished, for instance by inserting a second capacitor in parallel to the feedback resistance in Figure 4.12(a).

### Example 4.7 Design of a differentiator

We want a differentiator with transfer 1 at 10 Hz and a differentiating character from DC up to 100 Hz. Find proper values for the passive components.

According to (4.18) the cutoff frequency is located at $1/(2\pi \cdot R_1 C)$, so $2\pi \cdot R_1 C = 10^{-2}$. This is achieved by (for instance) $C = 100$ nF and $R_1 = 16$ k$\Omega$. To obtain a transfer 10 at 10 Hz, $R_2$ should be ten times $R_1$, so 160 k$\Omega$. Note that offset is not a serious problem in this design: the transfer at DC is zero. The input impedance is frequency dependent, but above the cutoff frequency (100 Hz) almost constant, and equal to $R_1$ (16 k$\Omega$).

When taking into account the frequency dependent transfer of the operational amplifier, a sharp peak in the characteristic is found, around 100 Hz. Since this is beyond the differentiation range, it is wise to further reduce the gain for frequencies higher than 100 Hz, for instance by introducing a second cutoff point at 500 Hz. This is realized by a capacitance $C_2$ in parallel to $R_2$. With $2\pi \cdot R_2 C_2 = 1/500$, the value of $C_2$ is found to be 2 nF.

### 4.2.3. Second order bandpass filter

A passive bandpass filter can be made by connecting a lowpass and a highpass filter in series (Figure 4.14(b)). When mutual loading of these sections is disregarded,

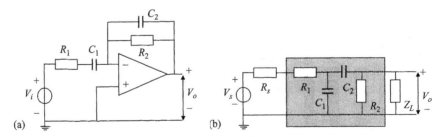

**Figure 4.14** *Second order bandpass filter: (a) active; (b) passive filter*

the total transfer is just the product of that of the low- and highpass filters, yielding a bandpass characteristic. Obviously, a load impedance may influence substantially the bandpass characteristic. This is not the case for the active bandpass filter shown in Figure 4.14(a). The transfer function of this filter with just one operational amplifier is found using the general expression in (4.15):

$$\frac{V_o}{V_i} = -\frac{Z_2}{Z_1} = \frac{j\omega R_2 C_1}{(1 + j\omega R_1 C_1)(1 + j\omega R_2 C_2)}$$
$$= -\frac{R_2}{R_1} \cdot \frac{j\omega R_1 C_1}{1 + j\omega R_1 C_1} \cdot \frac{1}{1 + j\omega R_2 C_2} \quad (4.19)$$

The second line of this equation clearly shows that this transfer can be written as the product of a highpass and a lowpass characteristic, with cutoff frequencies at $\omega = 1/R_1 C_1$ and $\omega = 1/R_2 C_2$, respectively. If these cutoff frequencies are far apart ($1/R_1 C_1 \ll 1/R_2 C_2$) the passband transfer is about $-R_2/R_1$.

**Example 4.8 Design of a bandpass filter**
Make a design of an active bandpass filter, with passband from 150 Hz to 20 kHz and a passband transfer 10.

The values of the components should satisfy:

- $R_2 = 10 \cdot R_1$ (passband transfer);
- $2\pi \cdot 150 \cdot R_1 C_1 = 1$ (highpass cutoff frequency);
- $2\pi \cdot 20000 \cdot R_2 C_2 = 1$ (lowpass cutoff frequency).

A possible design is: $R_1 = 10\,\text{k}\Omega$, $R_2 = 100\,\text{k}\Omega$, $C_1 = 100$ nF and $C_2 = 80$ pF.

This bandpass filter has a moderate selectivity: the slopes in the frequency characteristic are just 20 dB/decade up and down. Bandpass filters with a much better selectivity are based on resonance effects, like in an LC-filter. Inductorless resonance is achieved by frequency-selective positive feedback. One example from this category is the Wien filter, a single-operational amplifier circuit with only resistors and capacitors. This and all other second-order resonance filters have a transfer function that can be described by three parameters: the resonance frequency, the maximum amplitude transfer (at resonance) and the quality factor (Q) or damping factor. The Q-factor is set by the difference of two resistance values. The design of a high-Q bandpass filter is rather critical, because the sensitivity for parameter changes is proportional to the Q-factor itself. Furthermore, the finite and frequency dependent gain of the operational amplifier affects the filter performance significantly, in particular when the filter operates in a frequency range where this gain had dropped to much lower values [4.1, 4.2].

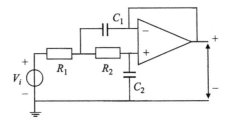

**Figure 4.15** *Second order Butterworth filter (unity gain Sallen-Key configuration)*

### 4.2.4. Second and third order lowpass filter with Butterworth characteristic

The filter in Figure 4.15 is a second order lowpass, according to the "Sallen and Key" configuration [4.3]. The slope in the stopband amounts to 40 dB per decade, so twice as steep as a first order lowpass.

The calculation of the transfer function is a little tedious. Applying the nullor properties of the (ideal) operational amplifier and Kirchhoff's laws, the transfer appears to be:

$$\frac{V_o}{V_i} = \frac{1}{1 + j\omega(R_1 + R_2)C_2 - \omega^2 R_1 R_2 C_1 C_2} \quad (4.20)$$

With proper component values the characteristic can be made such that it satisfies the Butterworth condition, which means a maximally flat transfer in the passband. Mathematically, the Butterworth condition is:

$$\left|\frac{V_o}{V_i}\right| = \frac{1}{\sqrt{1 + (\omega/\omega_k)^{2n}}} \quad (4.21)$$

in other words: the denominator of the transfer function contains only even powers of $\omega$. The Sallen-Key filter has Butterworth properties if $C_1/C_2 = (R_1 + R_2)^2/2R_1 R_2$. In a simple design, we choose $R_1 = R_2 = R$, hence $C_1 = 2C_2$.

The selectivity of the filter from Figure 4.15 can easily be increased by connecting a (passive) lowpass RC-filter at its output. To be more robust against loading errors a buffer amplifier or non-inverting amplifier could be added. The transfer of this third order filter is the product of the transfer given in equation (4.20) and the transfer of the lowpass network: $1/(1 + j\omega R_3 C_3)$, with $R_3$ and $C_3$ the component values of this RC-network. Again, the total configuration has a Butterworth character if the transfer satisfies the condition as given in equation (4.21). There is some freedom in choosing the component values. Taking $R_1$, $R_2$ and $R_3$ equal to $R$, then with $C_1 = 4C_2$ and $C_3 = 2C_2$ the filter is maximally flat in the passband.

The circuits in this section are just a few examples of analogue filters that are built up with operational amplifiers, resistors and capacitors only. There are many other possible designs, and the reader is referred to the literature for further orientation on this subject (Section 4.4).

## 4.3. Modulation and detection

Modulation is a particular type of signal conversion, that makes use of an auxiliary signal, the *carrier*. One of the parameters of this carrier signal is varied correspondingly to the input (or measurement) signal. The result is a shift of the complete signal frequency band to a position around the carrier frequency. Due to this property, modulation is also referred to as *frequency conversion*.

The modulated waveform offers various advantages over the original waveform and is, for that reason, widely used in electronic signal processing systems. One application of modulation is frequency multiplexing, a method for a more efficient way of signal transmission. A very important advantage of modulated signals is their better noise and interference immunity. In measurement systems modulation offers the possibility of bypassing offset and drift from amplifiers.

The carrier is a signal with a simple waveform, for instance a sine wave, square wave, or pulse shaped signal. Several parameters of the carrier can be modulated by the input signal, for instance the amplitude, the phase, the frequency, the pulse height or the pulse width. These types of modulation are referred to as amplitude modulation (AM), phase modulation, frequency modulation (FM), pulse height and pulse width modulation, respectively. Amplitude modulation is a powerful technique in instrumentation to suppress interference signals; therefore, this chapter is confined is to amplitude modulation only.

### 4.3.1. *Frequency spectrum of AM signals*

A general expression for an AM signal with a sinusoidal carrier modulated by an input signal $v_i(t)$ is:

$$v_m = \hat{v}_c\{1 + k \cdot v_i(t)\} \cos \omega_c t \qquad (4.22)$$

where $\omega_c$ is the frequency of the carrier and $k$ a coefficient determined by the modulator. Suppose the input signal is a pure sine wave:

$$v_i = \hat{v}_i \cos \omega_i t \qquad (4.23)$$

The modulated signal is:

$$\begin{aligned} v_m &= \hat{v}_c\{1 + k \cdot \hat{v}_i \cos \omega_i t\} \cos \omega_c t \\ &= \hat{v}_c \cos \omega_c t + \tfrac{1}{2} k \cdot \hat{v}_c \hat{v}_i \{\cos(\omega_c + \omega_i) + \cos(\omega_c - \omega_i) t\} \end{aligned} \qquad (4.24)$$

which shows that this modulated signal has three frequency components: one with the carrier frequency ($\omega_c$), one with a frequency equal to the sum of the carrier

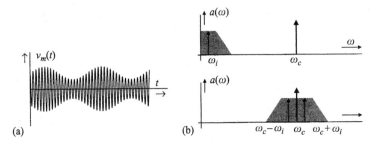

**Figure 4.16** *Amplitude modulated signal: (a) time signal; (b) frequency spectrum; top: input signal; bottom: modulated signal. Grey areas: signal spectrum; arrows: a single component showing the relative position in the spectrum*

frequency and the input frequency ($\omega_c + \omega_i$) and one with the difference between these two frequencies ($\omega_c - \omega_i$). Figure 4.16(a) shows an example of such an AM signal. The input signal is still recognized in the "envelope" of the modulated signal, although its frequency component is not present.

When the carrier is modulated by an arbitrary input signal, each of the frequency components produces two new components, with the sum and the difference frequency. Hence, the whole frequency band is shifted to a region around the carrier frequency (Figure 4.16(b)). These bands at either side of the carrier are called the *side bands* of the modulated signal. Each side band carries the full information content of the input signal. The AM signal does not contain low frequency components anymore. Therefore, it can be amplified without being affected by offset and drift. If such signals appear anyway, they can easily be removed from the amplified output by a highpass filter.

There are many ways to modulate the amplitude of a carrier signal. We will discuss three methods:

- multiplying modulator
- switch modulator
- bridge modulator.

*Multiplication* of two sinusoidal signals $v_c$ and $v_i$ (carrier and input) results in an output signal:

$$v_m = K \cdot \hat{v}_c \hat{v}_i \{\cos(\omega_c + \omega_i)t + \cos(\omega_c - \omega_i)t\} \tag{4.25}$$

with $K$ [$V^{-1}$] the scale factor of the multiplier. This signal contains only the two side band components and no carrier; it is called an AM signal with suppressed carrier. For arbitrary input signals the spectrum of the AM signal consists of two (identical) side bands without carrier. Figure 4.17(a) shows an example of such an AM signal. Note that the "envelope" is not identical to the original signal shape

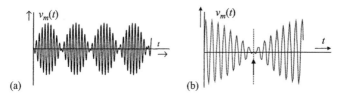

**Figure 4.17** *AM signal with suppressed carrier: (a) sinusoidal carrier modulated with a sine wave signal; (b) phase shift where the "envelope" crosses zero (at arrow)*

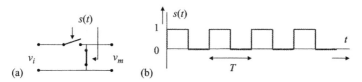

**Figure 4.18** *(a) series-shunt switch as modulator; (b) time representation of the switch signal*

anymore. Further, the AM signal shows a phase shift in the zero crossings of the original input signal (Figure 4.17(b)).

In the *switch modulator* the measurement signal is periodically switched on and off, a process that can be described by multiplying the input signal with a switch signal $s(t)$, being 1 when the switch is on, and 0 when it is off (Figure 4.18).

To show that this product is indeed a modulated signal with side bands, we expand $s(t)$ into its Fourier series:

$$s(t) = \frac{1}{2} + \frac{2}{\pi}\left\{\sin \omega t + \frac{1}{3}\sin 3\omega t + \frac{1}{5}\sin 5\omega t + \cdots\right\} \quad (4.26)$$

with $\omega = 2\pi/T$, and $T$ the period of the switching signal. For a sinusoidal input signal with frequency $\omega_i$, the output signal contains sums and differences of $\omega_i$ and each of the components of $s(t)$.

This modulation method produces a large number of side band pairs, positioned around odd multiples of the carrier frequency ($\omega_c$, $3\omega_c$, $5\omega_c$, ...), as shown in Figure 4.19. The low-frequency component originates from the multiplication by the mean of $s(t)$ (here 1/2). This low-frequency component and all components with frequencies $3\omega_c$ and higher can be removed by a filter. The resulting signal is just an AM signal with suppressed carrier. A similar modulator can be achieved by periodically changing the *polarity* of the input signal. This is equivalent to the multiplication by a switch signal with zero mean value; in that case there is no low-frequency band as in Figure 4.19.

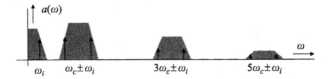

**Figure 4.19** *Spectrum of an AM signal from a switching modulator. Grey areas: the whole spectrum; arrows: a single component to show the relative position*

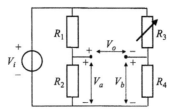

**Figure 4.20** *Wheatstone bridge as modulator*

Advantages of the switch modulator are its simplicity and accuracy: the side band amplitude is determined only by the quality of the switch. The absence of DC and low-frequency components considerably facilitates the amplification of modulated signals: offset, drift and low-frequency noise can be kept far from the new signal frequency band. When very low voltages must be measured it is recommendable to modulate these prior to any other analogue signal processing that might introduce DC errors. An application of this concept is encountered in the *measurement bridge*, that can be considered as a third modulation method. The principle is illustrated with the resistance measurement bridge or Wheatstone bridge of Figure 4.20.

The bridge is connected to an AC signal source $V_i$; the AC signal (usually a sine or square wave) acts as the carrier. In this example we consider a bridge with only one resistance ($R_3$) that is sensitive to the measurand. Assuming fixed, equal values of the three other resistances, the signal $V_a$ is just half the carrier, whereas $V_b$ is an AM signal: half the carrier modulated by $R_3$. The bridge output $V_o$ is the difference between these two signals, so an AM signal with suppressed carrier.

This output can be amplified by a differential amplifier with high gain; its low frequency properties are irrelevant; the only requirements are a sufficiently high bandwidth and a high CMRR for the carrier frequency to accurately amplify the difference $V_a - V_b$.

### Example 4.9 Design of a carrier frequency measurement system

The output of a measurement system is a voltage that varies linearly with the measurand. The maximum signal (full scale) is $10\,\mu$V and the required accuracy is 1% full scale. The signal bandwidth is DC – 10 Hz. Since offset and offset drift of available amplifiers is too high to meet the requirements, a carrier frequency system is designed. Find a proper carrier frequency for this system.

One percent accuracy of $10\,\mu V$ means $0.1\,\mu V$ stability of the input offset, which makes modulation necessary indeed. The carrier frequency should be substantially larger than the maximum signal frequency (10 Hz). A 1 kHz carrier frequency would be sufficiently high to allow suppression of additional bands in a switched modulator. For suppression of possible interference from the mains (50/60 Hz) it is recommended not to choose a multiple of this frequency.

### 4.3.2. Demodulation

The reverse process of modulation is *demodulation* (sometimes called *detection*). Looking at the AM signal with carrier (for instance in Figure 4.17) we observe the similarity between the envelope of the amplitude and the original signal shape. An obvious demodulation method would therefore be *envelope detection* or *peak detection*. Although this method is applied in radio receivers for AM signals and requires only a diode and a capacitor, it is not recommended for measurement signals because of the inaccuracy and noise sensitivity: each transient is considered as a new peak belonging to the signal. A better demodulation method is (single or double-sided) rectification followed by a lowpass filter. When properly designed, the output follows the envelope, and thus corresponds with the amplitude of the original measurement signal. The value of the cutoff frequency of the detector is a compromise between the remaining ripple at the output and the ability to follow a declining input amplitude.

Obviously, the envelope detectors operate only for AM signals *with* carrier. In an AM signal *without* carrier, the envelope is not a copy of the input anymore. Apparently, additional information is required with respect to the phase of the input, for a full recovery of the original waveform (Figure 4.17(b)).

A very effective method to solve this problem, and which has a number of additional advantages too, is *synchronous detection*. This method consists of multiplying the AM signal by a signal having the same frequency as the carrier. If the carrier signal is available (as is the case in most measurement systems) this synchronous signal can be the carrier itself or can be derived from it.

Assume a modulated sinusoidal input signal with suppressed carrier:

$$v_m = \hat{v}_m \{\cos(\omega_c + \omega_i)t + \cos(\omega_c - \omega_i)t\} \tag{4.27}$$

This signal is multiplied by a synchronous signal with a frequency equal to that of the original carrier, and a phase angle $\varphi$. The result is:

$$\begin{aligned} v_{dem} &= \hat{v}_m \hat{v}_s \{\cos(\omega_c + \omega_i)t + \cos(\omega_c - \omega_i)t\} \cdot \cos(\omega_c t + \varphi) \\ &= \hat{v}_m \hat{v}_s \left[\cos \omega_i t \cos \varphi + \tfrac{1}{2} \cos\{(2\omega_c + \omega_i)t + \varphi\} \right. \\ &\quad \left. + \tfrac{1}{2} \cos\{(2\omega_c - \omega_i)t + \varphi\}\right] \end{aligned} \tag{4.28}$$

With a lowpass filter, the components around $2\omega_c$ are removed, leaving the original component with frequency $\omega_i$. This component has a maximum value for $\varphi = 0$,

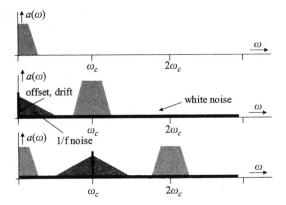

**Figure 4.21** *Signal spectra in a sequence of modulation and synchronous detection*

that is, when the synchronous signal has the same phase as the carrier. For $\varphi = \pi/2$ the demodulated signal is zero, and it has the opposite sign for $\varphi = \pi$. This phase sensitivity is an essential property of synchronous detection. Moreover, it allows phase measurements as well (Section 4.3.3).

Figure 4.21 reviews the whole measurement process, for a bridge type modulating system. The starting point is a low frequency, narrow band measurement signal (top). This signal is modulated and subsequently amplified. The spectrum of the resulting modulated signal is depicted in the middle. Also some additional error signals introduced by the amplifier are shown: offset (at DC), drift and 1/f noise (LF) and wide band thermal noise. Much of these interfering signals can possibly be filtered out using a bandpass filter around the spectrum of the modulated measurement signal: a *pre-detection filter*. The bottom part shows the spectrum of the demodulated signal. By multiplication with the synchronous signal, all frequency components are converted to a new position. The spectrum of the amplified measurement signal folds back to its original position and the LF error signals are converted to a higher frequency range. A lowpass filter removes all components with frequencies higher than that of the original band.

An important advantage of this detection method is the elimination of all error components that are not in the (small) band of the modulated measurement signal. If the measurement signal has a narrow band (slowly fluctuating measurement quantities), a low cutoff frequency of the filter can be chosen. Hence, most of the error signals are removed, and a tremendous improvement of the SNR is achieved, even with a simple first order lowpass filter in the demodulator.

**Example 4.10 Noise performance of a synchronous detector**
The modulated signal from Example 4.7 is amplified by an operational amplifier of type II in Table 4.1. We want to analyse the noise performance of the system.

The amplifier equivalent input voltage noise is $25\,\text{nV}/\sqrt{\text{Hz}}$. We assume that this is the contribution of thermal noise around the carrier frequency. Since the signal bandwidth is 10 Hz, the bandwidth of the LP filter after demodulation can be set to 10 Hz too. The remaining in-band noise covers 20 Hz (two sidebands), and amounts to $25 \cdot \sqrt{20} = 112\,\text{nV}$. This is just a little more than 1% full scale.

### 4.3.3. Measurement systems based on synchronous detection

The preceding section showed that synchronous detection is a powerful mechanism in instrumentation to measure small AC signals with a low signal-to-noise ratio. Synchronous detection is applied in many measurement instruments, for instance network analysers and impedance analysers (Chapter 7). In such instruments the measurement signals are all sinusoidal (the analysis takes place in the frequency domain). Synchronous detection is also used in special types of amplifiers, like the lock-in amplifier and the chopper amplifier. Their principles are explained in the next sections.

**Lock-in amplifiers**

A lock-in amplifier is an AC amplifier based on synchronous detection and intended to measure the amplitude and phase of small, noisy, narrow-band measurement signals. A simplified block diagram of a lock-in amplifier is depicted in Figure 4.22.

The amplifier has two input channels: the signal channel and the reference channel. The signal channel is composed of AC amplifiers and a bandpass filter. This filter is used to remove part of the input noise prior to synchronous detection: the *pre-detection filter*. Depending on the type of lock-in amplifier, it is a manually adjustable filter or an automatic filter. The reference channel is composed of an amplifier, an adjustable phase shifter and a comparator. The comparator converts the reference signal into a square wave for a proper control of the switches in the (switching) synchronous detector. The adjustable phase shifter allows maximizing of the amplifier's sensitivity when the input signal and the reference signal are out

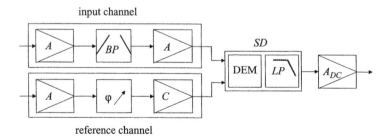

**Figure 4.22** Block diagram of a lock-in amplifier system: $A$ = AC-amplifier; BP = adjustable bandpass filter; C = comparator; SD = synchronous detector; DEM = demodulator, LP = lowpass filter, $A_{DC}$ = DC amplifier

of phase, see equation (4.28). With a calibrated phase shifter, the original phase difference between the input and reference signal can be measured.

An adjustable low-drift DC amplifier brings the output to the proper level for the displaying of the measurement value. The output signal is a DC or slowly varying signal proportional to the amplitude of the input signal.

The cutoff frequency of the lowpass filter at the output of the synchronous detector fixes the bandwidth and thus the measurement time of the whole system. This filter can be set manually, according to the user's wish. For a small SNR of the input signal, the filter should have a low cutoff frequency, to suppress as much noise as possible. However, it should be high enough to not attenuate the input signal itself. In other words: the bandwidth of the lowpass filter should match the bandwidth of the input signal. The operator should make a trade-off between measurement time and noise suppression. When the input signal has a very small bandwidth (a sine wave with an almost constant amplitude) a long measurement time can be chosen; in this way very small signals (down to 1 nV) can be measured.

A lock-in amplifier as described here is suitable for the measurement of AC signals of which a synchronous signal is available (like with a bridge measurement or a transfer measurement). If there is no such a synchronous signal, it must still be produced. This is explained by the diagram in Figure 4.23, where a PLL is used for the generation of the synchronous signal.

A PLL or *phase-locked-loop* is a control system with a synchronous detector and a VCO (*voltage controlled oscillator* with sinusoidal or square wave signal) in a feedback loop. Let the major frequency component of the measurement signal (a narrow band sinusoidal measurement signal) be $\omega_c$. Assume further the (fundamental) frequency of the VCO is $\omega_{VCO}$. When both these signals are multiplied and the product is lowpass filtered, the result is zero only when both frequency components are equal and are $\pi/2$ out of phase. This situation is achieved by feedback of the filtered product to the control input of the VCO. In the steady state, the output frequency of the VCO equals the frequency of the input signal. Since this output is now a noise-free sine or square wave, it is suitable as a reference signal for the lock-in amplifier.

### Chopper amplifiers

A chopper amplifier or chopper stabilized amplifier is a special type of amplifier for the measurement of very small DC voltages or low frequency signals. To get rid

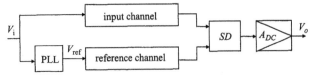

**Figure 4.23** *Lock-in amplifier with self-generating reference signal by applying a PLL*

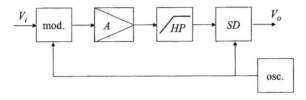

**Figure 4.24** *Block diagram of a chopper stabilized amplifier; HP = highpass filter*

of the offset and drift, as encountered in normal DC amplifiers, the measurement signal is first modulated using a switch modulator, then amplified and finally the signal is demodulated by synchronous detection (Figure 4.24).

Chopper amplifiers are also called indirect DC amplifiers for obvious reasons. Another name is *electrometer*, named after its predecessor, an electrometer tube, a special device that performed the modulation of the input signal. Chopper amplifiers are available as modules or integrated circuits. Table 4.1 (column "type V") shows for comparison some specifications of an integrated chopper stabilized operational amplifier. In this device the chopping is performed by CMOS transistors, with a switching frequency of about 500 Hz. The most striking advantage of the chopper amplifier is the combination of a low input offset voltage drift and a low bias current; their non-zero values are due mainly to imperfections of the switches.

## 4.4. Further reading

There is an overwhelming amount of literature on analogue signal conditioning, as text books and as papers in scientific journals. Many books on "instrumentation" discuss a wide spectrum of topics, including sensors, conditioning and conversion. Some of them are already cited in preceding chapters. A few more books containing one or more aspects of signal conditioning in the analogue domain are mentioned below, and a longer list about filters in particular.

R.G. Irvine, *Operational Amplifier Characteristics and Applications*, Prentice Hall, 1994, ISBN 0-13-606088-9.
A useful and well readable book on operational amplifiers. It shows how to design not only linear circuits, but also non-linear circuits; digital circuits; filters and oscillators using the operational amplifier as a building block. The book also contains a chapter on the specifications of operational amplifiers, and gives a view on the inside of such devices.

M.L. Meade, *Lock-in Amplifiers: Principles and Applications*, Peregrinus, on behalf of the IEE, 1983, ISBN 0-906048-94-x.
This work fully covers the topic of lock-in amplifiers and the underlying detection principles. Although over 20 years old, it is still useful to learn about the background of this powerful technique to detect small and noisy signals.

K.L. Su, *Analog Filters*, Chapman & Hall, 1996, ISBN 0-412-63840-1.
A book on active filters at the undergraduate level. Starting with a few chapters on approximations of filter characteristics and on passive LC ladder filters, it continues with operational amplifier based analogue filters. There is a separate chapter on sensitivity analysis and another on switched capacitance filters. Examples and solutions are included. Recommended when starting to study the subject.

G.H. Tomlinson, *Electrical Networks and Filters: Theory and Design*, Prentice Hall, 1991, ISBN 0-13-248261-4(3).
This work covers most relevant topics on network theory and filter design in less than 300 pages. Consequently the discussion is basic, but useful. Although the major part of the book is devoted to network analysis (poles and zero's, frequency responses, synthesis of passive networks) there are chapters on analogue and digital filters and operational amplifiers as well.

R. Schaumann, M.S. Ghausi and K.R. Laker, *Design of Analog Filters: Passive, Active RC, and Switched Capacitor*, Prentice Hall, 1990, ISBN 0-13-201591-9.
A more advanced treatment of filter design. The book is structured as follows: LC filters (ladder, allpass); sensitivity analysis; operational amplifiers and circuits; second order active filters; higher order active filters; fully integrated filters (continuous-time); and switched capacitance filters. Many end-of chapter problems, but no solutions.

J.T. Taylor and Q. Huang, *CRC Handbook of Electrical Filters*, CRC Press, 1997, ISBN 0-8493-8951-8.
This multi-authored book not only covers digital and continuous-time active filters, but also discusses a few particular topics: switched capacitance filters, electromechanical filters (SAW, crystal) and computer aided design methods. For more advanced study.

T. Deliyannis, Y. Sun and J. Kel Fidler, *Continuous-Time Active Filter Design*, CRC Press, 1999, ISBN: 0-8493-2573-0.
The first sentence of the preface to this book is: "In this digital age, who needs continuous-time filters?". It might be clear at the end of this chapter that measurement and instrumentation still need filtering in the analogue domain. The book deals with the analysis and design of numerous filter types and structures. It also provides useful information on transfer functions, sensitivity and stability analysis, and active electronic circuits (5 chapters on filters based on OTA's), underpinning the design of filters with a specified transfer characteristic. Many references are cited at the end of each chapter. The book can be considered as a standard work on analogue filters, recommended for both students and professionals in measurement and instrumentation.

## 4.5. Exercises

1. In the instrumentation amplifier of Figure 4.8 the resistance values are: $R_a = R_1 = R_3 = 1\,\text{k}\Omega$ and $R_b = R_2 = R_4 = 15\,\text{k}\Omega$. The resistance values have a tolerance $\pm 0.2\%$. Find the differential gain and the CMRR of this amplifier.

2. Both inputs of an oscilloscope with a differential voltage input are connected to a sine wave voltage with 10 V amplitude. The screen shows a sine wave with 20 μV amplitude, to be contributed to the CM input voltage. Find the CMRR of this oscilloscope.

3. An operational amplifier is configured as voltage buffer (Figure 4.4b). The DC gain of the operational amplifier is $10^5$, the CMRR at DC amounts 80 dB. If these are the only parameters that affect the transfer of the buffer, determine the deviation from the ideal transfer 1.

4. The cutoff frequency of a passive second order highpass filter is 3 kHz. Find the transfer at 150 Hz.

5. The bandwidth of a measurement signal is 0–1 Hz. There is a noise component of 50 Hz, which should be reduced by at least a factor of 100; the measurement signal may not be affected by more than 1%. Determine the cutoff frequency and the minimum order of a lowpass filter to meet these goals.

6. Consider the measurement bridge of Figure 4.20, supplied with a sine wave voltage of 1 kHz and 8 V rms. The measurement resistance varies linearly with the measurand, which is a signal with bandwidth 0–4 Hz. The bridge amplifier has a specified input noise voltage of $20\,\text{nV}/\sqrt{\text{Hz}}$. Calculate the relative resistance change $\Delta R/R$ that produces an output signal equal to the noise output.

# Chapter 5

# Digital Signal Conditioning

Although some signals are *digital* by nature, most digital signals in measurement processes occur as a result of AD-conversion of an analogue signal. This happens according to a general measurement set-up as described in Section 1.3. The advantage of digitising an analogue signal is that the resulting digital signal can be coded easily and that the code sequences can be processed by a digital computer. This chapter discusses AD-conversion and digital signal conditioning with the exception of the AD-converter itself, which is the topic of Chapter 6.

This chapter is organised as follows. In Section 5.1 digital signals and codes are described. This is followed in Section 5.2 by a discussion about sampling and quantization. In Section 5.3 we investigate signal sampling devices and multiplexers, which is necessary to obtain high quality digital signals. In Section 5.4 an inventory of some operations that can be performed on digital signals is given. This chapter concludes with Section 5.5, describing the hardware for these operations.

**5.1. Digital (binary) signals and codes**

A digital or binary signal has only two levels, denoted as "0" and "1". The relation between these codes and voltage or current values depends on the applied technology (bipolar transistors, MOSFET's and possibly others). In digital circuits with, for instance, bipolar transistors a "0" corresponds to a voltage below 0.8 V; a "1" is represented by a voltage above 2 V. Digital signals have a much lower interference sensitivity compared to analogue signals. This tolerance for interference and noise is paid for by a strong reduction in the information content. An analogue voltage of, say, 6.32 V contains much more information than a digital value "0", because in the latter case there are only two possible levels. The minimum amount of information (yes or no; high or low; "0" or "1"; on or off) is called a *bit*, an acronym for binary digit.

Usually a measurement signal contains much more information than only 1 bit. To adequately represent the information by a binary signal, a group of bits is necessary; such a group is called a *binary word*. A word consisting of 8 bits is called a *byte* (an impure acronym for "by eight"). The words byte and bit should not be confused: in this book kb means kbit and kB stands for kbyte. Because the SI prefixes strictly represent powers of 10, they should not be used to represent powers of 2. Thus, one kilobit, or 1 kbit, is 1000 bit and not $2^{10}$ bit=1024 bit.

In December 1998 the International Electrotechnical Commission (IEC) approved as an IEC International Standard names and symbols for prefixes for binary multiples for use in the fields of data processing and data transmission. They are listed in Appendix B3.

With $n$ bits, $2^n$ different words can be constructed, not more. The number of bits is bound to a maximum, not only for practical reasons, but also due to imperfections of the components in the AD-converter that generates the binary words. So, the analogue-to-digital conversion causes at least one extra error, the quantization error. Typical word lengths of AD-converters range from 8 to 16 bits, corresponding to a quantization error of $2^{-8}$ to $2^{-16}$ ($4 \cdot 10^{-3}$ to $4 \cdot 10^{-6}$). Obviously, it is useless to take an AD-converter with many more bits than in correspondence with the inaccuracy or the resolution of the measurement signal.

### Example 5.1 Resolution of a digital signal representation
The range of a measurement signal is 0–10 V. There are 10 bits available to represent this signal. The resolution of this representation is $2^{-10} \approx 0.1\%$ or about 10 mV.

Another measurement signal has an inaccuracy of 0.01%. The number of bits, required for a proper representation, is 14, because $2^{-14} < 10^{-4} < 2^{-13}$.

A binary word can be written as:

$$G = [a_n a_{n-1} \cdots a_2 a_1 a_0 a_{-1} a_{-2} \cdots a_{-m}] \tag{5.1}$$

where $a_i$ is either 0 or 1 (numbers) and $n$ and $m$ are positive numbers. The value of $G$ in the decimal number system is:

$$G = a_n 2^n + a_{n-1} 2^{n-1} + \cdots + a_2 2^2 + a_1 2 + a_0 + a_{-1} 2^{-1} + a_{-2} 2^{-2}$$
$$+ \cdots + a_{-m} 2^{-m} \tag{5.2}$$

The coefficient $a_n$ contributes most to $G$ and is therefore called the *most significant bit* or *MSB*. The coefficient $a_{-m}$ has the lowest weight and is called the *least significant bit* or *LSB*. In AD- and DA-converters the digital signals are binary coded fractions of a reference voltage $V_{ref}$. The relation between the analogue and digital signals of the converter is:

$$V_a = G \cdot V_{ref} = V_{ref}(a_{n-1} 2^{-1} + a_{n-2} 2^{-2} + \cdots + a_1 2^{-n+1} + a_0 2^{-n})$$
$$= V_{ref} \sum_{i=0}^{n-1} a_i 2^{i-n} \tag{5.3}$$

where now $a_{n-1}$ is the MSB and $a_0$ the LSB. So, $G$ is a binary number between 0 and 1. Consequently, the MSB of a converter corresponds to a value $\frac{1}{2} V_{ref}$, the next bit is $\frac{1}{4} V_{ref}$ and so on until, finally, the LSB with value $2^{-n} \cdot V_{ref}$.

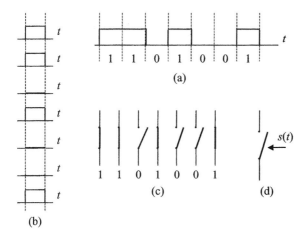

**Figure 5.1** *Various representations of a binary word: (a) serial, dynamic; (b) parallel, dynamic; (c) parallel, static; (d) serial, static*

Figure 5.1 shows several electronic representations of a binary word. Figures 5.1(a) and (b) are dynamic representations (voltages or currents as time signals). In Figure 5.1(a), the bits of a word are generated and transported one after another: it is a *serial word*. Such a serial word requires a relatively long time for the presentation of a measurement value, and this time grows in proportion to the number of bits. In Figure 5.1(b) the bits are generated simultaneously: one line for each bit. There are as many parallel lines as there are bits in a word: it is a *parallel word*. The information is available at one instant, but more hardware is required (cables, connectors, components). Figures 5.1(c) and (d) illustrate static representations of the same binary word, in Figure 5.1(c) with a set of switches. A "0" corresponds to a switch that is off and a "1" to a switch that is on. Finally, in Figure 5.1(d) the binary word is represented by just a single switch, set on or off by a binary control signal $s(t)$. If the time needed to represent one bit (the *bit-time*) is $t_b$, than the transport of a serial $n$-bit word takes $n \cdot t_b$; the transmission of a parallel word is substantially shorter, but takes at least one bit-time.

## 5.2. Sampling

Sampling an analogue signal $x_c(t)$ means to take *samples* of this signal at specific moments. Often the time interval $\Delta t$ between the sample moments is constant. Sampling a signal $x_c(t)$ results in a sequence of samples $\{x[n]\}$ with:

$$x[n] = x_c(n\Delta t) \tag{5.4}$$

To maintain the relation between the sequence of samples and its analogue original, the sample time $\Delta t$ must be known. In that case, from the sequence of samples a continuous time signal $x_s(t)$ can be constructed according to:

$$x_s(t) = x_c(t) \sum_{n=-\infty}^{\infty} \delta(t - n\Delta t) = \sum_{n=-\infty}^{\infty} x[n]\delta(t - n\Delta t)$$

$$= \sum_{n=-\infty}^{\infty} x_c(n\Delta t)\delta(t - n\Delta t) \qquad (5.5)$$

where $\delta(t)$ is the "Dirac pulse", which is zero everywhere, except in $t = 0$ and which has an area equal to 1:

$$\int_{-\infty}^{\infty} \delta(t)dt = 1 \qquad (5.6)$$

From the properties of the Dirac pulse it appears that $x_s(t)$ is zero everywhere, except at the sample moments $t = n\Delta t$, where the signal equals $x_c(n\Delta t)$, multiplied by the Dirac pulse $\delta(t - n\Delta t)$. Figure 5.2 illustrates this sampling model.

The use of this construct is that $x_s(t)$ is a continuous time signal, whose properties (e.g. information content) can be compared with the original signal $x_c(t)$.

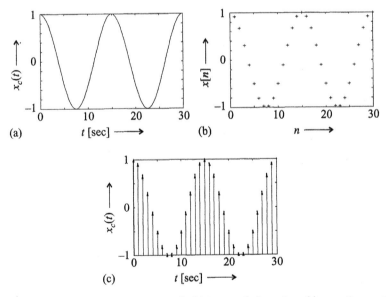

**Figure 5.2** *(a) Continuous time signal; (b) its sampled version; (c) a continuous time representation with pulses*

## 5.2.1. *Intrinsic information losses by sampling*

Sampling inevitably leads to loss of information. The time intervals between the samples may show fluctuations and the sample values contain errors. Moreover, even at perfect sampling, the signal values between the sample moments are lost and can only be recovered under strict conditions.

### Example 5.2  Consequences of sampling

Consider the *harmonic* signal $x_c(t) = \cos \pi t$. Figure 5.3(a) shows the analogue signal, while in Figure 5.3(b) the sampled signal is sketched with a sample period $\Delta t = 1$. In Figure 5.4(a) the signal $x_c(t) = \cos 3\pi t$ is shown. Figure 5.4(b) shows the sampled signal. It can be observed that both sampled signals are precisely the same, although the period times of the analogue signals differ by a factor of 3.

The information loss by sampling is analysed conveniently using the frequency content of the signals. Every continuous time signal $x_c(t)$ can be assembled from a set of *harmonic* base signals $\exp(j\omega t)$, according to:

$$x_c(t) = \int_{-\infty}^{\infty} X_c(\omega) e^{j\omega t} d\omega \qquad (5.7)$$

where $X_c(\omega)$ is the continuous time *Fourier transform* of $x_c(t)$.

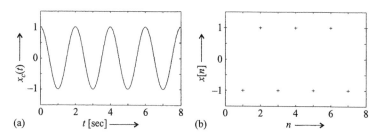

**Figure 5.3** *(a)* $x_c(t) = \cos \pi t$; *(b)* sampled signal with $\Delta t = 1$

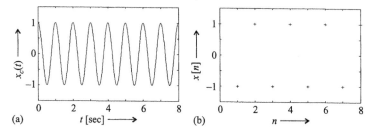

**Figure 5.4** *(a)* $x_c(t) = \cos 3\pi t$; *(b)* the sampled version is precisely the same as in Figure 5.3(b)

$X_c(\omega)$ is in general complex valued, even if $x_c(t)$ is real as in our case. In graphs, often the absolute value of $X_c(\omega)$ is shown, which is the amplitude of the corresponding harmonic. The inverse relation of (5.7) is:

$$X_c(\omega) = \int_{-\infty}^{\infty} x_c(t) e^{-j\omega t} dt \tag{5.8}$$

According to these two relations, $X_c(\omega)$ contains all the information about the signal $x_c(t)$. In other words, $x_c(t)$ can be computed from $X_c(\omega)$ using (5.7). Therefore the information loss due to sampling a signal $x_c(t)$ can be determined by comparing the differences between $X_c(\omega)$ and the Fourier transform of the sampled signal $x_s(t)$, which is $X_s(\omega)$.

It can be proven [1] that $X_s(\omega)$ shows a repetition of several frequency signals, $X_c(\omega)$, shifted over a specific frequency distance:

$$X_s(\omega) = \Delta t \sum_{k=-\infty}^{\infty} X_c(\omega - k\omega_s) \tag{5.9}$$

with $\omega_s = 2\pi/\Delta t$ the angular frequency in rad/s. This phenomenon is called *aliasing*. Figure 5.5 shows a signal $X_c(\omega)$ and its aliases for two cases: in Figure 5.5(a), $X_c(\omega)$ and its aliases are separated, in Figure 5.5(b) they show overlap.

### 5.2.2. Sampling theorem

From Figure 5.5(a) it follows how the original continuous time signal $x_c(t)$ can be recovered from its sampled version $x_s(t)$. To this, the Fourier transform of the

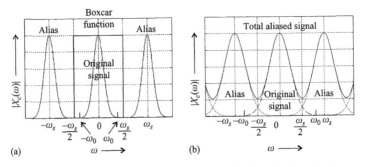

**Figure 5.5** *Aliases in the Fourier transform $X_c(\omega)$ of a signal $x_c(t)$; (a) aliases can be separated from the original; (b) aliases mix up with the original*

sampled signal, $X_s(\omega)$, is multiplied with the *boxcar* filter function:

$$H(\omega) = \begin{cases} 1 & |\omega| < \omega_s/2 \\ 0 & \text{elsewhere} \end{cases} \quad (5.10)$$

This filter removes the aliases and the Fourier transform of the original signal is left undisturbed. Back-transformation to the time domain results in a perfect *reconstruction* of the original signal $x_c(t)$. Thus it appears that no information is lost by the sampling process.

In Figure 5.5(b) $X_c(\omega)$ cannot be separated from its aliases by any filter and therefore no perfect reconstruction is possible. Comparing both graphs, a condition can be deduced, required to obtain a perfect reconstruction: $X_c(\omega)$ should be completely separable from its aliases. This means that there has to be an angular frequency $\omega_0$ such that:

$$X_c(\omega) = 0, \quad |\omega| > \omega_0 \quad (5.11)$$

Thus, the signal must be *band-limited*. In that case, $X_c(\omega)$ can be reconstructed perfectly if the angular frequency satisfies:

$$\omega_s \geq 2\omega_0 \quad (5.12)$$

This is the sampling theorem and $\omega_s$ is called the Nyquist rate. If the Nyquist criterion is not satisfied, different aliases show overlap (Figure 5.5(b)). During reconstruction of the continuous time signal two phenomena then occur:

1. The frequency content of $x_c(t)$ for $|\omega| > \omega_s/2$ is lost.
2. High frequencies from the first alias of the original signal appear in the reconstruction as low frequencies. The affected frequencies are between $|\omega| = \omega_s - \omega_0$ and $|\omega| = \omega_s/2$.

It is up to the user to decide if the loss of high frequency information is important or not. If yes, the sampling frequency has to be increased to reduce this loss. However, aliasing is always quite annoying, as it disturbs low frequency information in the signal. Aliasing effects can be reduced by low-pass filtering the continuous time signal, *prior* to sampling (Section 5.2.4). Thus, if a signal does not satisfy the Nyquist criterion, it can be made to do so by using a *pre-sampling* filter.

### Example 5.3 Sampling a harmonic signal
Consider again the harmonic signal of example 5.2: $x_c(t) = \cos \pi t = (\exp(j\omega t) + \exp(-j\omega t))/2$. It contains just two angular frequencies: $\omega = \pi$ and $\omega = -\pi$. Thus the angular frequency for sampling should satisfy $\omega_s \geq 2\pi$. The maximum sampling period is $\Delta t = 2\pi/\omega_s = 1$. In this case the sampling period is just short enough to have no information loss.

The signal $x_c(t) = \cos 3\pi t$ has angular frequencies: $|\omega| = 3\pi$ and for this signal a sampling period of 1 second is too long. The maximum sampling period is $\frac{1}{3}$ second. Now, the frequency $\omega = 2\pi$ appears in the sampled signal as a frequency $\omega = \pi$. This is what was observed in Example 5.2.

A real measurement signal will in general contain components over the whole frequency range; however, it often falls off with increasing frequency. The interesting part of the signal is determined by its bandwidth, defined by:

$$|X_c(\omega_B)| = \frac{1}{\sqrt{2}}|X_c(\omega_{\max})| \tag{5.13}$$

where $\omega_B$ is the frequency at the upper boundary of the passband and $\omega_{\max}$ is the frequency where the signal spectrum is at its maximum. Thus, frequencies above the upper boundary contain less than half of the power contained at $\omega_{\max}$ and are considered to be uninteresting. So, one might consider $\omega_B$ as the maximum frequency to obtain a sample frequency according to the Nyquist criterion ($\omega_s \geq 2\omega_B$).

### 5.2.3. Reconstruction of the analogue signal

An analogue signal can be reconstructed from its samples by connecting the sample values by line segments. This is called interpolation. We consider here only zero order, first order and "ideal" interpolation.

**Zero and first order interpolation**
If the reconstructed signal is $x_r(t)$, zero order interpolation is obtained by giving $x_r(t)$ the value of the nearest sample:

$$x_r(t) = x[n] \quad \text{if}: \left(n - \tfrac{1}{2}\right)\Delta t < t \leq \left(n + \tfrac{1}{2}\right)\Delta t \tag{5.14}$$

First order interpolation is obtained by connecting the samples with straight-line segments:

$$x_r(t) = \frac{(t - (n-1)\Delta t)x[n] - (t - n\Delta t)x[n-1]}{\Delta t} \quad \text{if}: (n-1)\Delta t < t \leq n\Delta t \tag{5.15}$$

Figure 5.6 illustrates these types of interpolation.

**Ideal interpolation**
The ideal interpolation can be found from the prescript in Section 5.2.2, stated in the frequency domain: remove the aliases by an ideal boxcar filter. It can be proved [1]

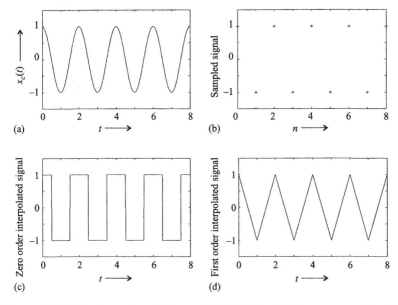

**Figure 5.6** *(a) Analogue signal; (b) sampled digital signal; (c) reconstructed analogue signal using zero order interpolation; (d) reconstructed analogue signal using first order interpolation*

that in the time domain the ideal reconstruction is:

$$x_r(t) = \sum_{n=-\infty}^{\infty} x[n]\, \text{sinc}\frac{\pi(t - n\Delta t)}{\Delta t} \qquad (5.16)$$

with: $\text{sinc}(x) = (\sin x)/x$

That is, the reconstructed signal is the summation of an infinite sequence of sync-pulses, multiplied with the sample values $x[n]$.

An AD-converter as will be described in Section 6.2 performs a zero order interpolation. Programs depicting graphs of digital signals often connect the sample points by straight lines, which is an example of first order interpolation. Both approximations are not ideal and lead to additional information losses even if the sampling process satisfies the Nyquist criterion. Therefore in practice one should use a sample frequency, which is much higher (at least a factor of 5) than the Nyquist rate.

### 5.2.4. Pre-sampling filter

Aliasing degrades the quality of a digitized signal. Aliasing errors can be reduced by low-pass filtering the analogue signal before digitization. However, the low-pass filter is in general not ideal and therefore some high frequency components of the signal nevertheless will remain. Suppose we succeed in keeping the level of the aliasing disturbance below the quantization noise of the ADC. Then the effects of quantization dominate. Obviously, the following quantities are related:

- the resolution of the ADC;
- the sampling frequency of the ADC;
- the shape and the cut-off frequency of the low-pass filter;
- the frequency content of the signal.

We now have a criterion for a sensible design of the AD-system.

### Example 5.4 Design of a pre-sampling filter
A measurement signal covers the full range of an N-bit AD-converter, running from $-A$ to $A$ [V]. The sample frequency of the ADC is $f_s$. About 20% of the signal power is contained in frequencies equal or above $f_s$. Discuss the design of a pre-sampling filter.

Frequencies above $f_s/2$ will be folded back and should therefore be removed. We choose an nth order Butterworth filter (Section 4.2.4). The quantization step interval of the ADC is:

$$q = \frac{2A}{2^N} \tag{5.17}$$

The quantisation noise has a uniform distribution with standard deviation: (see Section 3.2.1):

$$\sigma = \frac{q}{2\sqrt{3}} = \frac{A}{\sqrt{3 \cdot 2^N}} \tag{5.18}$$

The frequency response of an $n$th order Butterworth filter (Section 4.2.4) is:

$$|H(\omega)| = \frac{1}{\sqrt{1 + (\omega/\omega_c)^n}} \tag{5.19}$$

In worst case, the 20% of the signal power $P$ contained in frequencies equal or above $f_s$ is concentrated at half the sampling frequency:

$$P = \left|X\left(\tfrac{1}{2}\omega_s\right)\right|^2 = \tfrac{1}{5} A^2 \left|H\left(\tfrac{1}{2}\omega_s\right)\right|^2 \tag{5.20}$$

Equating $P$ to the variance of the noise $\sigma^2$ at half of the sample frequency (from (5.18)), we have:

$$\frac{1}{5} \frac{1}{1 + (\omega_s/2\omega_c)^n} = \frac{1}{3 \cdot 2^N} \tag{5.21}$$

Solving the cut-off frequency results in:

$$f_c = \frac{f_s}{2\sqrt[n]{(3/5) \cdot 2^N - 1}} \tag{5.22}$$

In the case of a fourth order filter, a sampling frequency of 50 kHz and a 12 bits ADC, the cut-off frequency of the filter should be no more than 3550 Hz. Using an eighth order filter, the cut-off frequency can be chosen to be 9400 Hz. Thus, compared with the fourth order filter, more low frequency components are passed without disturbance, increasing the signal quality.

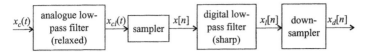

**Figure 5.7** *Digitization using over-sampling: the input signal $x_c(t)$ is low-pass filtered yielding an analogue low frequency signal $x_{cl}(t)$. This is digitized to the digital signal $x[n]$, which is sharply low-pass filtered yielding $x_l[n]$. Finally this signal is down-sampled to the final result $x_d[n]$*

### 5.2.5. Over-sampling

Low-pass filters with a sharp cut-off characteristic may be expensive and they have a non-linear phase response. An alternative is to apply a lower order pre-sampling filter and to sample the signal at a higher rate than actually required (Figure 5.7).

Using a sharp digital anti-aliasing filter, the aliased frequencies are removed, after which the signal is *down-sampled* to the final sample rate. This technique is called *over-sampling*. Finite impulse response (FIR) digital filters can be implemented having a linear phase response. The analogue low-pass filter can be designed as follows: The final sampling frequency is $\omega_s$, so signal components with a frequency below $\omega = \omega_s/2$ are preserved. The analogue filter should have a transfer of 1 for frequencies below $\omega_s/2$. Suppose the sampling frequency is $M\omega_s$, with $M$ a positive integer, the over-sampling factor. If $H(\omega) = 0$ for $\omega > \omega_0$, then aliasing will occur only between $M\omega_s - \omega_0$ and $M\omega_s/2$. To preserve signals with frequencies below $\omega_s/2$ the following condition should be fulfilled:

$$M\omega_s - \omega_0 > \frac{\omega_s}{2} \quad (5.23)$$

or

$$\omega_0 < \omega_s \left(M - \tfrac{1}{2}\right) \quad (5.24)$$

Thus, the higher the over-sampling frequency, the more relaxed are the requirements of the analogue filter.

**Example 5.5 Over-sampling with double sampling frequency**
Figure 5.8 shows the Fourier transform of a signal $X_c(\omega)$. If this signal is sampled with a sampling frequency $\omega_s$, the frequency content of the signal can be preserved between 0 and $\tfrac{1}{2}\omega_s$. This happens if an analogue boxcar filter $H_B(\omega)$ according to (5.10) is applied. If the signal is over-sampled with the double sampling frequency $2\omega_s$, we can apply a "relaxed" analogue filter $H_R(\omega)$, which only has to satisfy:

$$H_R(\omega) = \begin{cases} 1 & |\omega| < \tfrac{1}{2}\omega_s \\ 0 & |\omega| > \tfrac{3}{2}\omega_s \end{cases} \quad (5.25)$$

The frequencies between $\omega_s$ and $3/2\omega_s$ will be "folded back" to the range between $1/2\omega_s$ and $\omega_s$. However, the range between 0 and $1/2\omega_s$ will remain free of aliasing. Now a digital

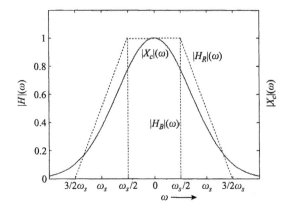

**Figure 5.8** *Over-sampling with the double sample frequency, for explanation see Example 5.5*

filter can be applied to remove the frequency content between $1/2\omega_s$ and $\omega_s$, and then the signal can be sub-sampled according to the original sample frequency $\omega_s$.

## 5.3. Sampling devices and multiplexers

Sampling an analogue signal is carried out by a *sample-and-hold* device. It converts the analogue signal to a signal containing a piecewise constant level during a hold time. This level is the sample level. During the hold time the signal is available for digitization by the AD converter, which is discussed in Chapter 6. In the simplest case, one analogue signal is digitized and stored in a microcomputer system. However, by *multiplexing*, more than one signal can be stored at the same time. Of course, using this technique all signals will have to share the available hold time so that the effective sampling rate for each separate signal will be lowered. The option of multiplexing has been discussed already in Section 1.3. There are two possibilities (Figure 5.9):

1. Sampling and digitizing each signal separately and then feeding the resulting digital signals to a digital multiplexer (space division multiplexing).
2. Multiplexing the analogue signals first and then sampling and digitizing the resulting multiplexed signal (time division multiplexing).

Advantages and drawbacks of both possibilities have been discussed already in Section 1.3.

In Section 5.3.1, sample and hold devices will be discussed, while Section 5.3.2 and Section 5.3.3 are dedicated to analogue and digital multiplexers, respectively.

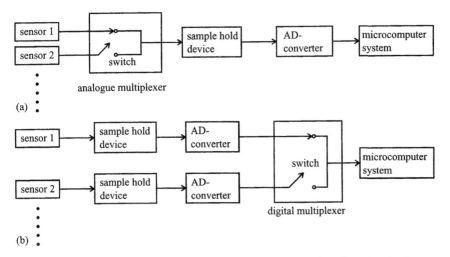

**Figure 5.9** *Multiplexing using: (a) an analogue multiplexer; (b) a digital multiplexer*

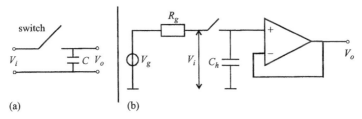

**Figure 5.10** *Sample-and-hold device: (a) equivalent scheme; (b) simple realisation with an operational amplifier*

### 5.3.1. Sample and hold devices

The aim of a sample-and-hold device is to sample a measurement signal and keep its output signal at the sample level during a specified time, required for the conversion process of the AD-converter. The scheme of Figure 5.10a does precisely that: when the switch is closed, the output follows the input (track phase). When the switch is open, the output is constant as long as no current flows from the capacitor (hold phase). Figure 5.10(b) shows a simple realisation with an operational amplifier to prevent loading of the output. Figure 5.11 shows a typical analogue input signal ($v_i$) and the resulting output signal $v_o$ from the sample-and-hold device.

A sample-and-hold device shows a number of errors:

1. The circuit is an operational amplifier circuit and the output signal suffers from all its deficiencies, like offset voltage, drift, bias current, noise, gain error and a limited bandwidth.

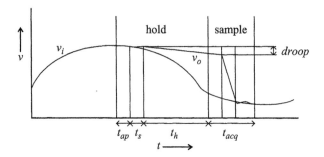

**Figure 5.11** *Input signals ($v_i$) and output signals ($v_o$) of a sample-and-hold device with various phases during the process*

**Table 5.1** *Typical characteristics of a sample-and-hold device*

| | |
|---|---|
| Aperture time ($t_a$) | 35 ns |
| Settling time ($t_s$) | 0.5 μs |
| Acquisition time ($t_h$) | 5 μs |
| Aperture jitter | 0.5 ns |
| Droop rate | 1 mV/ms |
| Offset voltage | 5 mV |
| Bias current | 2 nA |
| Input resistance | $10^{12}$ Ω |
| Input capacitance | 10 pF |

2. During the hold phase, the output voltage may get lower, due to the bias current. This is called *droop*.
3. Switching from the track to the hold phase or vice versa takes time. We distinguish (among others):

- The aperture time ($t_{ap}$), the time to switch from sample to hold mode.
- The settling time ($t_s$), the time the output signal needs to become stable after the device has started to hold the signal.
- The acquisition time ($t_{acq}$), which is the total time required for the sample and hold device to pick up the input signal again after switching from the hold to the sample phase.
- The hold time $t_h$, the time finally available for the AD-converter to perform the digitization.

These times have to be taken into account while operating the AD-converter. Specifically the aperture time shows random variations called jitter. Some specifications of the analog devices AD585 sample-and-hold device (with built in capacitor) are given in Table 5.1.

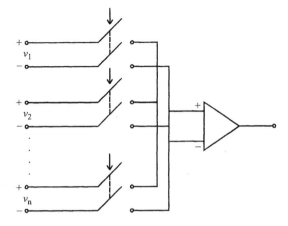

**Figure 5.12** *Multiplexer set-up in differential mode*

### 5.3.2. Analogue multiplexing

Typically an analogue multiplexer (MUX) contains a number of switches, amplifiers (to prevent the multiplexer from loading the input circuit), and additional circuitry for control of the switches. The "on-resistance", the resistance of the switch when it is closed, is in the order of a few hundred ohms. Therefore, measures may have to be taken to prevent loading effects of the output circuit, e.g. by terminating the output of the MUX with a high input-impedance amplifier. Such an amplifier may be part of a sample-and-hold circuit, following the MUX (see Figure 5.9(a)).

The signals to be multiplexed may be single-ended or differential. A differential signal requires two switches (see Figure 5.12), while for a single-ended signal only one switch is required. Different types of multiplexers are offered for either of the two applications, but also architectures are available able to handle both differential and single-ended signals.

In general the switching part of a multiplexer can also be used in the reversed direction, as a de-multiplexer.

Some important features characterising a multiplexer are given in Table 5.2. The OFF-isolation specifies how well the not transmitted signal(s), are suppressed, keeping the transmitted signal undisturbed. The settling time and the switching time influence the time it takes to acquire a sample and therefore they influence the maximum bandwidth of the acquired signals.

### 5.3.3. Digital multiplexing

A digital multiplexer performs the same function as an analogue multiplexer, except that now the input and output signals are digital. Thus, a number of parallel binary

**Table 5.2** *Some typical multiplexer specifications*

| Parameter | Value | Condition |
|---|---|---|
| Input voltage range | −10V to 10V | current < 200 μA |
| OFF isolation | 66 dB | @ 10 MHz |
| Settling time | 4.0 μs | 10 V step to 0.02% |
| Switching time between channels | 1.2 μs | |
| Input resistance | 100 Ω | |

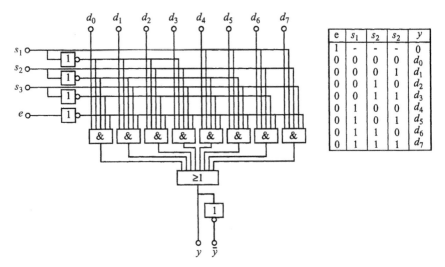

**Figure 5.13** *Digital multiplexer with truth table. Note that line crossings do not indicate connections.*

inputs are transferred to the output in a specified order, where each input occurs one at a time. A digital multiplexer can be realised with logical elements and switches. Figure 5.13 shows a scheme of a digital 8 bits multiplexer, including an optional enable input.

The input signal consists of eight parallel binary signals $d_1$ to $d_8$. According to the settings of switches $s_1$ to $s_3$, one of these signals is transferred to the output signal $y$. If the enable signal is zero, the output is kept zero, regardless of the input signal. The switching scheme will be kept such that the output signal consists of sequences of signals in which each input signal occurs once. The truth table (function table) of the multiplexer indicates its output $y$, as a function of its inputs $\{e, s_1, \ldots, s_3, d_1, \ldots, d_7\}$.

## 5.4. Digital functions

Using digital hardware, functions can be realised to manipulate digital signals. In this section a corollary is given together with references to literature for further details (Section 5.6). We use European symbols for the logic elements.

### 5.4.1. Decoding and encoding

As explained in Section 5.1, in $n$ bits $2^n$ binary values can be coded. For instance, with a binary word containing eight bits, 256 different values can be coded. A decoder forms a particular combination of values from the input. If the input contains $n$ digital signals, the output will contain maximally $2^n$ independent digital signals, one for each separate value. As an example, a one out of four decoder is shown in Figure 5.14. The truth table shows that indeed each possible binary value at the input has its own digital signal at the output.

There is another way to look at the output of the encoder. From Figure 5.14 it can be observed that the output is related to the input as:

$$y_0 = \bar{d}_0 \bar{d}_1 \quad y_1 = \bar{d}_0 d_1 \quad y_2 = d_0 \bar{d}_1 \quad y_3 = d_0 d_1 \qquad (5.26)$$

$y_0$ to $y_3$ are called the minterms of $\{d_0, d_1\}$ and they play an important role in combinational logic, which is discussed in Section 5.4.2.

An encoder performs the inverse operation of a decoder, which is to code a number of binary values into a set of digital signals. One obtains an encoder by reversing the signal direction in Figure 5.14.

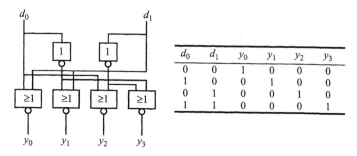

**Figure 5.14** *Logic scheme and truth table of a one out of four decoder*

### 5.4.2. *Combinational operations*

Logical operations can be divided into two types: combinational and sequential operations. A combinational operation combines the actual inputs to the output without memory function. The output of a sequential operation also depends on the past of the input and therefore contains a memory function. The output of a combinational operation between inputs $d_1$ to $d_n$ will be $F(d_1,\ldots,d_n)$ and it can be proven that the output can always be written in the standard form:

$$F(d_1,\ldots,d_n) = \sum_i g_i m_i \qquad (5.27)$$

where $g_i = 0 \vee 1$ and $m_i$ is a *minterm*, containing AND-operations between all inputs or their complimentary form (see Section 5.4.1).

**Example 5.6**
Consider the logic function: $F(d_1, d_2, d_3) = d_1 \bar{d}_2$

It is equal to: $F(d_1, d_2, d_3) = d_1 \bar{d}_2 d_3 + d_1 \bar{d}_2 \bar{d}_3$, which is of the required form (5.27).

In the same way, the output can always be written in the form:

$$F(d_1,\ldots,d_n) = \prod_i G_i M_i \qquad (5.28)$$

where now $M_i$ is a "*maxterm*", containing OR-operations between all inputs or their complementary form and $G_i = 0 \vee 1$.

**Example 5.7**
Consider the same logical function: $F(d_1, d_2, d_3) = d_1 \bar{d}_2$

It is equal to:

$$F(d_1, d_2, d_3) = (d_1 + d_2 + d_3)(d_1 + d_2 + \bar{d}_3)(\bar{d}_1 + \bar{d}_2 + d_3)(\bar{d}_1 + \bar{d}_2 + \bar{d}_3)$$
$$\times (d_1 + \bar{d}_2 + d_3)(d_1 + \bar{d}_2 + \bar{d}_3)$$

which is of the form (5.28).

Both these forms can be simply realised in programmable hardware.

### 5.4.3. *Sequential operations*

The output of a sequential operation not only depends on the current input, but also on the past. This requires a memory function, obtained by a feedback mechanism (Figure 5.15).

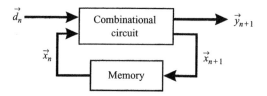

**Figure 5.15** *Set-up for sequential operations. The arrows above the symbols indicate vectors. Thus, the quantities $\vec{d}, \vec{x}$ and $\vec{y}$ may contain more than one component.*

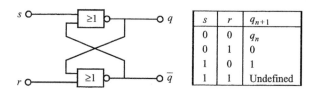

**Figure 5.16** *SR flip-flop with truth table*

### Example 5.8 SR flip-flop

A *flip-flop* has two inputs and two outputs. The outputs are each other's complements. An *SR flip-flop* with truth table is depicted in Figure 5.16. As can be observed from the truth table, if $s = r = 0$ the output $q$ does not change, reflecting the memory function. If $r = 0$, the output is set to one. If $r = 1$, it is reset to zero.

In general, a sequential operation will be implemented using a combinational circuit and a memory section. The combinational circuit is characterised by a state $\vec{x}$. The next state $\vec{x}_{n+1}$ and the output $\vec{y}_{n+1}$ are both determined by the current state $\vec{x}_n$ and the current input $\vec{d}_n$:

$$\vec{x}_{n+1} = \vec{f}(\vec{x}_n, \vec{d}_n)$$
$$\vec{y}_{n+1} = \vec{g}(\vec{x}_n, \vec{d}_n) \quad (5.29)$$

Note that input, state and output are vectors to indicate that they may contain more than one component. In total such a system is often called a *state system* or a *state machine*. Using (5.29), mathematical state systems can be implemented in hardware.

## 5.5. Hardware for digital signal conditioning

Circuits for digital signal conditioning purposes can be realised using several hardware options. In the first place, simple ICs are available containing several logical elements. However, for more complicated conditioning purposes, devices are on the market that can be configured once, or are reprogrammable.

### 5.5.1. Programmable logic devices

A programmable logic device consists of an array of logical elements, which is initially fully connected. Every connection contains a fuse, which can be blown by a large current. By blowing fuses selectively, the corresponding connections are broken, while others are preserved. This determines the program of the device. Of course this can only be done once, using a "programmer", which controls the currents to blow the fuses. A variety of architectures is available, varying in the logical elements and the programmable connections between them.

### 5.5.2. (Programmable) read only memory

In a *read only memory (ROM)* information is stored permanently. It is realised by combining a decoder with a set of OR gates. Figure 5.17 shows a 4 × 3 ROM. A cross (×) indicates a connection between a horizontal and a vertical line. The information in the ROM is defined by placing the crosses in the scheme, which will lead to a truth table, relating the indexes $i_0$ and $i_1$ to the outputs $y_0$ to $y_2$. Formally, a ROM has no signal input. The indexes are used to address the information in the ROM, defined by the places of the crosses (connections). That is, by using consecutively all combination of indexes as an input, one reads out the truth table. One may however use the indexes as a signal input. As the decoder produces all minterms of the index signals (see Section 5.4.2), it is possible to implement any logical function according to equation (5.27), using a (P)ROM.

Other architectures can be considered as variations on the (P)ROM concept. Using a complicated architecture requires much effort, programming the device. A simpler architecture may not be suitable for the logical function one has in mind. Table 5.3 mentions some architectures with their most important features.

### 5.5.3. Microcontrollers

A *microcontroller* is a chip, containing the functionality of a computer system, including a CPU, different kinds of memory, a system bus and digital IO hardware.

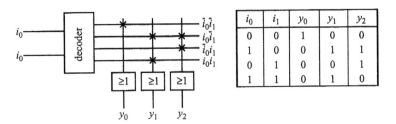

**Figure 5.17** *A ROM with truth table. The crosses indicate connections between horizontal and vertical lines*

**Table 5.3** *Some architectures for programmable logic devices with features*

| Name | Feature |
|---|---|
| Programmable Logic Array (PLA) | Programmable AND-gates instead of decoder |
| Programmable Array Logic Device (PAL) | Programmable AND-array, fixed OR-array |
| Field Programmable Gate Array (FPGA) | Only programmable AND functions, in large quantities |
| Field Programmable Logic Sequencer (FPLS) | Suitable for realising sequential circuits using added registers |

**Table 5.4** *Some selected specifications of the Atmel AT90S8535 microcontroller*

| CPU | | AD-converter | |
|---|---|---|---|
| MIPS | Up to 8 MIPS at 8 MHz | Resolution | 10 bits |
| Program Memory | 8 kB of In-System Programmable Flash | Sample rate | Up to 15 kSamples/sec at maximum resolution |
| EEPROM | 512 bytes | | |
| SRAM | 512 bytes | | |

For signal acquisition, an AD-converter is available on board. The CPU has a general-purpose instruction set like a microprocessor. Programs are developed using a PC and downloaded using a serial interface. Programming can happen in assembly language, but there are also (cross-)compilers for high-level languages, mainly C. The microcontroller is a very flexible instrument suitable for the development of intelligent measurement and control applications, because of its various interface options and because it can be programmed and reprogrammed freely. The following elements are part of a microcontroller:

- ALU and memory for processing and storing data.
- AD-converter for acquiring measurements.
- Analogue comparator, for fast detection algorithms.
- Serial peripheral interface (SPI) for control and synchronisation with other hardware and for programming the chip in place.
- Asynchronous receiver and transmitter (UART) for data transport to other hardware components.
- Watchdog timer for controlling the program flow.

Table 5.4 shows selected specifications of the Atmel AT90S8535 microcontroller.

### 5.5.4. DSPs

The architecture of a *digital signal processor* (DSP) has been optimised for processing digital signals with very high data rates. To this, separated processing

**Table 5.5** *Some specifications of the Analog Devices ADSP-21992*

| CPU | | AD-converter | |
|---|---|---|---|
| Speed | Up to 160 MIPS | Resolution | 14 bits |
| Program RAM | 32 kWords, 64 kB (32 bit data width) | Sample rate | 20 MSamples/sec |
| Data RAM | 16 kWords, 32 kB (16 bit data width) | | |

elements are used like multipliers, shifters and adders. Also data and address busses are separated according to a Harvard architecture.

Although originally a microcontroller and a DSP had different application areas, the difference between the two is fading away now: DSPs are equipped with a variety of peripheral devices, like an AD-converter and serial interfaces.

As an example, the ADSP-21992 (Analog Devices) contains the following interfaces:

- SPI and SPORT serial interfaces
- CAN-bus interface
- AD-converter
- pulse width modulation (PWM) generation unit.

On the other hand, the CPUs of microcontrollers are becoming faster, allowing more mathematical computations and because of their simpler architecture, they are easier to program. Depending on the specific application the designer will have to make a choice between both options:

- A microcontroller can be used if the maths is not to heavy and the required speeds do not exceed the capabilities of the microcontroller.
- A DSP can be used if the connections with the external world are made possible by some DSP and if the real time requirements of the computations are not hampered by the control actions required by the application.
- A hybrid solution, consisting of a microcontroller *and* a DSP is possible and most powerful, but in general also most expensive.

Table 5.5 shows some specifications of the Analog Devices ADSP-21992.

### 5.6. Further reading

J.H. McClellan et al., *Signal Processing First*, Pearson Education International, 2003, ISBN 0-13-120265-0.
General textbook about digital signal processing, contains an extensive treatment about sampling.

A. Bateman, *The DSP Handbook*, Pearson Education, 2002, ISBN 0-201-39851-6.
Algorithms for digital signal processing plus the implementation on DSPs.

A.F.P. van Putten, *Electric Measurement Systems*, Institute of Physics Publishing, Bristol, UK, ISBN 0-7503-0340-9 (2nd ed.).
Contains some worked out examples of digital measurement systems.

M. Mano M, Ch. Kime, *Logic and Computer Design Fundamentals*, 2nd ed., Prentice-Hall, Upper Saddle River NJ, USA, 2001, ISBN 0-13-031486-2.
General textbook about digital hardware.

## 5.7. Exercises

1. A signal is sampled using a sampling frequency of 2000 Hz. Give the Nyquist rate.

2. A signal is sampled using a sampling frequency of 2000 Hz. Prior to sampling the signal is low-pass filtered according to:

$$H(f) = \begin{cases} 1 & f < 500 \text{ Hz} \\ 0 & f > f_c \text{ Hz} \end{cases}$$

Here $f_c$ is the cut-off frequency of the filter: Above $f_c$ the signal is blocked; between 500 Hz and $f_c$ the signal is passed partially.

Give $f_c$, such that frequencies below 500 Hz are passed without aliasing effects.

3. Given the typical specs for a sample and hold device from Table 5.1 and a conversion time of an AD-converter, $t_{\text{conv}}$, of 15 μs, what is the bandwith of the signal that can be digitised without information loss?

# Chapter 6

# Analogue to Digital and Digital to Analogue Conversion

An analogue signal is not suitable for processing by a digital computer. Only after conversion into a digital signal, can it be handled by the computer. Conversely, many actuators and other output devices require analogue signals, hence the digital signals from a computer have to be converted into an analogue format. Analogue-to-digital converters (AD-converters or ADC's) and digital-to-analogue converters (DA-converters or DAC's) are available as modules or integrated circuits. In Chapter 5 digital signals, binary codes and some digital processing elements have been introduced. This chapter starts with an overview of the main causes of conversion errors. The remaining sections are devoted to particular types of converters and their major characteristics.

## 6.1. Conversion errors

In this section we discuss the main errors occurring in the conversion from the analogue to the digital signal domain. These errors include the quantization and sampling errors (which are independent of the converter type) and some converter-specific errors, due to shortcomings of the electronic components of the converter.

### 6.1.1. Quantization noise

Figure 6.1 represents the transfer characteristic of an AD converter. The range of the converter is set by the reference voltage $V_{ref}$. Usually, the range is from 0 to $V_{ref}$, but some converters allow a user-specific range setting, for instance from $-\frac{1}{2}V_{ref}$ to $+\frac{1}{2}V_{ref}$. An ideal converter performs a perfect rounding off to the nearest quantization level. Obviously, the maximum error is $\frac{1}{2}$LSB or $-\frac{1}{2}$LSB. For a fixed input signal and range the quantization error decreases with an increasing number of bits $n$. It is possible to express the quantization error in terms of noise power or in rms value.

Suppose the intervals between two successive levels are all the same and equal to $q$ (uniform quantization, see Chapter 3). In the case of a random input signal $V_i(t)$, the error $e(t)$ has a uniform probability density within each interval $(-q/2, +q/2)$.

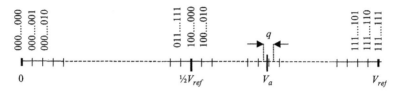

**Figure 6.1** *Quantization; the analogue voltage $V_a$ is rounded off to the nearest code*

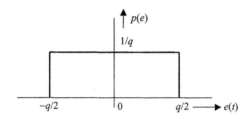

**Figure 6.2** *Uniform probability density of the quantization error*

The probability density is shown in Figure 6.2 (compare Figure 3.1); note that

$$\int_{-\infty}^{\infty} p(e)\,de = 1 \quad (6.1)$$

The mean square of the error follows from:

$$\bar{e}^2 = \int_{-\infty}^{\infty} p(e)e^2\,de = \frac{q^2}{12} \quad (6.2)$$

We can also express the quantization noise in terms of signal-to-noise ratio (SNR), see Section 3.5.2. As the number of distinct levels is $2^n$ over the full range (0 to $V_{ref}$ or $-\frac{1}{2}V_{ref}$ to $\frac{1}{2}V_{ref}$), the quantization step $q = V_{ref}/(2^n - 1) \approx 2^{-n}V_{ref}$. Evidently, the signal-to-noise ratio of the quantized signal depends on the input voltage. For a full-scale sinusoidal input the SNR is:

$$\mathrm{SNR} = 20\log\frac{V_{i,rms}}{e_{rms}} = 20\log\frac{V_{ref}/2\sqrt{2}}{q/2\sqrt{3}} = 20\log\frac{\sqrt{3}}{\sqrt{2}}2^n = 1.78 + 6n \text{ dB} \quad (6.3)$$

This is the maximum signal-to-noise ratio of a quantized noiseless input signal.

### Example 6.1 Number of required bits
The *SNR* of a quantized sinusoidal signal with a full scale amplitude with respect to the range of the ADC should be limited to 80 dB. Find the minimum number of bits.

This minimum number of bits is found to be from $1.78 + 6n > 80$ or $n(\min) = 14$.

Analogue to Digital and Digital to Analogue Conversion   143

When a static signal is applied to an ADC, the measurement error is constant, and lies within the interval $[-\frac{1}{2}\text{LSB}, +\frac{1}{2}\text{LSB}]$. This static error can be reduced by a technique called *dithering*. A wide band noise signal or a high frequency ramp signal is superimposed on the input signal. The amplitude of the dithering signal should be larger than 1 LSB. Now the ADC produces a series of codes, the average of which represents the analogue value. In this way the resolution can be better than 1 LSB. Moreover, non-linearity is also improved by dithering [1]. Non-linearity introduces harmonic distortion: when a pure sine wave is applied to the analogue input, the non-equal quantization intervals produce harmonics in the (reconstructed) output wave form. Dithering reduces this effect.

AD conversion comprises discretisation in the time domain too: the analogue signal is sampled, usually at equally spaced time intervals. Errors due to the sampling process have been discussed in Section 5.2. Here we will discuss briefly the main characteristics of AD- and DA-converters, as specified by most manufacturers.

## Reference
The reference (voltage) determines the full scale and hence the range of the converter. The reference either is an internal voltage (integrated in the converter IC) or should be applied to a connecting pin. The range is fixed or variable; unipolar (from 0 to $V_{ref}$) or bipolar (from $-\frac{1}{2}V_{ref}$ to $+\frac{1}{2}V_{ref}$).

## Resolution
The resolution of an AD converter can be defined in various ways. It is related to the number of bits, hence the number of distinct quantization levels. An $n$-bit converter has a resolution expressed as $n$ (bits), $2^n$ (number of different codes) or $V_{ref}/(2^n - 1)$ (minimum change in the analogue value corresponding to a digital step of 1 LSB).

### Example 6.2 Resolution of a DA converter
Find the resolution of an 8-bit DA converter with a range from 0 to 8V.

The resolution is expressed as 8 bit, $2^8 = 256$ (levels), or $8/(2^8 - 1) \approx 31.4\,\text{mV}$.

## LSB (least significant bit)
The term LSB covers two different quantities:

- the bit of the binary numerical representation that carries the smallest weight;
- the change in the analogue value corresponding to two consecutive binary codes, hence $V_{ref}/(2^n - 1)$.

## Integral non-linearity
The integral non-linearity is the maximal deviation of the analogue value in a plot of the actual (measured) transfer characteristic, from a specified straight line. This line may be between the end points or the "best fit". Non-linearity is expressed as a fraction of full-scale, or in multiples or fractions of 1 LSB (Figure 6.3(a)).

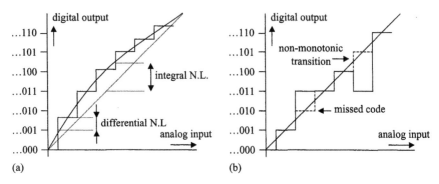

**Figure 6.3** *Illustration of some conversion error specifications: (a) integral and differential non-linearity (N.L.); (b) monotonicity and missed code*

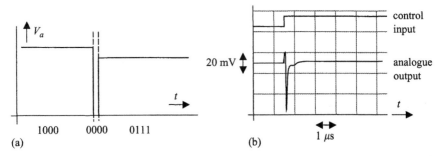

**Figure 6.4** *(a) Illustration of a half scale glitch occurring if the MSB switches prior to the other bits; (b) example of a specified glitch for a 1 LSB code change*

**Differential non-linearity**

This is the maximum deviation in a 1 LSB analogue step from the nominal step ($V_{ref}/2^n - 1$), expressed in multiples (or fractions) of 1 LSB (Figure 6.3(a)).

**Monotonicity**

A DA-converter is said to be monotonic if the output increases or remains the same when the digital input increases, over its full range (Figure 6.3(b)).

**Glitch errors; deglitching**

When the input of a DA converter is increased or decreased by 1 LSB, it passes through a transition. The largest or major transition is at half-scale, where all switches change state (for instance in an 8-bit converter, from 1000 0000 to 0111 1111 or vice versa). Unequal switching times of the bit switches introduce incorrect codes during such a transition, resulting in transient peaks. Figure 6.4 shows a pure half-scale glitch, and an example of a real glitch for a 1 LSB code change.

Generally, the transients are very fast, and can be quite large. Fortunately, the slew rate of the output amplifier reduces their effect at the output of the converter.

Glitching may further be reduced (called *deglitching*) by adding a sample-hold circuit (Section 5.3.2) at the output of the DA converter. Glitching errors are specified in the unit V·s (or V·ns), representing the area of the pulse in the voltage-time plot).

**Settling time**
The settling time is the time required for a prescribed input change of a DA converter to reach the final output state and remain at that output within a specified fraction. This fraction is usually expressed in LSB; the input change is usually full scale.

**Conversion time**
The conversion time is the time required for a complete conversion from the start conversion command up to the valid final code.

**Conversion rate**
Normally, the conversion rate is the inverse of the conversion time. Very fast converters use *pipelining*, that is, a new conversion is initiated before the previous conversion is terminated (Section 6.5). The conversion process is performed in a number of phases, controlled by clock pulses. For instance, in the first phase the analogue value is compared to a set of references resulting in a set of comparator outputs. In the second phase these digital signals are converted into a binary word, possibly followed by a third phase in which the digital word is clocked to the output. Obviously, each of these phases can handle different analogue values simultaneously, increasing the conversion *rate* (but not the conversion *time*) by a factor of three. So, for such converters the conversion rate is specified separately.

## 6.2. Parallel DA-converters

This section deals with only one type of DA-converter: the parallel converter with *ladder network*. It is the most widely used type for general applications, available as integrated circuit for a very low price.

The first step of the DA-conversion is the transfer from the $n$ parallel signal bits to a set of $n$ parallel switches. The binary signals must satisfy certain conditions in order to be able to activate the electronic switches. Once the digital parallel input signal is copied to the switches, the proper weighing factors must be assigned to each of the (equal) switches: half the reference voltage to the switch for the MSB ($a_{n-1}$), a quarter of the reference to the next switch $a_{n-2}$ and so on. Figure 6.5 gives an example of how this could be realized.

To each switch a current is allotted corresponding to the weighing factor: $\frac{1}{2}I$ for the MSB, $\frac{1}{4}I$ for the next bit and so on. The current for the last switch (LSB) is $I \cdot 2^{-n}$. In this circuit, the switches have two positions: left and right ("0" and "1", respectively). When in the left-hand position, the current flows directly to ground.

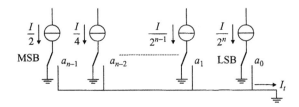

**Figure 6.5** *Assignment of weighing factors by current sources*

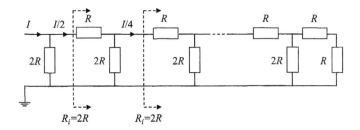

**Figure 6.6** *Ladder network*

In the right-hand position, the current flows to a summing point. The sum of the currents $I_t$ is just the analogue value that corresponds to the binary input code.

The weighted currents $I/2$, $I/4$ etc. should be derived from the reference voltage in such a way that the accuracy is better than half an LSB. This is achieved by a special resistance network, a so called *ladder network*. A particular property of this network is the input resistance which is independent of the number of sections, as can easily be seen from Figure 6.6.

Consider the currents through this network. At the first (leftmost) node the input current $I$ splits up into two equal parts: one half flows through the resistor $2R$, the other half towards the rest of the network, having an input resistance of $2R$ as well. The latter current, $\frac{1}{2}I$, splits up again at the second left node in two equal parts, and so on. Hence, the currents through the "rungs" of the ladder have values $I/2$, $I/4$, $I/8$ and so on. The network performs a successive division by two, using only two different resistance values, $R$ and $2R$, considerably simplifying an accurate design for large numbers of bits. Figure 6.7 shows how the ladder network is applied in a DA-converter circuit.

The weighted currents flow either to ground (switch position "0") or to the summing point of the current-to-voltage converter (switch position "1"). Note that this point is at zero potential, because the non-inverting input of the summing amplifier is fixed to ground. So the resistances $2R$ are always grounded, irrespective of the switch state. The largest current $I/2$ flows through the switch for the MSB ($a_{n-1}$), the smallest current $I/2^n$ through the switch for the LSB ($a_0$). The following equations

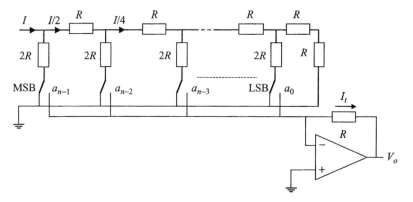

**Figure 6.7** *DAC with ladder network*

apply for this circuit:

$$I = \frac{V_{ref}}{R} \qquad (6.4)$$

($R$ being the input resistance of the ladder network) and

$$I_t = a_{n-1}\frac{I}{2} + a_{n-2}\frac{I}{2^2} + \cdots + a_1\frac{I}{2^{n-1}} + a_0\frac{I}{2^n} = \sum_{i=0}^{n-1} a_i \frac{I}{2^{n-i}} \qquad (6.5)$$

The output voltage is:

$$V_o = -R \cdot I_t = -R \sum_{i=0}^{n-1} a_i \frac{I}{2^{n-i}} = -V_{ref} \sum_{i=0}^{n-1} a_i 2^{-n+i} \qquad (6.6)$$

which is, apart from the minus sign, equivalent to equation (5.3).

Most general purpose, integrated DA-converters contain an in-built reference voltage (with Zener diode). The ladder network is composed of laser-trimmed SiCr resistors. The switches are made of CMOS transistors. Obviously, the non-zero tolerances of these resistances cause differential non-linearity. Moreover, the on-resistance of the electronic switches and more particularly the spread in these values contribute to the non-linearity as well. To improve this property of a DAC manufacturers try to apply self-calibration techniques in integrated converters [2].

Table 6.1 lists typical specifications of a general purpose 12 bit DAC, with internal reference source and output amplifier.

**Table 6.1** *Selected specifications of a general purpose integrated DAC (CMOS technology; internal reference source 4V)*

| Characteristic | Value | Unit |
|---|---|---|
| Resolution | 12 | bit |
| Temperature range | −40 to +85 | °C |
| Integral non-linearity | ±2 | LSB |
| Differential non-linearity | ±$\frac{1}{2}$ | LSB |
| Monotonicity | guaranteed over specified temperature range | |
| Settling time (full step) | 7 | μs |
| Glitch | 5 | nV.s |
| Digital "0" | max. 1.5 | V |
| Digital "1" | min. 3.5 | V |
| Full scale output | 4 | V |
| Power supply voltage | 5 | V |
| Power consumption | 2.5 | mW |

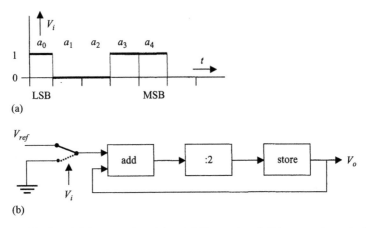

**Figure 6.8** *Serial DA conversion: (a) serial binary signal; (b) conversion principle*

## 6.3. Serial DA-converters

The weighing factors of a binary word follow from the position of the bits relative to the "binary" point. In an ADC or DAC these weighing factors are $\frac{1}{2}$ (for the MSB), $\frac{1}{4}$, $\frac{1}{8}$ and so on up to $\frac{1}{2}^n$ (for the LSB). In a parallel DAC the allocation of the weighing factors is based on the spatial arrangement of the bit lines or the switches. In a serial word (Figure 6.8), the allocation is based on the time order of the bits.

The first step is the conversion of the active binary quantity (a voltage or a current) into the corresponding switch state (on or off). The switch is controlled by the bits.

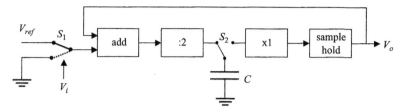

**Figure 6.9** *A serial DAC according to the protocol of Figure 6.8*

It connects either a reference voltage $V_{ref}$ (switch on) or ground (switch off) to the converter circuit (Figure 6.8(b)). The conversion process takes place in a number of sequential phases.

Suppose the LSB is in front (the leading bit). The first phase consists of the following actions: division of the voltage $a_0 \cdot V_{ref}$ by 2, and storage of this value in a memory device. When the second bit arrives, the value $a_1 \cdot V_{ref}$ is added to the contents of the memory, the result is divided by 2 and stored again (the old value is deleted). The memory now contains the value $(a_0/4) \cdot V_{ref} + (a_1/2) \cdot V_{ref}$. This process is repeated as long as bits arrive at the input of the converter. At the advent of the MSB (the $n$th bit), the contents of the memory is:

$$(a_0 2^n + a_1 2^{n-1} + \cdots + a_{n-2} 2^{-2} + a_{n-1} 2^{-1}) V_{ref} \tag{6.7}$$

which is just the desired analogue value $GV_{ref}$ (Chapter 5). Figure 6.9 shows a circuit that works exactly in agreement with this procedure. The memory device is a capacitor, which is charged when the switch $S_2$ is in the left position; when the switch is in the right position, the charge is maintained and can be measured via a buffer amplifier.

Typical resolution of a general purpose serial DAC is 16 bit. The conversion time for a 16 bit serial DAC is about 1 μs.

## 6.4. Parallel AD-converters

The output of an AD-converter is a binary code representing a fraction of the reference voltage (or current) that corresponds to the analogue input. Evidently, this input signal must range between zero and the reference: the full scale of the converter. This section is restricted to only one type of AD-converter, the successive approximation AD-converter. This type of converter belongs to the class of compensating AD-converters, that all use a DA-converter in a feedback loop. The basic principle is given in Figure 6.10.

Besides a DA-converter, the AD-converter contains a comparator, a clock generator and a word generator. The word generator produces a binary code that is applied to the input of the DA-converter; it is also the output of the AD-converter. The word

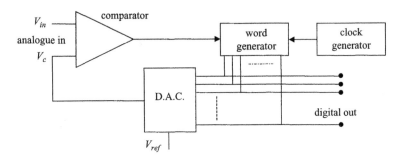

**Figure 6.10** Basic structure of a compensating ADC

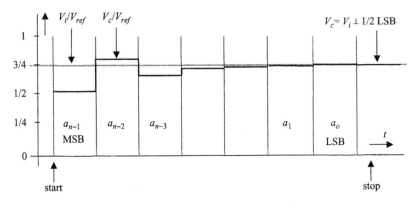

**Figure 6.11** Example of the compensating process in a SAR ADC

generator is controlled by the comparator whose output is "1" for $V_{in} > V_c$ and "0" for $V_{in} < V_c$, where $V_{in}$ is the input of the AD-converter and $V_c$ is the compensation voltage, identical to the output of the DA-converter. When the comparator output is "1", the word generator produces a next code that corresponds to a higher compensation voltage $V_c$; when the output is "0", a new code is generated that corresponds to a lower value of $V_c$. This process continues until the compensation voltage equals the input voltage $V_{in}$ (up to a difference of less than 1 LSB). As $V_c$ is the output of the DA-converter, its input code is just the binary representation of the analogue input voltage. In the final state, the (average) input of the comparator is zero: the input voltage $V_{in}$ is compensated by the voltage $V_c$.

The clock generator controls the word generator circuit: at each (positive or negative) transition of the clock, a new word is generated.

Now the question arises of how to reach the final state as quickly as possible, that is with the minimum number of clock pulses. The input signal is not known beforehand, so the best estimation is halfway the allowable range, that is $\frac{1}{2} V_{ref}$. So, the first step of the conversion process is the generation of a code that corresponds to a compensation voltage of $\frac{1}{2} V_{ref}$ (Figure 6.11).

The output of the comparator (high or low) indicates whether the input voltage is in the upper half of the range ($\frac{1}{2}V_{ref} < V_{in} < V_{ref}$) or in the lower half ($0 < V_{in} < \frac{1}{2}V_{ref}$). So, the output of the comparator is just equal to the MSB $a_{n-1}$ of the digital code we are looking for. If $a_{n-1} = 1$, the compensation voltage is increased by $\frac{1}{4}V_{ref}$, followed by a check during the next comparison whether the input is in the upper or in the lower half of the remaining range. For $a_{n-1} = 0$, the compensation voltage is decreased by that amount. The second comparison results in the generation of the second bit $a_{n-2}$. The successive comparisons are governed by the clock generator: usually one comparison (so one output bit) per clock pulse. After $n$ comparisons the LSB $a_0$ is known and the conversion process is finished.

The word generator that functions according to the principle explained here is called a *successive approximation register* or *SAR*. Most low-cost AD-converters are based on this principle. The bits are generated successively in time, but are stored in a digital memory (integrated with the converter), to have them available as a parallel binary word. Some converter types have a serial output as well, with the bits available as a series word only during the conversion process.

Figure 6.12 shows the architecture of an integrated general purpose ADC. This converter contains a clock generator, a reference voltage source $V_{ref}$, a comparator and buffer amplifiers at each output. The functions of the terminals are as follows. The analogue input voltage is connected between terminals 1 (ground) and 2 (note the relatively low input resistance of this particular architecture). At floating terminal 3, the input voltage range runs from 0 to +10 V; at grounded terminal 3, an additional current source is connected to the input of the internal DA-converter, resulting in a shift of the input range, which becomes now from −5 to +5 V. Terminal 4 is for the supply voltage (with respect to ground). Terminals 5 through 14 form the

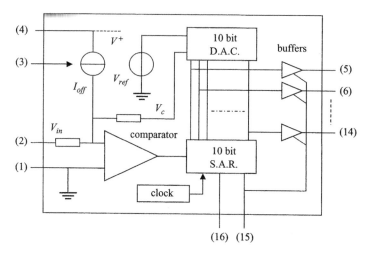

**Figure 6.12** *Internal structure of an integrated SAR ADC*

10 bit parallel digital output. Between the DA-converter and the output terminals, ten buffer amplifiers are connected, called *tri-state buffers*. A tri-state buffer has three states: "0", "1" and "off"; in the off-state there is no connection between the converter and the output terminals. This third state is controlled via an extra input to each buffer. The tri-state buffers allow the complete circuit to be electronically disconnected from the rest of the system. This offers the possibility of connecting several devices with their corresponding terminals in parallel, without the danger of mutually short-circuiting the circuits.

As soon as the conversion process is finished, the SAR generates a "0" at terminal 15 (during the conversion this output is "1"). The buffers connect the binary code to the corresponding output terminals. Further, this output signal can be used to let the processor know that the conversion is finished and that the data at the outputs are valid (this terminal is also called the *data ready* output). Finally, a binary signal at terminal 16 starts the converter. As long as this input is "1", the converter is in a wait state; as soon as the input is made "0" (for instance by a signal from a computer), the conversion starts.

This is just one example out of many types of AD-converters based on successive approximation. They differ with respect to number of bits, architecture details, the definition of control pins and the character of the output (current or voltage, buffered or unbuffered etc.).

## 6.5. Direct AD-converter

The structure of a direct AD-converter is very straightforward (Figure 6.13). The input voltage is compared simultaneously to all possible binary fractions of the reference. For an $n$-bit converter there are $2^n$ distinct levels. The reference voltage is subdivided into $2^n$ equally spaced voltages. With the same number of comparators the input voltage $V_{in}$ is compared to each of these levels simultaneously. A number of comparator outputs, counted from the top is low, the rest is high, like the temperature indication of a liquid-in-glass thermometer. A digital decoder combines these $2^n$ values and generates the required binary code of $n$ bits. This AD-converter is also known as *flash converter*, because of the high conversion speed (limited by the time delay of the comparators and the decoder only). A disadvantage is the large number of components (and thus a high price). With special architectures and high speed technology this conversion principle enables the realization of an 8-bit converter with a conversion rate of over 300 MHz ($3 \cdot 10^8$ samples per second).

It is possible to significantly reduce the number of components, at the price of a somewhat lower speed (Figure 6.14). Here, the input voltage $V_{in}$ is compared to $\frac{1}{2}V_{ref}$, resulting in the MSB. If $a_{n-1} = 1$ (hence $V_{in} > \frac{1}{2}V_{ref}$), $V_i$ is reduced with an amount $\frac{1}{2}V_{ref}$; otherwise it remains the same. To determine the next bit, $a_{n-2}$, the corrected input voltage should be compared to $\frac{1}{4}V_{ref}$. However, it is easier to

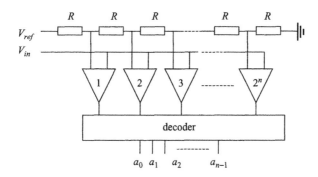

**Figure 6.13** *Structure of an n-bit direct ADC*

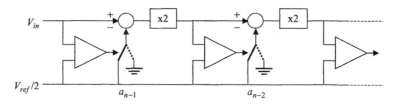

**Figure 6.14** *Principle of the cascaded ADC*

compare twice the value with $\frac{1}{2}V_{ref}$ which is the same. To that end, the voltage $V_{in} - a_{n-1} \cdot \frac{1}{2}V_{ref}$ is multiplied by 2 and compared with half the reference voltage by a second comparator. This procedure is repeated up to the last bit (LSB).

As the comparators are now connected in series, this converter is called a *cascaded* AD-converter. It is somewhat slower than the preceding type, because the delay times are in series; however, the number of components is reduced considerably.

Table 6.2 shows a specification example of the two AD-converters described before. Note that the conversion frequency of the cascaded converter is higher than the inverse of its conversion time, due to pipelining (see also Section 6.2): the next bit is applied to the first comparator when the preceding bit is in the second comparator and so on.

Another method to reduce the number of components, but without much loss of speed, is the two-stage flash converter, depicted in Figure 6.15. In this type of converter, the conversion is split up into two successive parts. First, the first 3 or 4 most-significant bits are converted. Next, the obtained binary code is converted back into an analogue signal using a DAC. The output of this DAC is subtracted from the original analogue input. Finally, the difference is converted to obtain the remaining group of bits. This principle is applicable to all type of converters, but is in particular applied in flash converters, as it reduces significantly the number of components.

## 154 Measurement Science for Engineers

**Table 6.2** *Specification example of a direct and a cascaded DAC*

| Property | Parallel | Cascaded |
|---|---|---|
| Resolution | 8 bit | 8 bit |
| Inaccuracy | ±0.1% ±½LSB | ±0.1% ±½LSB |
| Monotony | guaranteed | guaranteed |
| Diff. non-linearity | 0.01% | 0.01% |
| Conversion time | 35 ns | 150 ns |
| Conversion frequency | 20 MHz | 11 MHz |

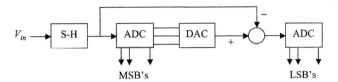

**Figure 6.15** *Two-stage flash converter*

### Example 6.3 Reducing the number of components
Suppose the first conversion results in the first 4 most significant bits of an 8 bit word. This requires $2^4 = 16$ comparators. The four last bits require also a converter with 16 comparators, hence a total of 32 comparators, instead of $2^8 = 256$ in a full flash converter. This reduction is paid by an extra 4-bit DAC and a sample-hold circuit, to keep the sampled input during both conversion phases.

In practical converters the discretization intervals are unequal, due to resistance tolerances and comparator offsets. This results in non-linearity, and consequently the generation of harmonic distortion. A possible way to reduce the inaccuracy and non-linearity is dynamic element matching. Like dithering, this method introduces random components too in the signal, but now by randomly rearranging the elements (resistors, capacitors) that determine the quantization levels of the converter [3]. Evidently, randomizing the elements' position in the circuit is performed by electronic switches (the components themselves remain in their position on the chip). The effect is a more accurate quantization (based on the average of the element values) and a spread of the noise components over the frequency range (instead of discrete harmonic frequencies).

### 6.6. Integrating AD-converters

In an *integrating* AD converter the input signal is integrated during conversion: the digital output is proportional to the averaged value of the input signal. As also noise and other interference signals are integrated, their contribution to the output is reduced (presuming zero average value of these signal components). For this reason, integrating AD-converters are widely used in precision DC current and voltage

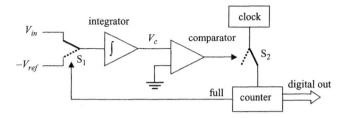

**Figure 6.16** *Principle of a dual-slope integrating ADC*

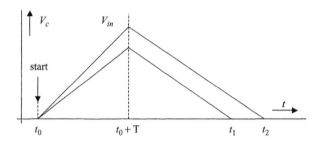

**Figure 6.17** *Integrator output voltage in the dual-slope ADC of Figure 6.16, for two different values of the analogue input*

measurement instruments. Due to the integration of the input, these converters have a slower response compared to the other AD-converters. Figure 6.16 shows the principle of a double integrating AD-converter, which is called the *dual-slope* or *dual-ramp* converter.

The conversion is performed in two phases. First, the input signal is integrated and then the reference voltage is integrated. The conversion starts with the switch $S_1$ in the upper position: the input signal is connected to the integrator, which integrates $V_{in}$ during a fixed time period $T$. At positive constant input and a positive transfer of the integrator, the output increases linearly in time. The comparator output is positive and keeps switch $S_2$ on. This switch lets pass a series of pulses with frequency $f_0$ to a digital counter. In digital voltmeters, this is usually a decimal counter. Upon each pulse, the counter is incremented by one count, as long as $S_2$ is on. When the counter is "full" (it has reached its maximum value), it gives a command to switch $S_1$ to disconnect $V_{in}$, and to connect the (negative) reference voltage to the integrator. At the same moment the counter starts again counting from zero. Since $V_{ref}$ is negative and constant, the output of the integrator decreases linearly with time. As soon as the output is zero (detected by the comparator), $S_2$ switches off and the counter stops counting. The content of the counter at that moment is a measure for the integrated input voltage.

Figure 6.17 depicts the integrator output for one conversion period and two different (constant) input voltages. The counter is reset to zero; $S_2$ closes as soon as $V_c$ is positive: at that time, $t_0$, the counter starts counting. At $t = t_0 + T$ the counter is full; at that moment the integrator output voltage is

$$V_c(t_0 + T) = \frac{1}{\tau} \int_{t_0}^{t_0+T} V_{in} \, dt \qquad (6.8)$$

where $\tau$ is the proportionality factor ($RC$-time) of the integrator. At constant input voltage, the slope of the output is $V_{in}/\tau$. From $t = t_0 + T$ the (negative) reference is integrated, hence $V_c$ decreases with a rate $V_{ref}/\tau$. When $V_c = 0$ (at $t = t_1$), $S_2$ opens and the counter stops. The voltage rise during the first integration phase (of $V_{in}$) is fully compensated by the voltage drop during the second integration phase (of $V_{ref}$), so:

$$\frac{V_{in}}{\tau} T = \frac{V_{ref}}{\tau}(t_1 - t_0 - T) \qquad (6.9)$$

from which follows:

$$V_{in} = V_{ref} \frac{t_1 - (t_0 + T)}{T} \qquad (6.10)$$

The voltage $V_{in}$ is transferred to a time ratio, and is measured by the decimal counter. Note that the parameter $\tau$ of the integrator does not appear in this ratio. The first integration period $T$ is made such that the counter is full at a power of 10, so $10^n$. As the counter starts again with the second integration, at $t = t_0 + T$, the number of pulses during the second integration is equal to $(t_1 - t_0 - T) \cdot 10^n / T = (V_{in}/V_{ref}) \cdot 10^n$. When $V_{in}$ is not constant during the integration period, the number of pulses at the end of the second integration period equals:

$$counts = \frac{10^n}{V_{ref}} \int_{t_0}^{t_0+T} V_{in} \, dt \qquad (6.11)$$

At first sight this conversion method seems rather laborious. However, the resulting output code depends solely on the reference voltage (and the input voltage of course) and not on other component values. The only requirement for the integrator and the frequency $f_0$ is that they should be constant during the integration period.

The method also allows a strong reduction of interference from the mains (50 or 60 Hz spurious signals). To suppress 50 Hz interference, the integration time is chosen a multiple of 20 ms. The average of a 50 Hz signal is just zero in that case. This choice results in a minimum conversion time of 40 ms (full scale), or a maximum conversion rate of 25 per second.

Dual-slope converters are available as a module or as an integrated circuit. Some integrated converters go with special microprocessor chips performing the control and data handling; with special data processing the conversion rate can be increased at the same noise suppression capability. Advances in integration technology have

# Analogue to Digital and Digital to Analogue Conversion

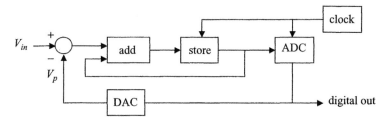

**Figure 6.18** *Concept of a $\Sigma\Delta$ ADC; $V_p$ is the predicted analogue value*

resulted in the design of complete measurement systems for electrical parameters, based on the dual slope principle, on one or a few chips. They have a typical inaccuracy of less than $10^{-4}$ to $10^{-5}$.

## 6.7. Sigma-delta AD-converters

The general concept of a $\Sigma\Delta$ converter (also called delta-sigma converter) is based on a series of consecutive AD-conversions with a low resolution, followed by averaging the outcomes of these conversions. The procedure is shown in Figure 6.18.

The conversion consists of a number of phases. In the first phase, the analogue input signal is converted into an $n$-bit binary word, resulting in the first prediction, with an error that is within the quantization interval $[0, q]$. Meanwhile, the input value is stored in an analogue memory (for instance a sample-hold circuit). The binary coded first prediction is converted back to the analogue domain, and subtracted from the input signal (the *delta* process). This results in an analogue error signal $E(1)$ (still within the quantization interval). In the next phase of the conversion, the error signal is added to the stored input signal (the summation or *sigma* process). This sum is converted by the same ADC into a binary coded second prediction, again with an error within the quantization interval. The second prediction is converted back to an analogue signal, and subtracted from the input, resulting in a new error signal $E(2)$. This process is repeated a number of times.

The result of this procedure is a series of output codes, all of them representing the analogue value with a quantization error corresponding to the number of bits of the ADC. However, the mean value of this series is a better approximation of the analogue input.

### Example 6.4
The ADC and DAC in Figure 6.18 have both 4 bits resolution. The reference voltage is 10 V, the input voltage is 3.00 V. This value lies between 2.500 (code 0100) and 3.125 (code 0101). In the initial state the output code is 0000, hence the error equals the input voltage. The table shows the first 8 conversion steps.

**158**  Measurement Science for Engineers

| $E(n)$ (V) | $V_s$ (V) | Code | $V_p$ (V) |
|---|---|---|---|
| 3.000 | 3.000 | 0100 | 2.500 |
| 0.500 | 3.500 | 0101 | 3.125 |
| −0.125 | 3.375 | 0101 | 3.125 |
| −0.125 | 3.250 | 0101 | 3.125 |
| −0.125 | 3.125 | 0101 | 3.125 |
| −0.125 | 3.000 | 0100 | 2.500 |
| 0.500 | 3.500 | 0101 | 3.125 |
| −0.125 | 3.375 | 0101 | 3.125 |

Apparently, after 5 steps the sequence is repeated in this typical example. The mean of the predicted values is just 3.000 V.

The length of the sequence depends on the input value, so does the accuracy of the converter. Obviously, the calculation of the average is performed in the digital domain, using the outputs of the "code" column.

An important feature of the $\Sigma\Delta$ ADC is its simplicity. The achievable resolution is much better than the resolution of the applied AD converter itself. This allows the use of the most simple converter: a 1-bit ADC, consisting of just a single comparator. A 1-bit DA converter is very simple too: a single switch that connects the output either to zero (0) or a reference voltage (1). In fact, circuit complexity is exchanged for speed requirements.

## Example 6.5
Assume a 1 bit converter with reference voltage 10 V. The next table gives the first 16 converter steps for an input voltage 3.100 V. Code = 0 for $V_{sum} < 5$ V; code = 1 for $5 \leq V_{sum} < 10$ V.

| $E$ (V) | $V_s$ (V) | Code | $V_p$ (V) | $Av(n)$ |
|---|---|---|---|---|
| 3.1 | 3.1 | 0 | 0 | |
| 3.1 | 6.2 | 1 | 10 | |
| −6.9 | −0.5 | 0 | 0 | |
| 3.1 | 2.6 | 0 | 0 | 0.25 |
| 3.1 | 5.7 | 1 | 10 | |
| −6.9 | −1.2 | 0 | 0 | |
| 3.1 | 1.9 | 0 | 0 | |
| 3.1 | 5.0 | 1 | 10 | 0.375 |
| −6.9 | −1.9 | 0 | 0 | |
| 3.1 | 1.2 | 0 | 0 | |
| 3.1 | 4.3 | 0 | 0 | |
| 3.1 | 7.4 | 1 | 10 | 0.333 |
| −6.9 | 0.5 | 0 | 0 | |
| 3.1 | 3.6 | 0 | 0 | |
| 3.1 | 6.7 | 1 | 10 | |
| −6.9 | −0.2 | 0 | 0 | 0.3125 |

Analogue to Digital and Digital to Analogue Conversion    159

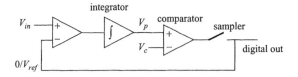

**Figure 6.19** *Layout of a sigma-delta converter with integrator*

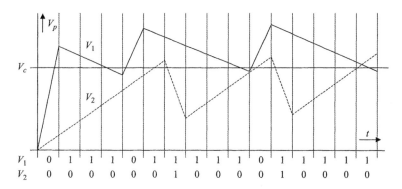

**Figure 6.20** *Time diagram of an integrating sigma-delta process, for two different input signals: $V_{in,1} > V_{in,2}$. The generated one-bit codes are given too*

In the last column the average over the first 4, 8, 12 and 16 predictions are listed. Obviously, with an increasing number of steps, the mean value comes closer to the actual analogue input.

The noise performance of the converter is substantially improved when the summing device is replaced by an integrator. Its effect is similar to that in a dual ramp ADC: noise in the input is averaged out. This also holds for the quantization noise originating from the (1-bit) quantizer in this converter. Figure 6.19 shows a realization of such a 1-bit sigma-delta modulator with integrator.

The comparator acts as a 1-bit ADC; its output is either high ("1") or low ("0"). This output is sampled, and serves also as the analogue version of the 1 bit DAC ($V_{ref}$ or 0), so no additional DA-converter is required.

Figure 6.20 illustrates the operation principle. The vertical lines indicate sample moments. In the initial state the output of the integrator is zero. When a (constant) input signal is applied, the output of the integrator $V_p$ increases linearly with time, at a rate proportional to $V_{in}$. During this state the sampler takes over the output of the comparator, which is 0 in this example; these are the first "predictions". This process goes on until $V_p$ crosses the comparator voltage $V_c$, and the output of the comparator changes state. The new value is taken over at the next first sample moment. From that moment the reference voltage $V_{ref}$ is subtracted from the input voltage, and since $V_{in} < V_{ref}$ the input of the integrator is negative.

So its output decreases with a rate equal to the difference between $V_{in}$ and $V_{ref}$. Again, the sampler takes over the resulting predictions, which are now 1. For the case $V_p = V_2$ in Figure 6.20, this takes only one sample because $V_p$ has already decreased below $V_c$ before the next sample moment. The integrator continues with integration of $V_{in}$. This process continues for some time, resulting in a number of bits. The average of these bits is proportional to the input voltage, just as shown in the table of Example 6.5. In the same figure, the process is shown for a larger value of $V_{in}$, resulting in the voltage $V_1$, and a smaller average number of zeros.

Similar to the dual slope ADC, the input is integrated during one or more sample periods, so noise is (partially) averaged out.

Sigma-delta converters use essentially oversampling: to obtain high accuracy in a reasonable conversion time, the sample rate should be high (MHz range), even for low-frequency input signals. Besides the required averaging, which is done in the digital part of the converter, noise performance can further be improved by additional digital filtering, in particular noise shaping and decimation filtering. Details of these techniques are beyond the scope of this book.

Due to their simplicity, one-bit converters are very suitable for integration with other signal conditioning circuits on a single chip.

## 6.8. Further reading

B. Loriferne, *Analog-digital and Digital-analog Conversion*, Heyden, 1982, ISBN 0-85501-497-0.
Basics on ADCs and DACs, mainly for undergraduate students. Apart from the conversion principles as presented before, Loriferne's book contains a chapter on digital-synchro and synchro-digital converters, a type of converter in particular used for angular encoders.

A. Moscovici, *High Speed A/D Converter: Understanding Data Converters Through SPICE*, Kluwer Academic Publishers, 2001, ISBN 0-7923-7276-X.
This book focuses on the speed of AD converters. Since comparators and sample-hold circuits are part of most types of converters, these circuits are discussed in more detail in separate chapters. Not only the standard flash type converters are discussed, but also some special configurations for high speed applications, such as the folding ADC and pipeline ADC's.

K.M. Daugherty, *Analogue-to-digital Conversion: a Practical Approach*, McGraw-Hill, 1995, ISBN 0-07-015675-1.
A general text book, with many practical indications for the designer of data acquisition systems. Only a little mathematics is used, except for the section on delta-sigma converters where the z-transform is introduced to illustrate noise shaping.

David F. Hoeschele Jr., *Analogue-to-digital and Digital-to-analogue Conversion Techniques*, Wiley, 1994, ISBN 0-471-57147-4.

A general textbook on AD and DA converters. As well as common topics such as ADC principles, DAC principles, ADC testing and converter specifications, the book also discusses supporting circuits (comparators, reference voltages, multiplexers), both on the system level and the transistor level.

R.J. van de Plassche, *Analogue-to-digital and Digital-to-analogue Converters*, Kluwer Academic Publ., 1994, ISBN 0-7923-9436-4.
The major focus of this book is on speed and accuracy of converters and associated circuits. It contains many examples of various converters, with detailed structures on a transistor level. Design tricks and technological implications are also included. A suitable book for the starting designer of integrated converters.

## 6.9. Exercises

1. Determine the signal-to-noise ratio of a 10 bit ADC, for a full scale sinusoidal input signal.

2. Calculate the minimum number of bits that is required to limit the SNR (due to quantization noise) to 60 dB.

3. A dual slope converter should be able to fully suppress both 50 Hz and 60 Hz interference signals. What should be the minimum integration time?

4. Give the number of comparators required for the following types of 10 bits AD-converters:

direct ADC (according to Figure 6.13);
cascaded ADC (according to Figure 6.14);
two-stage flash converter (according to Figure 6.15);
SAR converter (according to Figure 6.11).

5. Make a table as in Example 6.3, but now for an input voltage $V_i = 7.3$ V. Determine the average over the first 5, the first 10 and the first 15 outputs.

# Chapter 7

# Measurement of Electrical, Thermal and Optical Quantities

According to the definition of measurement, the purpose of a measurement system is to map the values of a measurand to a specified set of numbers or symbols. In an electrical measurement system, the physical quantity is converted into an electrical signal using a sensor or transducer (except for a pure electrical measurand), while the measurement result, the numbers obtained, are represented again in the physical domain, using an actuator (a display, a memory device).

Between 1970 and 1980 the development of sensors received a strong technology push because of the successful silicon IC-technology. Silicon possesses many physical effects applicable to the construction of sensors. This allows the integration of sensors together with the electronic circuits on a single chip. Such sensor systems are not only small in dimension, but they can also be manufactured in large volumes at a very low price per device. The technology to realize such sensor systems is based on either bulk micromachining or surface micromachining. This technology has grown mature and silicon based sensors are commercially available now. At the end of the 20th century, the use of silicon technology was extended from electrical circuits to complete mechano-electrical microstructures (MEMS), resulting in the construction of silicon actuators and complete measurement systems on a silicon chip.

In this chapter we discuss various methods for the measurement of quantities in the time, electrical, magnetic, thermal and optical domains. In these cases, emphasis is on the sensor, where the primary conversion from the physical into the electrical domain takes place. Measurements for mechanical and chemical quantities will be discussed in separate chapters (Chapters 8 and 9).

## 7.1. Measurement of electrical quantities and parameters

Since most measurement systems ultimately process electrical signals, it is important to study the measurement of quantities in the electrical domain. We focus on the measurement of voltage, current and impedance. Resistance, capacitance and inductance are special cases of impedance, and electric charge is just the time integral of current. Further, we limit the discussion to ranges that are normally encountered in practical situations. For measurement of extremely small or large values or under special conditions the reader is referred to the literature on these issues.

### 7.1.1. DC voltage

General purpose (digital) voltmeters consist of an analogue voltage amplifier with adjustable gain and a display. The gain is manually or automatically set to a value for which the signal matches the input range of the ADC. The gain can be smaller than 1, to allow the measurement of voltages higher than the maximum range of the ADC. Such electronic voltmeters have a high input impedance, minimizing loading errors due to the (unknown) source impedance (Section 3.5.3).

The overall specifications of a general purpose electronic voltmeter are related to the quality of the built-in amplifier, as discussed in Section 4.1. Similar considerations apply to oscilloscopes, which too have (adjustable) amplifiers at the input. Most oscilloscopes have an optional offset adjustment, allowing the adjustment of the baseline on the screen. Oscilloscopes are, however, inaccurate measurement instruments since the output is read from a display with limited resolution. They have particular value when measuring AC voltages with unknown time behaviour.

There are many possible causes of measurement errors, even in this simple case. The model of Figure 7.1 is helpful to identify particular classes of errors, due to differential and common mode resistances of both the signal source and the measuring instrument, as well as common mode voltages.

When the voltage to be measured is relative to ground, the corresponding ground terminal of the instrument can be used, but care should be taken not to create ground loops. Further, ground terminals of both source and measuring instrument should have the same potential ($V_g$ = zero in Figure 7.1), otherwise it may cause substantial errors when measuring very small voltages. When measuring the voltage

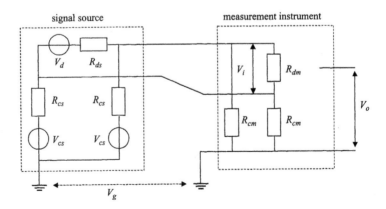

**Figure 7.1** *Resistances and interference signals in a voltage measurement system*

between two points that are not grounded (differential mode or DM) the instrument must have a floating input. Furthermore, its CMRR (Section 3.5.4) should be high to avoid measurement errors caused by the common mode signals.

In summary, before connecting a DC-voltmeter to a signal source, we must know about:

- the voltage, being DC indeed
- the expected range
- the source resistances (common and differential)
- the instrument resistances
- the CM voltage when measuring DM voltages
- the CMRR of the instrument
- the accuracy requirements.

**Example 7.1 Differential mode voltage measurement**
A (floating) voltage source is characterized by:

- source resistance $<100\ \Omega$
- CM voltage $<5$ V
- expected DM voltage in the order of 100 mV.

The DM voltage should be measured with an accuracy better than $10^{-3}$ of the indication. The measurement is performed with an instrument having the following specifications:

- Input resistance $>1\ M\Omega$
- CMRR (common mode rejection ratio): $>10^5$
- FS (full scale) inaccuracy at 100 mV: $<10^{-4}$.

Discuss the performance of this measurement, for an indicated value of 60 mV.

The required inaccuracy is $10^{-3}$ relative or 60 $\mu$V absolute. The error budget is:

- due to the input resistance: $<10^{-4}$ (relative) or 6 $\mu$V absolute
- due to the CMRR: output voltage $<50\ \mu$V (absolute)
- due to full scale inaccuracy (100 mV): $<10\ \mu$V (absolute).

So, the total measurement error amounts to 66 $\mu$V, which is still too large. Apparently the error is mainly determined by the CMRR, and hence an instrument with a better CMRR should be taken or the CM voltage of the source should be reduced in some way.

Since there are no perfect DC signals, we must consider their time behaviour as well. On the one hand the system should be fast enough to follow (slow) signal variations. On the other hand, the system should be able to average out noise on the signal. The SNR improvement by averaging or filtering is proportional to the averaging time (hence the measurement time) and inversely proportional to the bandwidth of the system. So in selecting the proper instrument, we should consider:

- the required part of the signal spectrum
- the type and level of the source noise
- the required SNR
- the noise parameters of the measurement system
- the measurement time and response time.

**Example 7.2 Measurement of the intensity of a reflected laser beam**
The intensity of a reflected laser beam is measured using a photodiode, the current of which is to be converted into a voltage using an operational amplifier configured as a current-to-voltage converter (see Chapter 4). The intensity changes slowly, with a bandwidth of 0.1 Hz. The photo current is in the order of 0.1 nA. Give a possible design and discuss its noise performance.

We select a type III operational amplifier from Table 4.1. The resistance $R$ in the current-to-voltage converter is chosen to be 10 M$\Omega$, so the output voltage is about 1 mV. The (thermal) current noise amounts to $\sqrt{(4\,kT\,B/R)} = 4.1 \cdot 10^{-14}\sqrt{B}$ with $B$ the system bandwidth. Further, the spectral current noise of the operational amplifier is specified to be 2 pA/$\sqrt{Hz}$, hence the noise contribution is $2 \cdot 10^{-12}\sqrt{B}$ ampere. Apparently, the amplifier noise dominates the resistance noise.

Since the signal bandwidth is only 0.1 Hz, the noise can be reduced significantly by a first order low-pass filter, for instance at 4 Hz. In that case, the noise current is 4 pA, and the (max) SNR with respect to current is 25 (or 28 dB). The response time corresponds to a time constant of 0.25 s.

### 7.1.2. AC voltage

All precautions that need to be taken when measuring a DC voltage measurement apply to AC measurements as well. Moreover, not only must the input resistance of the instrument now be large compared to the source resistance, also the input capacitance should be sufficiently low to prevent load at higher frequencies. Obviously, the instrument's bandwidth should fully cover that of the measurement signal.

Another point of attention is the signal shape. When using an RMS measuring instrument, the user must be aware of the fact that possibly such an instrument gives a correct output only for sinusoidal inputs. The rms value of a sinusoidal signal with amplitude $A$ is

$$\sqrt{\frac{1}{T}\int_0^T (A\sin\omega t)^2 dt} = \frac{1}{2}A\sqrt{2} \qquad (7.1)$$

The use of an instrument that measures the mean of the rectified signal is much easier than a real rms measurement. The mean of a rectified sine wave with amplitude $A$ equals:

$$\frac{1}{T}\int_0^T |A \sin \omega t| dt = \frac{2A}{\pi} \tag{7.2}$$

Therefore, an rms instrument that operates according to the rectifying principle has a built-in scale correction of a factor $(1/4)\pi\sqrt{2} \approx 1.11$. Obviously, such an instrument is calibrated correctly only for sine waves. If the wave shape is not known, it should be identified first.

So-called "true rms" meters measure the rms value according to the definition. These instruments are thus suitable for all kinds of wave shapes. One method to measure true rms is based on the heat generated by the signal when applied to a dissipative element. The temperature rise is measured using a temperature sensor, for instance a thermocouple (Section 7.4.2). The transfer of this system is rather non-linear, and depends on various system parameters. The system performance is improved considerably by using a second, identical dissipator and thermal sensor in a control loop (Figure 7.2). The control amplifier supplies a DC current to the second dissipator such that in the steady state the temperature difference is zero. In that situation the rms value of the DC signal equals that of the AC signal, and can be measured by a DC voltmeter. This thermal principle allows the measurement of rms voltage (and current) of signals with frequencies up to 100 MHz.

It should be noted that rms or true rms instruments generally should not measure the DC component in the signal, which is accomplished by a coupling capacitor in series with the input.

Even the use of an oscilloscope to measure or identify signal shapes must be performed with care. A typical value for the input resistance of an oscilloscope is 1 MΩ. When this value appears to be insufficiently high, an attenuating probe might be used. This probe attenuates the input signal by a factor of 10, accomplished by a simple resistive voltage divider. However, due to the input capacitance (typically 10 pF), the attenuation is frequency dependent. Hence, the instrument must be calibrated prior to an AC measurement. The procedure is illustrated in Figure 7.3.

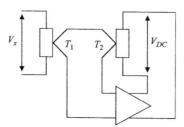

**Figure 7.2** *Thermal measurement of rms voltage*

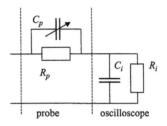

**Figure 7.3** *Model of the probe attenuation network*

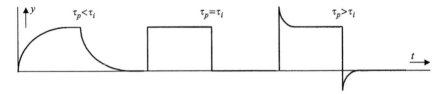

**Figure 7.4** *Probe adjustment: various situations*

Here, $R_i$ and $C_i$ are the input resistance and input capacitance of the oscilloscope. $R_p$ is a fixed resistance, included in the probe housing, and has a value 9 times the input resistance $R_i$. The DC attenuation in this case is just 10. Further, $C_p$ is an adjustable capacitance.

The general expression for the transfer of the network is:

$$H(j\omega) = \frac{R_i}{R_i + R_p} \cdot \frac{1 + j\omega R_p C_p}{1 + (j\omega R_i R_p / R_i + R_p)(C_i + C_p)} \qquad (7.3)$$

This transfer is frequency independent only when the condition

$$R_i C_i = R_p C_p \qquad (7.4)$$

is fulfilled. In other words: the time constant of the oscilloscope input impedance should be equal to that of the probe impedance. In that case, the transfer of the probe attenuator is just equal to

$$H = \frac{R_i}{R_i + R_p} \qquad (7.5)$$

The probe capacitance $C_p$ can be adjusted manually. To facilitate adjustment of the probe, a square shaped test signal is available on the oscilloscope. By connecting the probe to this test point, the time response is displayed on the scope screen. Only when the screen displays a pure square shape is the transfer of the oscilloscope frequency independent. Depending on the type of deviation (Figure 7.4) the probe capacitor is raised or lowered until a condition of frequency independence is reached.

Summarizing, next to the basic preparations for the measurement of a DC signal as outlined before, an AC measurement requires at least the following precautions:

- gain knowledge about the signal shape and its frequency range
- check instrument with respect to:
  - input impedance (including capacitance)
  - rms or true rms
  - proper probe adjustment (when applicable).

### 7.1.3. Current

Current is a typical "through" or flow variable. Measurement of such variables requires the circuit to be "opened" in order to insert the measurement device. "Across" or effort variables do not require insertion of the measurement device, but just a connection to the points that define the measurand. For this practical reason current measurements are less popular. In the case of an occasional current measurement, the circuit is opened to insert an ammeter. When somewhere in the system a current has to be measured continuously, this is often done by inserting a small resistance in the corresponding network branch and a subsequent measurement of the voltage across this proof resistance. This is for instance the case in battery management systems, where not only the voltage but also the current through the battery is continuously monitored. Obviously, the insertion of the resistance affects the original circuit, thus a loading error is introduced. Such test resistances should therefore be as small as possible, that is, small compared to the circuit resistance $R_s$. Similarly, the resistance $R_v$ of the voltmeter should be large compared to the proof resistance.

The measurement error caused by $R_v$ and $R_s$ can be calculated using the model of Figure 7.5. The measured voltage equals

$$V_o = I \cdot (R_s || R || R_v)$$
$$= I \cdot \frac{R_s \cdot R \cdot R_v}{R_s R + R_s R_v + R_v R}$$
$$= \frac{I \cdot R}{1 + R/R_v + R/R_s} \approx I \cdot R \cdot \left(1 - \frac{R}{R_v} - \frac{R}{R_s}\right) \qquad (7.6)$$

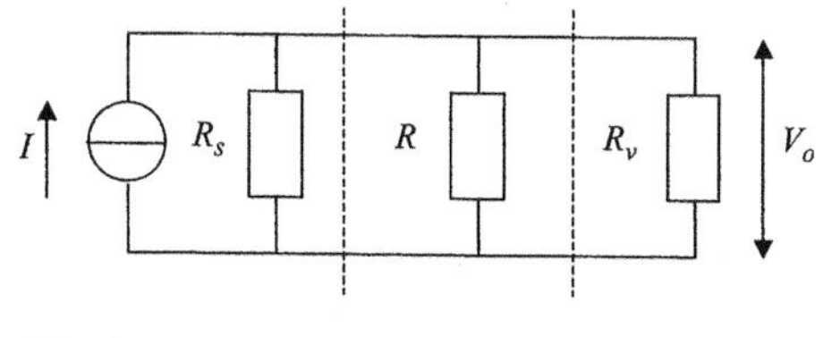

**Figure 7.5** *Model of a current measuring configuration with source resistance and load resistance*

The approximations hold for the conditions $R \ll R_s$ and $R \ll R_v$, stressing the need for a low value of the proof resistance $R$. The relative error due to the proof resistance equals $-R/R_v - R/R_s$.

### Example 7.3 Current measurement

In the circuit below all resistances have the value 10 kΩ. The current $I$ is measured by connecting a proof resistance $R$ in the branch with $R_3$. The voltage across $R$ is measured by a voltmeter with input resistance 1 MΩ. Discuss the requirements for a total measurement uncertainty less than 0.03%.

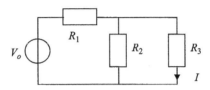

There are three major error sources: due to the circuit resistance, the voltmeter resistance and the uncertainty of the voltmeter. Suppose first that these errors contribute equally to the total measurement error. Then the proof resistance should be less than 0.01% of both the voltmeter input resistance and the circuit resistance. The latter equals $(R_1 \| R_2 + R_3) = 15\,k\Omega$ (Thevenin current source model) so $R < 1.5\,\Omega$. The input resistance of the voltmeter is much larger and its effect on the total error is negligible. So the voltmeter error should be less than the remnant of the maximum error, that is 0.02%.

In cases where the circuit resistance is relatively low, a very small proof resistance is required. This may result in a small voltage and consequently a large measurement error. If the measurement current flows to ground, a proof resistance can be avoided by applying an active current-to-voltage converter (Section 4.1). Such a circuit has ideally zero input impedance. This solution is applied for instance in optical measurements using photodiodes (see Section 7.5).

When measuring very small currents, care should be taken to avoid spurious current leakage. In such cases, *active guarding* is an effective way to reduce leakage currents through, for instance, the impedance of connecting cables. The principle of active guarding is illustrated in Figure 7.6.

In Figure 7.6(a) the current source (modelling some sensor with current output) is connected directly to the input of a current meter by a coaxial cable. To avoid capacitively induced interference, the shield of the cable should be grounded. However, part of the signal current flows through the cable impedance $Z_c$ to ground, introducing a measurement error. With active guarding (Figure 7.6(b)) the voltage across the cable is kept zero by applying the input voltage (via a buffer amplifier) to the cable shield. Hence the total signal current flows through the input of the measurement system (having a low impedance $Z_i$) and possible interference signals to the cable disappear in the output of the buffer amplifier.

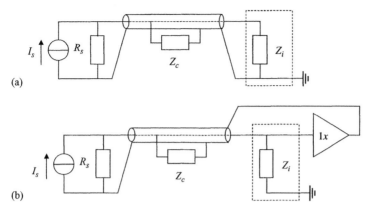

**Figure 7.6** *Active guarding to reduce leakage currents: (a) cable impedance affects the current measurement; (b) cable impedance is kept at zero potential difference by active guarding*

Finally, we mention the possibility for a current measurement without the need to open the circuit. It is performed with a *current transformer*. The current-carrying wire acts as the primary "coil". The ring shaped secondary coil (the current probe) can be opened like a pair of pincers, and closed around the wire carrying the current. In the secondary coil a current is induced that is proportional to the primary current, and to the number of turns. This method is used when it is not possible or allowed to open the circuit to insert a proof resistance (or ammeter), for instance in power lines.

### 7.1.4. Impedance

Traditionally, accurate measurements of resistance, inductance and capacitance are accomplished by a compensation or nulling method, using a bridge circuit. The principle is explained with the Wheatstone bridge[1] from Figure 3.16, which is repeated in Figure 7.7. This configuration, used for the accurate determination of resistance values, consists of two fixed resistances $R_1$ and $R_2$, the unknown resistance $R_x$ and an adjustable resistance $R_a$. The resistance $R_a$ is adjusted (manually) such that the output voltage $V_o$ of the bridge is zero, to be detected by a sensitive null detector.

When the output voltage is zero, the bridge is said to be in equilibrium; the equilibrium condition is:

$$R_x = \frac{R_1}{R_2} \cdot R_a \tag{7.7}$$

So, the value of the unknown resistance follows from the known values of the three other resistances constituting the bridge, independently of the supply voltage.

---

[1] Sir Charles Wheatstone (1802–1875), British physicist and inventor.

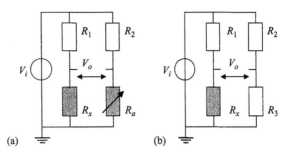

**Figure 7.7** *Measurement bridge: (a) compensation mode; (b) deflection mode*

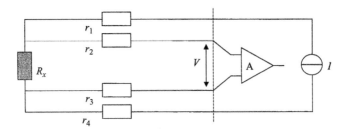

**Figure 7.8** *Four-wire configuration eliminating the resistance of the connecting wires*

The only requirements are a high sensitivity and low offset of the null detector. The compensation method requires further an accurate, adjustable resistance. These are expensive devices, and therefore only used for occasional precision measurements, such as calibration.

Figure 7.7(b) shows the bridge in deflection mode; it is used to accurately measure small resistance *changes*, and is therefore applied as an interface for all kind of resistive sensors (for instance strain gauges and resistive temperature sensors). In this application, the bridge is not nulled, and the output $V_o$ is just a measure for the (small) resistance change relative to a reference or nominal value. This is further explained in Section 8.2.

When the resistance value is not large compared with the connecting wires of the resistor to be measured, the wire resistances may introduce a substantial error. This is the case at very low resistance values (1 Ω or less) or when the measurement resistor is positioned at a large distance from the instrument (for instance a temperature sensor in an oven). The value $R_x$ in equation (7.7) now includes the resistances of the two wires, so the measurement result should be corrected for the sum of these wire resistances. When changing the measurement configuration (for instance another wire type or wire length) the correction factor should be changed accordingly. This recalibration is avoided in the four-wire configuration of Figure 7.8. It is essentially a current-voltage measurement. When the source resistance of the current source and the input resistance of the voltmeter are both infinite, the current $I$ flows completely through the resistance $R_x$. No current flows through the wires connecting the resistor with the voltmeter. So, $R_x$ just equals $V/I$.

Similarly to the Wheatstone bridge for measuring resistances, other bridges can be configured for the measurement of a capacitance and an inductance. Moreover, the unknown impedance can also be composed of a combination of these elements, for instance a resistance and capacitance in parallel. Obviously, this requires at least two adjustable components to be able to null the bridge output.

Nowadays, impedances are measured using an impedance analyser. With such instruments the real and imaginary parts of the unknown impedance, or its modulus and phase, are measured as a function of frequency. An impedance analyser provides a means to automatically plot such diagrams, for a frequency range up to several GHz. Obviously, the measurement principle of such an instrument is not based on an adjustable bridge, but on phase sensitive detection (Section 4.3). A sine wave (with fixed amplitude and variable frequency) from a VCO is applied to the unknown impedance $Z_x$. The current through this impedance is converted to a voltage of which the amplitude and phase (relative to the input voltage) is measured by synchronous detection. By digital processing other parameters of the unknown impedance can be calculated and displayed, over a wide frequency range, for instance the real and imaginary part of $Z_x$, the modulus and argument (amplitude and phase) of $Z_x$, or the parameters of a particular model, for instance a resistance, a capacitance and an inductance in series or in parallel.

## 7.2. Measurement of time and time-related quantities

Instruments for the measurement of frequency and time intervals are based on counting signal transitions. The input signal is applied to a comparator or a Schmitt trigger which converts the analogue signal into a binary form, with preservation of the frequency. The comparator output is applied to the counter only during an accurately known time interval, set by an oscillator (the time base oscillator) and an adjustable frequency divider (Figure 7.9). The counter counts the number of pulses or zero crossings per unit of time, hence the frequency. The result is displayed directly in Hz.

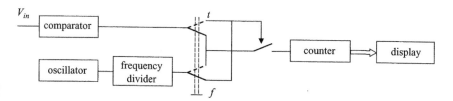

**Figure 7.9** *Structure of a digital counter switched to frequency mode; turning the double switch (period) time is measured*

174  Measurement Science for Engineers

The frequency range of the instrument depends on that of the counter and on the time interval. The latter can be varied stepwise with the adjustable frequency divider. By changing the range by factors of 10, the display can simply change accordingly by shifting the decimal point.

The measurement time of low frequencies would become too long when using this principle. In such cases it is more convenient to measure the time period rather than the frequency. This can simply be done by interchanging the position of the comparator and the time base oscillator in Figure 7.9. The counter counts the number of pulses from the reference oscillator during (half) the period of the input signal. Similarly, the range can be changed by using the frequency divider.

The phase difference between two periodic signals can be measured in the same way. A counter starts on the upgoing transition of the first signal (acting as the reference); the next upgoing transition of the second signal stops the counter; the number of counts determines the time delay between the signals, and the phase follows from a division by the total period time. Again, the resolution of the phase measurement is determined mainly by the counter frequency.

**Example 7.4 Resolution of a phase measurement**
The phase difference between two periodic signals with frequency 1 kHz is measured using a 100 MHz counter. Find the resolution of the phase measurement.

One period of the signal contains $10^5$ counts. The phase resolution is 1 count, so the resolution is $360/10^5$ is $0.036°$ or about 2 minutes of a degree.

The measurement range of frequency meters can be as high as several GHz. The input impedance has a high value (typically 1 M$\Omega$) for low frequency types and 50 $\Omega$ (characteristic impedance) for the high frequency instruments. The input signal should always exceed a specified minimum value, otherwise a correct conversion to a binary signal is not guaranteed.

The accuracy of a frequency measurement and a time period measurement is limited by the instability of the time base oscillator (the time base error) and noise in the input signal. Moreover, since the signal and time base oscillator are not synchronized, a trigger error occurs. The maximum error is plus or minus 1 count; hence the relative error in a frequency measurement increases with decreasing frequency. This is another reason to switch over to the time mode when the frequency is low. Obviously, the trigger error has a uniform distribution.

## 7.3. Measurement of magnetic quantities

In this section we describe four methods for the measurement of magnetic field strength $H$ (A/m) or magnetic induction $B$ (Tesla). The measurand is essentially a vector, so the sensor must consequently be direction sensitive. The four methods are:

- Coil
- Hall sensor
- Flux gate sensor
- Magnetoresistive sensor.

For other methods, for instance SQUIDs (Superconducting QUantum Interference Devices), the reader is referred to the literature.

### 7.3.1. Coil

Alternating magnetic fields can easily be measured using a small coil placed in the field. According to the Faraday[2] induction law voltage is induced in the coil satisfying:

$$V_{ind} = -\frac{d\Phi}{dt} \qquad (7.8)$$

The induction voltage is proportional to the flux variation, hence to the cross section of the coil. For a high sensitivity a large coil is required; on the other hand, a small coil is recommended to allow measurement of the local field strength in an inhomogeneous field. Furthermore, the sensitivity is proportional to the frequency of the alternating field, so for a proper measurement of the field strength this frequency should be a known parameter. Anyhow, the method is straightforward, but is restricted to AC fields.

The measurement of a DC field with a coil is possible by rotating the coil: in that case also a voltage is induced, which is proportional to the induction field $B$, the cross section of the coil and the rotation speed of the coil. The moving coil makes the method less attractive compared with other methods without moving parts.

### 7.3.2. Hall sensor

The Hall sensor[3] is based on Lorentz forces that act on moving charge carriers in a solid conductor or semiconductor, when placed in a magnetic field (Figure 7.10).

The force $F_l$ on a particle with charge $q$ and velocity $v$ equals

$$\bar{F}_l = q(\bar{v} \times \bar{B}) \qquad (7.9)$$

The direction of this force is perpendicular to both $B$ and $v$ (right-hand rule).

---

[2] Michael Faraday (1791–1867), British physicist.
[3] Edwin Herbert Hall (1855–1938), American physicist.

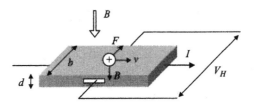

**Figure 7.10** *Principle of the Hall plate*

The flow of charges is obtained by applying a current $I$ through the device. As a result, charges are deflected and an electric field $E$ is built up in the plate. The charge carriers experience an electric force $F_e = qE$ that in the steady state counterbalances the Lorentz force: $F_e = F_l$. So we conclude that

$$\bar{E} = \bar{v} \times \bar{B} \quad (7.10)$$

From this equation the output voltage can be derived. Assuming all charge carriers have the same velocity, the current density $J$ equals $n \cdot q \cdot v$, with $n$ the charge density. In case $B$ is perpendicular to $v$ (as in Figure 7.10), the electric field equals simply $E = JB/nq$. Finally, with $I = b \cdot d \cdot J$ and $V = E \cdot b$, the voltage across the Hall plate is:

$$V = \frac{1}{nq} \cdot \frac{IB}{d} = R_H \cdot \frac{IB}{d} \quad (7.11)$$

The factor $1/nq$ is called the *Hall coefficient*, symbolized by $R_H$. The Hall effect can only be exploited in semiconducting materials: the factor $nq$ is small enough to obtain sufficient sensitivity. Earlier materials are semiconductor compounds such as GaAs, InAs and InSb. These devices had invariably a plate-like shape, because the sensitivity is inversely proportional to $d$ in equation (7.11). Nowadays, silicon is mainly used, because of the higher sensitivity and the compatibility with IC technology.

The assumption of equal velocity for the charge carriers is only an approximation: the velocity distribution within the material in not completely homogeneous. This results in a deviation of the Hall coefficient as given in equation (7.11). Practical values range from 0.8 to 1.2 times the theoretical values.

Hall plates suffer from offset: the construction is never fully symmetric: the voltage contacts are not precisely positioned perpendicularly to the main axis of the plate (that is, the direction of the control current). It can be compensated for using a compensation voltage derived from the current.

Silicon Hall devices, too, suffer from offset, raising from lay-out tolerances and material inhomogeneity. Various methods have been proposed to eliminate this offset, of which the spinning method gets much attention [1]. In a spinning Hall device the control current $I$ is rotated stepwise in the plane of the device, using electronic switches (for instance 8 steps). The output voltage is averaged over one full spinning period, resulting in a strong reduction of the offset. Typical offset values range from 10 to 50 mT for standard Hall-effect devices down to 10 μT for offset-compensated designs. Modern silicon Hall sensors are two-dimensional: they provide the two components of the magnetic field vector in the plane of the wafer. Current research is on the fabrication of single device 3D Hall sensors, in which the third direction of the magnetic field vector is measured as well, using the so called vertical Hall effect [2].

### 7.3.3. Fluxgate sensors

Like the Hall sensor, a fluxgate sensor measures both static and alternating magnetic fields. Basically the fluxgate sensor (or saturable-core magnetometer) consists of a core from soft magnetic material and two coils: an excitation coil and a sense coil (Figure 7.11(a)).

The excitation coil supplies an AC current that periodically brings the core into saturation. Hence, the permeability of the core material changes with twice the excitation frequency, between values corresponding to the unsaturated and the saturated state. An external magnetic field $H$ induces an additional induction field $B$ in the core of the sensor. Since the permeability varies periodically, so does the induction: it is modulated by the varying permeability. In the sense coil a voltage is induced according to equation (7.8). Assuming homogeneous and parallel fields, $\Phi = n \cdot B \cdot A$ with $A$ the cross section of the core and $n$ the number of turns. The voltage of the sense coil now becomes:

$$V_{ind} = -\frac{d\Phi}{dt} = -\frac{d\{nA\mu_0\mu_r(t)H\}}{dt} \qquad (7.12)$$

Hence the output voltage of the sensor is periodic, with an amplitude proportional to the magnetic field strength to be measured. Since the relation between $B$ and $H$ is strongly non-linear, the shape of the output voltage is a distorted sine wave, almost pulse like. Frequency selective detection of the output (synchronous demodulation on the second harmonic of the excitation frequency) makes the measurement highly insensitive to interference.

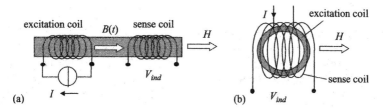

**Figure 7.11** *Structure of a flux gate sensor: (a) basic structure; (b) ring shaped*

Practical fluxgate sensors have a ring shaped core (Figure 7.11(b)). The excitation coil is wound toroidally around the ring whereas the sense coil fully encloses the ring, with its radial axis in the plane of the ring. At $H = 0$, the induced voltage is zero because of ring symmetry: the sensor is in balance. A field along the axis of the sense coil disturbs this symmetry, resulting in an output voltage proportional to the field strength.

A further improvement of the sensor performance is obtained by the feedback principle (Section 3.6.3). The output of the phase sensitive detector is amplified and activates a feedback coil around the sense coil. When the loop gain is sufficiently large the resulting induced field is zero, while the amplified sense voltage serves as the sensor output.

Flux gate sensors have a higher sensitivity than Hall sensors. Both types are currently subject to intensive research to enhance the sensitivity [3].

### 7.3.4. Magnetoresistive sensors

Some ferromagnetic alloys show anisotropic resistivity, which means that the resistivity depends on the orientation of the current through the material. This property arises from the interaction between the charge carriers and the magnetic moments in the material. As a consequence, the resistivity depends on the magnetization of the material. This effect, called anisotropic magnetoresistivity (AMR) was discovered in 1856, and is manifest in materials as iron, nickel and alloys of those metals (Permalloy). The effect is rather small, and requires very thin layers to be useful for sensing applications.

The effect can be described in various ways, for instance with the equation [4]:

$$\rho(\varphi) = \rho_0(1 + \beta \cos^2 \varphi) \tag{7.13}$$

Here, $\varphi$ is the angle between the current density vector $J$ through the device and the magnetization vector $M$ (Figure 7.12). In absence of an external magnetic field, this angle is zero. With an external in-plane field perpendicular to the current, the vector $M$ turns, according to equation (7.13). The material parameters $\rho_0$ and $\beta$ determine the sensitivity. The angle $\varphi$ runs from 0° (where the resistivity is maximal) to 90° (minimum resistivity). The maximum change in resistivity amounts a few percent only.

More recently, around 1980, magnetoresistive devices have been realized that exhibit a much larger sensitivity, according to the so called *giant magnetic resistance* effect. A GMR device consists of a multilayer of ferromagnetic thin films, sandwiched between conductive but nonferromagnetic interlayers. The axial impedance of an amorphous wire may change as much as 50% due to an applied magnetic field [5]. Combined with a magnetic source, GMR sensors are useful devices for the measurement of a variety of quantities in the mechanical domain, for instance (angular) position [6, 7] and torque [8].

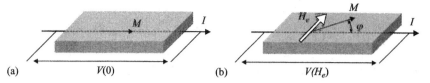

**Figure 7.12** *Angle between magnetization and current in a magnetoresistive material: (a) zero external field; (b) with external field*

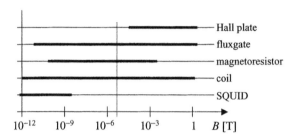

**Figure 7.13** *Comparison of the measurement range of various magnetic field sensors (after [1]). The vertical line at 50 µT represents the earth magnetic field*

In 1993 another group of materials, manganese based perovskites, has been discovered exhibiting an even stronger magnetoresistive effect, at least at low temperatures and high magnetic fields, where this effect is as high as 100%. For this reason the effect is called colossal magnetoresistivity (CMR). Unfortunately, at room temperature the effect is much smaller, but nevertheless can be employed for sensor purposes [9].

Finally, in Figure 7.13, the measurement range of various magnetic field sensors is shown for comparison.

## 7.4. Measurement of thermal quantities

This section is restricted to a discussion on temperature measurements only; other thermal quantities, for instance thermal conductivity and heat flow, will not be considered here. Two categories of temperature measurements are distinguished: contact thermometry and radiation thermometry. In the first category the object of which the temperature is measured should make a proper contact with a temperature sensor. This has the immediate consequence of (thermally) loading the measurement object. Furthermore, it takes time to heat up the sensor, roughly according to a first order response, the time constant being determined by the heat capacity of the sensor and the heat resistance between the sensor and the object.

The second category is based on the measurement of (thermal) radiation of the body whose temperature has to be measured. The radiation is directed to the temperature sensitive device, the temperature of which raises accordingly. The thermal load in this method is negligible.

Four different temperature measuring methods are presented here: resistive sensors, thermocouples, bandgap sensors and pyroelectric sensors.

### 7.4.1. *Resistive temperature sensors*

The resistivity of a conductive material depends on the concentration of free charge carriers and their mobility. The mobility is a parameter that accounts for the ability of charge carriers to move more or less freely throughout the atom lattice; their movement is constantly hampered by collisions. Both concentration and mobility vary with temperature, at a rate that depends strongly on the material.

In intrinsic (or pure) semiconductors, the electrons are bound quite strongly to their atoms; only a very few have enough energy (at room temperature) to move freely. At increasing temperature more electrons will gain sufficient energy to be freed from their atom, so the concentration of free charge carriers increases with increasing temperature. As the temperature has much less effect on the mobility of the charge carriers, the resistivity of a semiconductor decreases with increasing temperature: its resistance has a *negative temperature coefficient*.

In metals, all available charge carriers can move freely throughout the lattice, even at room temperature. Increasing the temperature will not affect the concentration. However, at elevated temperatures the lattice vibrations become stronger, increasing the chance of the electrons to collide and hamper a free movement throughout the material. Hence, the resistivity of a metal increases at higher temperature: their resistivity has a *positive temperature coefficient*.

The temperature coefficient of the resistivity is used to construct temperature sensors. Both metals and semiconductors are applied. They are called (metal) resistance thermometers and thermistors, respectively.

The construction of a resistance thermometer of high quality requires a material (metal) with a resistivity temperature coefficient that is stable and reproducible over a wide temperature range. By far the best material is platinum, due to a number of favourable properties. Platinum has a high melting point (1769°C), is chemically very stable, resistant against oxidation and available with high purity. Platinum resistance thermometers are used as an international temperature standard for temperatures between the boiling point of oxygen ($-182.97°C$) and the melting point of antimony ($+680.5°C$) but can be used up to 1000°C.

A platinum thermometer has a high linearity. Its temperature characteristic is given by:

$$R(T) = R_0(1 + aT + bT^2 + cT^3 + dT^4 + \cdots) \tag{7.14}$$

with $R_0$ the resistance at 0°C. A common value for $R_0$ is 100 Ω; such a temperature sensor is called a Pt-100. The normalized values of $R_0$ and the first three coefficients are:

$$R_0 = 100.00 \text{ Ω}$$

$$a = 3.90802 \cdot 10^{-3} \text{ K}^{-1}$$

$$b = -5.8020 \cdot 10^{-7} \text{ K}^{-2}$$

$$c = 4.2735 \cdot 10^{-10} \text{ K}^{-3}$$

according to the European norm DIN-IEC 751. The temperature coefficient is thus almost 0.4% per K.

The material of a thermistor (contraction of the words *therm*ally sensitive res*istor*) should have a stable and reproducible temperature coefficient too. Commonly used materials are sintered oxides from the iron group (chromium, manganese, nickel, cobalt, iron); these oxides are doped with elements of different valence to obtain a lower resistivity. Several other oxides are added to improve the reproducibility.

*Thermistors* cover a temperature range from −100°C to +350°C. Their sensitivity is much larger than that of resistance thermometers. Furthermore, the size of thermistors can be very small, so that they are applicable for temperature measurements in or on small objects. Compared to resistance thermometers, a thermistor is less stable in time and shows a much larger non-linearity.

As explained before, the resistance of most semiconductors has a negative temperature coefficient. That is why a thermistor is also called an NTC-thermistor or just an NTC. The temperature characteristic of an NTC satisfies the equation:

$$R(T) = R_0 \exp B \left( \frac{1}{T} - \frac{1}{T_0} \right) \qquad (7.15)$$

with $R_0$ the resistance at $T_0$ (0°C) and $T$ the (absolute) temperature (in K). The temperature coefficient (or sensitivity) of an NTC is:

$$\alpha = \frac{1}{R} \frac{dR}{dT} = -\frac{B}{T^2} [\text{K}^{-1}] \qquad (7.16)$$

The parameter $B$ is in the order of 2000 to 5000 K. For instance, at $B = 3600$ K and room temperature ($T = 300$ K), the sensitivity amounts −4% per K. To obtain a stable sensitivity, thermistors are aged by a special heat treatment. A typical value of the stability after aging is +0.2% per year.

The non-linearity of an NTC may be a disadvantage. In a computer-based measurement system the characteristic can be put into a look-up table. When this is not an option, the thermistor can be incorporated into a linearizing circuit; several of such circuits have been reported in literature, for example [10, 11].

Next to NTC thermistors there are PTC thermistors too. The temperature effect differs essentially from that of a thermistor. PTC's have a positive temperature coefficient over a rather restricted temperature range. Within the range of a positive temperature coefficient, the characteristic is approximated by

$$R(T) = R_0 \exp BT \qquad T_1 < T < T_2 \qquad (7.17)$$

The sensitivity in that range is $B$ (K$^{-1}$) and can be as high as 60% per K. PTC thermistors are rarely used for temperature measurements, because of lack of reproducibility. They are mainly applied as safety components to prevent overheating at short-circuits or overload.

### 7.4.2. Thermoelectric sensors

The (free) charge carriers in different materials have different energy levels. When two different materials are connected to each other, the charge carriers at the junction will rearrange due to diffusion, resulting in a voltage difference across this junction. Of course, neutrality is maintained for the whole construction. The value of this junction potential depends on the type of materials and the temperature.

By connecting one material between two other but equal materials, as shown in Figure 7.14(a), two junctions are created. At both junctions a junction potential appears, with opposite polarity. The voltage across the end points of the couple is zero as long as the junction temperatures are equal. If, on the contrary, the two junctions have different temperature, the thermal voltages do not cancel, hence there

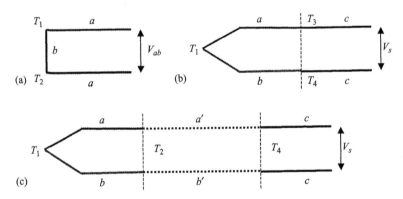

**Figure 7.14** *Thermoelectric sensors: (a) thermocouple; (b) thermocouple with connection wires; (c) with compensation cable*

is a net voltage across the end points of the couple, which satisfies the expression:

$$V_{ab} = \beta_1(T_1 - T_2) + \tfrac{1}{2}\beta_2(T_1 - T_2)^2 + \cdots \qquad (7.18)$$

This phenomenon is called the *Seebeck effect*, $V_{ab}$ is the Seebeck[4] voltage. The coefficients $\beta_i$ depend on the materials, and somewhat on the temperature. The derivative of the Seebeck voltage to the variable $T_1$ is:

$$\frac{dV_{ab}}{dT_1} = \beta_1 + \beta_2(T_1 - T_2) + \cdots = \alpha_{ab} \qquad (7.19)$$

and is called the *Seebeck coefficient*, which depends on the materials and the temperature as well. The Seebeck coefficient can always be written as the difference between two other coefficients: $\alpha_{ab} = \alpha_{ar} - \alpha_{br}$, with $\alpha_{ar}$ and $\alpha_{br}$ the Seebeck coefficients of the couples of materials $a$ and $r$ (a reference material) and $b$ and $r$, respectively. Usually, the reference material is lead.

The Seebeck effect is the basis for the thermocouple, a thermoelectric temperature sensor. One of the junctions, the reference junction or *cold junction*, is kept at a constant, well known temperature (for instance 0°C); the other junction (the *hot junction*) is connected to the object of which the temperature has to be measured. Actually, thermocouples measure only a temperature *difference*, not an absolute temperature.

To measure the Seebeck voltage of a junction, the end points of that junction have to be connected to a voltage measuring device, by electric wires of material $c$ (for instance copper) (Figure 7.14(b)). Now two new junctions are created, and also two new thermoelectric voltages. The total voltage equals

$$\begin{aligned} V_s &= V_{ca}(T_3) + V_{ab}(T_1) + V_{bc}(T_4) \\ &= V_{ca}(T_3) + V_{ab}(T_1) + V_{ba}(T_4) + V_{ac}(T_4) \\ &= V_{ab}(T_1) - V_{ab}(T_4) + V_{ca}(T_3) - V_{ca}(T_4) \end{aligned} \qquad (7.20)$$

When both junctions $a$-$c$ and $b$-$c$ have the same temperature (that is: $T_3 = T_4$) the last two terms cancel, and the remaining voltage only depends on the materials $a$ and $b$ of the thermocouple (*law of the third material*). This voltage is a measure of the temperature difference $T_1 - T_4$, with $T_4$ the temperature at the connection place with material $c$.

Unequal temperatures at the nodes of the connecting wires introduce a measurement error. This may happen when the distance between the measurement junction and the measurement instrument is large, and long connecting wires are needed. The best way to avoid such errors is the extension of the thermocouple wires up to the reading instrument. In general, the wires are too thin and fragile to ascertain proper operation in an industrial application. Making them thicker will increase

---

[4] Thomas Johann Seebeck (1770–1831), Estonian-German physicist.

the costs. A way out is the insertion of what is called a *compensation wire*: these are wires with different composition, diameter (and lower cost and quality), but having the same thermoelectrical characteristics as the couple itself. They only serve as an electrical connection between the open ends of the couple and the reference junction (see Figure 7.14(c)). Obviously, the cable does not introduce errors if:

the junctions $a - a'$ and $b - b'$ have the same temperature, and

the junctions $a - b$ and $a' - b'$ have the same Seebeck voltage.

It means that compensation wires should match the type of thermocouple used in the application.

Thermocouples cover a temperature range from almost 0 K to over 2900 K (however not by a single device). The various types are denoted by the characters K, E, J, N, B, R, S and T, an indication referring to the temperature range. Table 7.1 shows some properties of these types of thermocouples. The type indication refers to the application range.

The sensitivity of most couples from Table 7.1 increases somewhat with temperature. For each type the characteristic is standardized, including the non-linearity in the sensitivity.

Metal thermocouples have a relatively low sensitivity. To obtain a sensor with a higher sensitivity, a number of couples are electrically connected in series; all cold junctions are thermally connected to each other, as well as all hot junctions. The sensitivity of such a *thermopile* is $n$ times that of a single junction where $n$ is the number of couples.

Not only junctions of different metals generate thermovoltages, it also happens at a junction of different semiconductors and a junction of a metal and a semiconductor.

Table 7.1 *Properties of some popular thermocouples*

| Type indication | Composition [a; b] | Sensitivity [μV/K] | Range [°C] |
| --- | --- | --- | --- |
| K (chromel-alumel) | 90% Ni + 10% Cr; 94% Ni + 2% Al+rests | 39 (at 0°C) | −184 to 1260 |
| J (iron-constantan) | Fe, 60% Cu + 40% Ni | 45 (at 0°C) | −210 to 760 |
| E (chromel-constantan) | 90% Ni + 10% Cr; 60% Cu + 40% Ni | 60 (at 0°C) | −200 to 900 |
| S (platinum-rhodium) | Pt; 90% Pt + 10% Rh | 10 (at 1000°C) | −50 to 1600 |
| R (platinum-rhodium) | Pt; 87% Pt + 13% Rh | 14 (at 1600°C) | −50 to 1600 |
| B (platinum-rhodium) | 70% Pt + 30% Rh; 94% Pt + 6% Rh | 10 (at 1600°C) | 0 to 1800 |
| T (copper-constantan) | Cu; 60% Cu + 40% Ni | 40 (at 0°C) | −200 to 400 |

In particular, p-type silicon and n-type silicon are used for temperature measurements, because of the compatibility of these materials with integrated circuit technology. The Seebeck coefficient of silicon strongly depends on the doping level of the p or n materials, on the temperature and on the structure (monocrystalline silicon, polysilicon, amorphous silicon). Most integrated silicon thermocouples consist of junctions from single-crystal p- or n-doped silicon and aluminium, because these materials are present in standard IC technology. Typical values of the absolute Seebeck coefficient of such junctions at 300 K are around 1 mV/K, so much higher than metal-metal junctions.

As with metal thermocouples, the sensitivity of semiconductor thermoelectric sensors can be increased by a thermopile configuration. The planar technology allows the construction of several tens of junctions on a single silicon chip. Such silicon thermopiles are used for instance in infrared detectors, pyrometers (see Section 7.4.4) and other thermal sensing instruments.

### 7.4.3. Integrated temperature sensors

The relation between the current through a pn-junction and the voltage across it is given by:

$$V = \frac{kT}{q} \ln \frac{I}{I_s} \tag{7.21}$$

where $I_s$ is the saturation current. This equation holds for a pn-diode, and also for a bipolar transistor, for which $I = I_C$ (the collector current) and $V = V_{BE}$ (the base-emitter voltage) applies. The saturation current varies with temperature according to:

$$I_s \propto T^m e^{-V_{g0}/kT} \tag{7.22}$$

with $V_{g0}$ the extrapolated bandgap voltage at 0 K [12]. To use these properties for a temperature measurement, the current through the device is kept at a fixed value. Applying this to a bipolar transistor with zero base-collector voltage yields:

$$V_{BE}(T) = V_{BE0} + \lambda T + R(T) \tag{7.23}$$

with $V_{BE0}$ the extrapolated base-emitter voltage at 0 K, $\lambda$ the thermal sensitivity and $R(T)$ a small temperature dependent non-linearity term (of the second order). In integrated temperature sensors a combination of two transistors is used, having a fixed emitter area ratio $a$ and carrying different currents with a fixed ratio $r$, to cancel the non-linearity and other common effects. Then, the difference between the two base-emitter voltages satisfies the relation

$$\Delta V_{BE}(T) = \frac{kT}{q} \ln a \cdot r \tag{7.24}$$

This voltage is proportional to the absolute temperature, and a temperature sensor based on this property is called a *PTAT sensor*. The market offers several types

**Table 7.2** *Typical specifications of an integrated temperature sensor*

| Property | Value/range | Unit |
|---|---|---|
| Operating range | −25 ...+105 | °C |
| Non-linearity (full range) | 0.2 | °C |
| Output sensitivity | 1 | μA/K |

of integrated circuits containing a PTAT circuit. Table 7.2 shows some typical specifications of such a device. Obviously, the temperature range is limited by the technology used. Advantages are the easy interfacing and the low price.

### 7.4.4. Radiation thermometers

With a radiation type thermometer the temperature of a body can be measured contact free: the instrument responds to the thermal radiation of the body of which the temperature has to be measured. The total heat flux $W$ of a radiating body is proportional to the fourth power of temperature:

$$W = \sigma \cdot \varepsilon_r \cdot T_r^4 \tag{7.25}$$

with $\sigma$ the Stefan-Boltzmann constant[5,6] ($5.669 \cdot 10^{-8}$ W · m$^{-2}$K$^{-4}$) and $\varepsilon_r$ the coefficient of emission or the *emissivity* of the radiating surface. In a radiation thermometer, a thermal sensor is exposed to this radiation, and the temperature rise is measured accordingly. The sensor can be any of the devices discussed before. Since the detector itself also radiates heat, the net heat flux as detected by the instrument is

$$W = \sigma \left( \varepsilon_r T_r^4 - \varepsilon_d T_d^4 \right) \tag{7.26}$$

where $T_d$ is the detector temperature and $\varepsilon_d$ is the emissivity of the detector.

The heat flux radiated by a body covers a wave length range that depends on the temperature. The wave length where the radiation is maximum decreases with increasing temperature (Wien's law of radiation[7]). For instance, at room temperature this maximum occurs at 9.6 μm (IR). A body at 1000 K has its maximum emission at 3.5 μm, but part of the emission is within the visible range of the spectrum.

Since radiation thermometers operate contact free, they allow measurement of very high temperatures, up to 3500°C. The lower boundary of the range is about −50°C.

---

[5] Josef Stefan (1835–1893), Austrian Physicist.
[6] Ludwig Boltzmann (1844–1906), Austrian mathematical physicist.
[7] Wilhelm Wien (1864–1928), German physicist. Nobel Prize in Physics, 1911.

Another important advantage is the possibility to measure temperatures of materials having a low thermal conductivity, for instance stone.

The emissivity of a metal surface ranges from 0.03 (highly polished) to 0.8 (rough surface), and amounts 0.96 for graphitized surfaces. A proper thermal detector should have an emissivity close to 1.

A correct temperature measurement using the radiation method requires knowledge of the emissivity. The emissivity is the ratio of the emitted radiation of the actual radiating surface to that of a black body. Its value depends on the material, the surface condition and the wavelength. Usually, a radiation thermometer is calibrated for a surface with emissivity 1. If the emissivity differs from 1, the measurement result should be corrected. Since the heat is proportional to $T^4$ the relative measurement error due to an unknown emissivity is $\varepsilon^{1/4}$.

**Example 7.5 Correction for emissivity**
A radiation thermometer measures the surface temperature of a body with emissivity 0.6. The indicated temperature is 1000 K. Correct this value for the emissivity of the object.

The indicated value should be divided by $\varepsilon^{1/4} = (0.6)^{1/4} = 0.88$. So the object temperature is $1000/0.88 = 1135$ K.

The temperature sensor in a radiation thermometer can be any of the temperature sensors discussed before. If the sensor is a resistive temperature sensor, such as a platinum sensor or a thermistor, the instrument was called a *bolometer*. When the heat is measured by a thermoelectric sensor (like a thermocouple, a thermopile or a pyroelectric sensor), it was called a *pyrometer*. Nowadays, a pyrometer is more or less synonym for a radiation thermometer.

Two types of pyrometers are distinguished – the radiation pyrometer and the optical pyrometer. In a radiation pyrometer the radiation is focussed on a temperature sensitive sensor that is heated up by the radiation. The output of the sensor is a measure of the temperature. To increase the accuracy of the measurement, many pyrometers use some kind of feedback. An optical pyrometer contains a filament, which can be electronically heated to a known temperature. The filament is viewed with the hot body in the background. The temperature of the filament is adjusted manually until it appears to vanish (it has the same colour as the background). So, at this point the filament temperature is the same as the temperature to be measured. This instrument is also called a *disappearing filament pyrometer*. In the two-colour pyrometer or ratio pyrometer a similar approach is applied: the incoming radiation is adjusted automatically (for instance by a rotating wedge with graded transparency) until the transmitted intensity is a fixed fraction of the intensity of a heated filament. In all these instruments, the feedback principle reduces all kind of errors (see Chapter 3).

To further increase the accuracy, most pyrometers are provided with a modulator. The incident radiation is chopped by a rotating vane (chopper wheel), creating an alternating temperature (from ambient to exposed temperature). The resulting AC voltage has an amplitude proportional to the incident radiation, and hence to the temperature of the radiating body. The AC voltage is demodulated using synchronous detection (Section 4.3), to achieve high sensitivity and noise immunity.

## 7.5. Measurement of optical quantities

Most instruments for optical measurements respond to irradiance, the incident light power per unit area (see Table 2.11). For the measurement of photometric quantities instruments are used that have a spectral sensitivity matching that of the human eye. For instance, a lux meter measures the illuminance (that is the luminous flux per unit of area). The output is related to the human perception of light. In engineering practice, mainly radiometric quantities are considered. Since optical sensors have a wavelength-dependent sensitivity, for a correct measurement of an optical quantity the spectral sensitivity of the instrument should be taken into account.

Light is used in many instruments as an auxiliary energy source, to measure all kinds of other quantities. A particular property of the emitted light intensity is modulated by the measurand, for instance by the displacement of an object intercepting the light path. The light as received by a sensor contains the information about that measurand. Such modulating types of optical measurements outnumber by far the measurements on an independent light source, and will be discussed in more detail in later sections. In this section we briefly review some types of light sensors:

- LDR or light dependent resistor
- photodiode
- PSD (position sensitive diode)

Cameras make up a special class of optical sensors. They are used to extract more complex information about objects and processes, for instance in product inspection. CCDs and CMOS cameras will be discussed in Chapter 10 on imaging.

### 7.5.1. *Light dependent resistor (LDR)*

The resistivity of some materials, for instance cadmium sulphide (CdS) and cadmium selenide (CdSe), depends on the intensity of incident light. This is called the *photo-resistive effect*. A resistor made up of such a material is the LDR (light dependent resistor or *photoresistor*). In the absence of light the concentration of free charge carriers is low, hence the resistance of the LDR is high. When light falls on the material, free charge carriers are generated; the concentration increases and thus the resistance decreases with increasing intensity. The resistance of the LDR varies as $aE_d^{-b}$ where $E_d$ is the incident power per unit area, $a$ and $b$ being constants that depend on the material and the shape of the resistor.

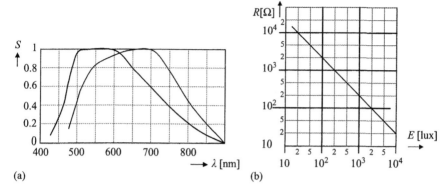

**Figure 7.15** *Characteristics of an LDR: (a) spectral sensitivity; (b) sensitivity characteristic*

The sensitivity of an LDR depends on the wavelength of the light, so it is expressed in terms of spectral sensitivity ($\Delta R/E_d$ per unit of wavelength). Figure 7.15(a) (left) shows the relative spectral sensitivity for two types of CdS resistors, one typical, and one with high sensitivity for red light. The sensitivity is normalized to the maximum sensitivity within the bandwidth. Apparently, below 400 nm and above 850 nm the LDR is not sensitive. The spectral sensitivity of this material matches rather well that of the human eye. Therefore, the sensitivity is also expressed in terms of lux. For other wavelengths other materials are used: in the near infrared region (1–3 μm) PbS and PbSe, and in the medium and far IR (up to 1000 μm) InSb, InAs and many other alloys. Figure 7.15(b) shows the resistance-illuminance characteristic of a typical CdS LDR. Note the strongly non-linear response of the device.

### Example 7.6 Sensitivity parameters of an LDR
Find the sensitivity parameters $a$ and $b$ of the LDR in Figure 7.15(b).

The resistance of the LDR is expressed as $R = aE^{-b}$ from which follows $\log R = \log a - b \log E$. Substitution of two values for R results in: $a = 2 \cdot 10^5$ and $b = 1.0$, hence $R = 2 \cdot 10^5 E^{-1}$ Ω (with $E$ in lux).

Even in complete darkness, the resistance appears to be finite; this is the *dark resistance* of the LDR, which can be more than 10 MΩ. The light-resistance is usually defined as the resistance at an intensity of 1000 lux; it may vary from 30 Ω to 300 Ω for different types. Photoresistors change their resistance value rather slowly: the response time from dark to light is about 10 ms; from light to dark, the resistance varies only by about 200 kΩ/s, resulting in a response time of about 1.5 s.

LDR's are used in situations where accuracy is not an important issue (dark-light detection only), for example in alarm systems, in the exposure-meter circuit of an automatic camera and in detectors for switching on and off streetlights.

### 7.5.2. Photodiode

The leakage current (reverse current) of a pn-diode originates from the thermal generation of free charge carriers (electron-hole pairs). Charge carriers that are produced within the depletion layer of the diode will drift away due to the electric field: electrons drift to the n-side, holes to the p-side of the junction. Both currents contribute to an external leakage current, $I_0$. The generation rate of such electron-hole pairs depends on the energy of the charge carriers: the more energy, the more free electron-hole pairs are produced. It is possible to generate extra electron-hole pairs by adding optical energy: if light is allowed to fall on the pn-junction, free charge carriers are created, increasing the leakage current of the diode. Photodiodes employ this effect: they have a reverse current that is about proportional to the intensity of the incident light. The photo diode can also act as a photovoltaic cell, when taking the voltage at zero current as output signal (Figure 7.16).

In a system where light is an auxiliary quantity, such as in optical encoders and non-contact displacement sensors (Section 8.8), the sensitivity of the optical sensor should match the spectrum of the light source. Therefore, when designing such a measurement system, the optical properties of both the source and the sensor should be considered carefully.

The major characteristics of a photodiode are:

- *spectral response* or *responsivity*: expressed in A/W or A/lm. Silicon photodiodes have a maximal response for light with a wavelength of about 800 nm;
- *noise equivalent power* or *NEP* (see 2.3.5);
- *optical bandwidth*: the wavelength interval at 50% of the relative spectral sensitivity. The optical bandwidth of a silicon photodiode is in general much wider than that of a LED or a laser diode. The bandwidth of a typical photodiode runs from 600 nm to 1000 nm approximately. LEDs have optical bandwidths in the order of 20 to 40 nm; a laser diode is much more monochromatic, with an optical bandwidth of only 1 nm;

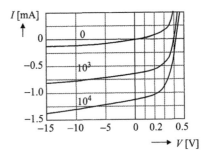

**Figure 7.16** *Typical current-voltage characteristic of a photodiode, for three values of the light input (in lux). The upper curve displays the dark current*

- *dark current*: the reverse current in the absence of light. As can be expected from the nature of the reverse current, the dark current of a photodiode increases strongly with increasing temperature; usually it is large compared with the reverse current of normal diodes, and ranges from several nA to μA, depending on the surface area of the device;
- *quantum efficiency*: the ratio between the number of optically generated electron-hole pairs and the number of incident photons; this efficiency is better than 90% at the peak wave length;
- *directivity*: specifies the sensitivity as a function of direction relative to the optical axis of the device. It is sometimes specified with the *half-angle* or beam width, defined as the angle spanned by the points in the directivity diagram where the sensitivity has dropped to half the value at the optical axis. The half-angle depends on the construction and a possibly built-in lens; values range from 5 (with lens) to 70° (without lens). Such values of half-angles also apply for LED's and laser diodes.

The detection limit of a photodiode is set by two main noise sources: a shot noise component (for a biased diode) and a thermal noise component. The first one equals

$$i_{n1} = \sqrt{2q I_d B} \tag{7.27}$$

where $I_d$ is the bias current through the diode and $B$ the bandwidth. The thermal noise is expressed as:

$$i_{n2} = \sqrt{\frac{4kTB}{R}} \tag{7.28}$$

Note that the biasing of the diode determines which noise component predominates: for an unbiased diode the thermal noise dominates over the shot noise.

### 7.5.3. Position sensitive diode (PSD)

A PSD or position sensitive diode is a light sensitive diode which is not only sensitive to incident light but also to the position where the incoming light beam hits the diode surface. It is an optical device (diode) which is particularly developed for the measurement of distances. Its operation and use will be detailed in Section 8.8. Here only the physical background is discussed. Figure 7.17(a) shows, schematically, the configuration of a PSD.

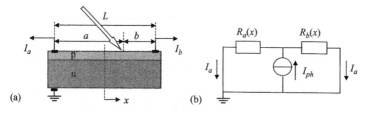

**Figure 7.17** *PSD: (a) principle of operation; (b) current division*

The device consists of an elongated silicon p-n photodiode with two connections at its extremities and one common connection at the substrate. An incident light beam can penetrate through the top layer, down to the depleted junction, where it generates electron-hole pairs. As in a normal photodiode, the junction is reversed biased, by connecting the substrate to the most positive voltage in the circuit. Due to the electric field across the junction, the free electrons are driven to the positive side, the free holes move to the negative side of the junction. The electrons arriving in the n-region flow through the common contact to ground; the holes in the p-layer split up between the two upper contacts, producing two external photo currents.

Figure 7.17(b) shows an electric model of the PSD. It is used to explain that the current division depends directly on the position of the light beam. The photo current (modelled as a current source $I_{ph}$) originates from the point where the beam hits the diode. The current ratio is determined by the resistances from this point to the two end points ($R_a$ and $R_b$). Applying Kirchhoff's current law[8] yields for the two output currents:

$$I_a = \frac{R_b}{R_a + R_b} \cdot I_{ph}$$
$$I_a = \frac{R_a}{R_a + R_b} \cdot I_{ph}$$
(7.29)

Assuming a homogeneous resistance over the whole length of the PSD, equation (7.29) can be rewritten as:

$$I_a = \frac{b}{a+b} \cdot I_{ph}$$
$$I_a = \frac{a}{a+b} \cdot I_{ph}$$
(7.30)

Generally, the origin of the coordinate system of the PSD is chosen at its centre. Thus, $x$ is the position of the light beam relative to the centre ($x = 0$). Further, when the total length of the PSD is $L$, the parameter $a$ equals $L/2 + x$ and $b$ equals $L - a$, so:

$$I_a = \left(\frac{1}{2} - \frac{x}{L}\right) \cdot I_{ph}$$
$$I_b = \left(\frac{1}{2} + \frac{x}{L}\right) \cdot I_{ph}$$
(7.31)

Apparently, the difference between the two output currents, $I_1 - I_2$ equals $2xI_{ph}/L$, and is linearly dependent of $x$ and $I_{ph}$. However, it is preferred to have an output signal that only depends on the position $x$, and not on the light intensity of the incident beam. In order to obtain such an output, $I_1 - I_2$ is divided by $I_1 + I_2 = I_{ph}$,

---

[8] Gustav Robert Kirchhoff (1824–1887), German Mathematician and physicist.

resulting in:

$$\frac{I_1 - I_2}{I_1 + I_2} = \frac{2x}{L} \tag{7.32}$$

So, to perform an intensity-independent position measurement, the PSD should be connected to an electronic interface for three signal operations: summation, subtraction and division.

With the PSD in Figure 7.17(a) the position of the light beam can be determined only in one direction. For a two-dimensional measurement, a 2D-PSD is required: this type of PSD has a square shaped sensitive surface and two pairs of upper contacts, one pair for each direction. Its operating principle is essentially the same as for the one-dimensional PSD.

## 7.6. Further reading

Most books on sensors and sensing systems discuss a wide variety of sensors. Some books focus on a limited class of sensors, for instance thermal sensors or acoustic sensors, or on a particular application area, for instance time measurement. The list below contains both general books on sensors and books on the measurement of electrical, magnetic, thermal and optical quantities.

S. Middelhoek, S.A. Audet, *Silicon Sensors*, Academic Press, 1989, ISBN 0-12-495051-5.
This is one of the first books giving a broad overview of the development of silicon sensors, with emphasis on the physics of sensors made from silicon. The introductory chapter reviews physical effects in silicon that can be used as a basis for signal conversion. Next, in subsequent chapters, the physical background and the technological aspects of silicon sensors are discussed: for radiant, mechanical, thermal, magnetic and chemical signals. The book ends with two chapters on sensor technology and sensor interfacing. The material is well documented with numerous references to original research papers.

S.M. Sze (ed.), *Semiconductor Sensors*, Wiley, 1994, ISBN 0-471-54609-7.
The organization of this book is similar to the previous one by Middelhoek; starting with two chapters on classification and technology, it proceeds with chapters on acoustic, mechanical, magnetic, radiation, thermal, chemical and biosensors; and concludes with a discussion on integrated sensor. Each chapter is written by one or two researchers in the field of semiconductor sensors. Most chapters focus on the technological aspects of each sensor group, but provide also information on the applications of commercialized semiconductor sensors.

R. Pallas-Areny, J.G. Webster, *Sensors and Signal Conditioning*, Wiley, 2001 (2nd edition), ISBN 0-471-33232-1.
This book discusses sensors and the associated interfaces, according to a different categorization from the one followed in this chapter: resistive sensors, reactive sensors, self-generating sensors, digital sensors. In separate chapters, it presents detailed information on all these sensors, their physical background, and a variety of constructions, and, for each sensor category, a chapter with detailed information on how to interface these sensors. End-of chapter

problems and end-of-book solutions are helpful for self-tuition. With respect to sensors and signal conditioning it is one level above the corresponding chapters in this book.

J. Fraden, *AIP Handbook of Modern Sensors – Physics, Designs and Applications*, American Institute of Physics, New York, 1993, ISBN 1-56396-108-3.
This book consists of three parts. The first part provides an overview on general sensor characteristics and a discussion of (most) physical topics encountered in subsequent chapters. Part two describes major properties of a selection of electronic interface circuits, including signal generators. The third part deals with a wide variety of sensors, some less common types as well. Since the physics has been treated in the first part, in this part the emphasis is on the construction of sensors, the measurement configuration and applications. The last (short) chapter is on chemical sensors.

L. Michalski, K. Eckersdorf, J. McGhee, *Temperature Measurement*, John Wiley 1991, ISBN 0-471-92229-3.
An application oriented, standard text book covering all relevant aspects of temperature measurements. The first half of the book discusses various types of thermometers: non-electric, thermoelectric, resistance and semiconductor thermometers and optical pyrometers. The second half deals with various other subjects of thermometry: temperature measurements of solid bodies, gases and liquids, the dynamic aspects of temperature measurements, transmitters and recorders, and some other particular applications.

G.C.M. Meijer, A.W. van Herwaarden (ed.), *Thermal Sensors*, IOP Publ., London, 1994, ISBN 0-7503-0220-8.
This book addresses the design of silicon sensors for the measurement of temperature and other thermal or related quantities. It contains overviews of thermal properties and relations, technological aspects, examples of sensing elements, and interface circuits. The material is illustrated with a number of case studies regarding various applications (wind meter, smart toaster, RMS converter, detection of micro-organisms etc.). The last chapter presents a number of design problems, mainly as the calculation of thermal quantities and how they are related to device characteristics.

R.S. Popovic, *Hall Effect Devices – Magnetic Sensors and Characterization of Semiconductors*, Adam Hilger, 1991, ISBN 0.7503.0096.5.
An excellent overview of Hall effect devices and all associated aspects, for both gradate and post graduate students. After a general introduction and a chapter on semiconductor physics, the physics of the Hall effect is described in much detail in Chapter 3 (covering about one third of the book). General characteristics are given in Chapter 4, where also interfacing of the devices is discussed. The remainder of the book deals with various Hall devices (not only just magnetic sensors but also magnetotransistors, MOS-Hall effect devices and carrier domain Hall effect devices) and the use of the Hall effect to characterize semiconductor materials.

P. Ripka, *Magnetic Sensors and Magnetometers*, Artech House Publ., 2000, ISBN 1-58053-057-5.
A recommendable book for any researcher working in a field related to magnetic measurements. The book covers almost every aspect of magnetic sensors, and is well written and

illustrated. As usual, the book starts with the basics of magnetism. In subsequent chapters, induction sensors, fluxgate sensors, magnetoresistors, Hall-effect sensors, magneto-optical sensors, resonance magnetosensors and SQUIDS are discussed. It follows an interesting Chapter 9 on "other principles", by the authors called "unusual types". Applications and calibration are covered in the next couple of chapters. The book finishes with a chapter on sensors for non-magnetic variables (force and torque, displacement and proximity), some of which are also described in the next chapter of this book.

## 7.7. Exercises

1. For the determination of the remaining talk time in a mobile phone, the voltage and current should be measured with high accuracy. Current is measured by a proof resistor in series with the battery. The measured voltages are converted to a digital format by an ADC.
   a. Determine the maximum value of the test resistance.
   b. What is the required resolution of the ADC (in bits) if:
      the maximum error in the measurement of the voltage is 0.1%;
      the maximum error in the current measurement is 1% over the range 10 mA to 2.2 A; the available voltage is 4.5 V?

2. A measurement bridge according to Figure 7.7(b) consists of three fixed resistances $R$ and one resistive sensor responding to the quantity $x$ as $R(x) = R(1+k \cdot x)$. The bridge supply voltage is $V_i = 10$ V. The numerical value of $k$ is 10 (units in $[x^{-1}]$). Determine the bridge output voltage for $x = 10^{-5}$ (units $[x]$).

3. The temperature in an oven is measured by a Pt-100. The resistance value appears to be 180.00 $\Omega$.
   a. Find the temperature, assuming the second order approximation of equation (7.14) and the standardized values for $R_0$, $a$ and $b$.
   b. What is the non-linearity error if the linear approximation (first order term in (7.14) only) is applied?
   c. What is the non-linearity error due to neglecting the third order term, at a temperature of 500°C?

4. The relative uncertainty in each of the parameters $V_R$, $V_o$ and $R_1$ in the simple interface circuit given below is 0.5%. What is the maximum contribution to the measurement uncertainty due to these errors, at 100°C?

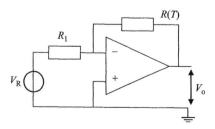

5. In the measurement system of exercise 3 $V_R = -10$ V and $R_1 = 1$ k$\Omega$. The uncertainty in the measurement result due to the offset voltage of the operational amplifier should not exceed 0.5 K at 100°C. What is the maximum allowable offset?

6. A temperature difference $T_x - T_y$ is measured using the thermocouple configuration given in the next figure. The material of the connecting wires is $a$, the thermocouples are of type K. Find the output voltage $V_o$ when $T_x = 20°C$, $T_y = 100°C$ and $T_z = 0°C$.

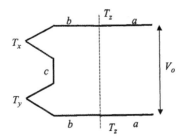

7. The next figure shows a PSD with interface circuit. The position of the light beam on the PSD is $x$, where $x = 0$ is the center of the PSD. The output quantity is $y = V_D/V_S$. All components have ideal properties. Find the position $x$ of the light beam for $y = -0.6$.

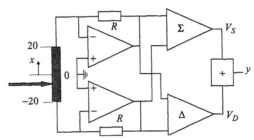

Chapter 8

# Measurement of Mechanical Quantities

There is an overwhelming number of sensor types for the mechanical domain available, operating on various physical principles. Some of them are based on electrical, optical and magnetic sensors, as discussed in the previous sections. A particular mechanical construction converts the displacement or force to be measured into an electrical, optical or magnetic quantity that can then be measured by one of the methods discussed before.

In this chapter we discuss sensors for measuring linear and rotational displacement, and force. Principally, a force sensor consists of a displacement sensor with a spring element. An accelerometer can be derived from a force sensor using a seismic mass.

The following classes of sensors for the measurement of displacement and force will be reviewed: resistive sensors (potentiometers and strain gauges), capacitive displacement sensors, magnetic and inductive displacement sensors, optical sensors, piezoelectric force sensors and accelerometers, and finally ultrasonic displacement sensors.

## 8.1. Potentiometric sensors

Potentiometric displacement sensors can be divided into linear and angular types, according to their construction. A potentiometric sensor consists of a (linear or circular) body which is either wire-wound or covered with a conductive film. A slider (wiper) can move along the wire or film, acting as a movable electrical contact. Usually, one of the end contacts is grounded and the other is connected to a voltage source. Then the slider voltage varies with the displacement, roughly from zero to the supply voltage. The ratio of the slider voltage and the supply voltage is called the *voltage ratio* or *VR*.

There is a wide variety of potentiometric displacement sensors on the market; linear types vary in length from several mm up to a few m, angular types for a range up to multiples of $2\pi$ (multiturn potentiometers, with helical windings). Built-in gears enlarge the range of single turn angular potentiometers up to a factor of 10. Combined with a spring, potentiometers act as a force sensor. With a proof mass fixed to the slider the construction is sensitive to acceleration.

The resolution of the wire-wound type potentiometer is limited by the wire diameter of the turns. When the wiper is moved continuously along the resistor body, it steps from one turn to the other, and the slider voltage changes leap-wise. So the resolution of a wire wound potentiometer is limited by the wire diameter. Actually, the wiper may short-circuit two turns when positioned just between them, resulting in a slight decrease of the total resistance, about midway between two quantization levels. The resolution can be increased (at the same outer dimension) by reducing the wire thickness. However, this reduces the reliability, because a thinner wire is less wear resistant.

The resolution of a film potentiometer is limited by the size of the carbon or silver grains that are impregnated into the plastic layer. The grain size is about 0.01 μm; the resolution is about 0.1 μm at best, but this is still more than two orders better compared to wire-wound types.

The specification of potentiometric sensors is standardized by the Instrument Society of America [1]. We list here the major items, in a short formulation:

- range (linear distance or angle)
- linearity (in % $VR$ over total range)
- hysteresis (in % $VR$ over specified range)
- resolution (average and maximum)
- mechanical travel (movement from one stop to the other)
- electrical travel (portion of mechanical travel during which an output change occurs)
- operating force or torque (break-out force or torque) to initialise movement
- dynamic force or torque (to continuously move the shaft after the first motion has occurred)
- temperature error (in % $VR$ per °C or in % $VR$ over a quarter of the full range).

The interfacing of a potentiometric sensor is essentially simple. To measure the position of the wiper, the sensor is connected to a voltage source $V_i$ with source resistance $R_s$; the output voltage on the wiper, $V_o$, is measured by an instrument with input resistance $R_i$ (Figure 8.1(a)). Ideally, the voltage transfer $V_o/V_i$ equals the voltage ratio ($VR$), and the transfer is proportional to the displacement of the wiper. Due to the source resistance $R_s$ and the load resistance $R_i$, the transfer will differ from $VR$. In particular, $R_i$ causes a non-linearity error.

In the ideal case, $R_s = 0$ and $R_i \to \infty$, the output voltage of the sensor satisfies

$$V_o = \frac{x}{L} V_i$$
$$V_o = \frac{\alpha}{\alpha_{max}} V_i \tag{8.1}$$

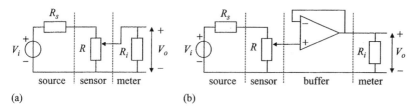

**Figure 8.1** *Interface circuits for a potentiometer: (a) direct connection; (b) with voltage buffer*

where $L$ is the total electrical length and $\alpha_{max}$ the maximum electrical angle.

A source resistance unequal to 0 causes a scale error. For $R_s \neq 0$ and $R_i \to \infty$ the sensor transfer is

$$V_o = \frac{R}{R+R_s} \cdot \frac{x}{L} \cdot V_i \approx \left(1 - \frac{R_s}{R}\right) \cdot \frac{x}{L} \cdot V_i \qquad (8.2)$$

The sensitivity is reduced by an amount $R_s/R$ with respect to the case of an ideal voltage source (zero source resistance).

A finite load resistance (the input resistance $R_i$ of the measurement instrument) results in an additional non-linearity error; assuming $R_s = 0$, the voltage transfer is:

$$\frac{V_o}{V_i} = \frac{x/L}{1 + (x/L)(1 - (x/L)) R/R_i} \approx \frac{x}{L}\left\{1 - \frac{x}{L}\left(1 - \frac{x}{L}\right)\frac{R}{R_i}\right\} \qquad (8.3)$$

The end-point linearity error (Section 3.5.4) with the end points defined by the ideal transfer $x/L$ amounts to $(x/L)^2(x/L-1)(R/R_i)$. The maximal error is $\frac{-4}{27}(R/R_i)$, occurring at $x/L = \frac{2}{3}$. This error adds up to the non-linearity error of the sensor itself, due to uneven winding or inhomogeneity of the wire or film material. Note that the maximum *relative* deviation occurs at the position $x/L = 0.5$, so midway along the potentiometer.

**Example 8.1 Non-linearity error of a potentiometric displacement sensor**
The intrinsic non-linearity of a 10 kΩ potentiometer is 0.1%. Find the minimum load resistance for an equal additional non-linearity error due to loading.

Using the value of the maximum non-linearity error as given above, we find

$$\frac{4}{27} \cdot R/R_i < 10^{-3} \text{ or } R_i > \frac{4}{27} \cdot R \cdot 10^3 = 1.5 \text{ M}\Omega$$

If the input resistance of the measurement instrument introduces an unacceptable additional non-linearity, a voltage buffer should be included (Figure 8.1(b)). Further, the sensitivity for the input voltage can be eliminated by connecting an ADC with external reference voltage. In this mode, the converter is used as a multiplying

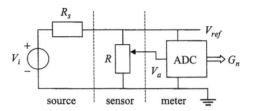

**Figure 8.2** *Buffering with a multiplying ADC*

ADC (Figure 8.2). When the reference voltage is made equal to the supply voltage of the potentiometer, the digital output is independent of $V_i$.

### 8.2. Strain gauges

Strain gauges are wire or film resistors deposited on a thin, flexible carrier material. The wire or film is very thin, so it can easily be stressed (strain is limited to about $10^{-3}$). An applied strain will change the length $l$ of the gauge, and this results in a resistance change, according to the relation:

$$R = \rho \frac{l}{A} \tag{8.4}$$

Here, $\rho$ is the resistivity of the material, $l$ the length and $A$ the cross section area of the wire or film. The resistivity of metals does not change much with strain; the parameters $l$ and $A$ change simultaneously when strained. Semiconductor strain gauges show a large variation of the resistivity with strain. This is called the *piezoresistive effect*, and is applied in for instance silicon pressure sensors. We will consider here only metal strain gauges.

There are two ways strain gauges are applied in practice:

- mounted directly on the object whose strain and stress behaviour have to be measured; when cemented properly, the strain of the object is transferred ideally to the strain gauge;
- mounted on a specially designed spring element (a bar, ring or yoke) to which the force to be measured can be applied (for instance measuring the stress in cables).

Strain gauges are suitable for the measurement of many mechanical quantities, for example force, pressure, torsion, bending and stress. Strain gauges respond primarily to strain (relative change in length, $\Delta l/l$). The applied force is found by using Hooke's law (equations (2.10) and (2.11)) and the value of the compliance or elasticity of the material on which the strain gauge is fixed.

## Example 8.2 Measurement of force in a bar

A steel bar with rectangular cross section area $A = 1$ cm$^2$ is subjected to tensile stress. The elasticity $c$ (or Young's modulus) of steel is $2 \cdot 10^{11}$ Pa. Find the strain at a load $F$ of $10^4$ N.

The specified load results in a stress of $T = F/A = 10^8$ Pa, producing a strain $S$ in the bar equal to $T/c = 2 \cdot 10^{-4}$. A strain gauge mounted on this bar experiences the same strain.

The sensitivity of a strain gauge is expressed in relative resistance change per unit of strain:

$$K = \frac{\Delta R/R}{\Delta l/l} \tag{8.5}$$

$K$ is called the *gauge factor* of the strain gauge. Actually, the change of the resistance is caused by three parameters: the resistivity, the length and the cross section area of the wire or film, cf. equation (8.4). So, the change of the resistance can be approximated by:

$$\frac{dR}{R} = \frac{d\rho}{\rho} + \frac{dl}{l} - \frac{dA}{A} \tag{8.6}$$

In general, all three parameters change simultaneously upon applying strain.

The gauge factor for metals can be calculated as follows. When an object is stressed in one direction, it experiences strain not only in this direction but also in perpendicular directions due to the Poisson effect[1]: when a wire or thin film is stressed in the principal plane (longitudinal direction) it becomes longer but also thinner: its diameter shrinks. The ratio of change in length and change in diameter is the *Poisson ratio*:

$$\mu = -\frac{dr/r}{dl/l} \tag{8.7}$$

where $l$ is the length of the wire and $r$ the radius of the circular cross section. The area of a wire with cylindrical cross section is $A = \pi r^2$, so

$$\frac{dA}{A} = 2\frac{dr}{r} \tag{8.8}$$

from which follows

$$\frac{dR}{R} = \frac{d\rho}{\rho} + \frac{dl}{l} - 2\frac{dr}{r} \tag{8.9}$$

or, using (8.7):

$$\frac{dR}{R} = \frac{d\rho}{\rho} + (1 + 2\mu)\frac{dl}{l} \tag{8.10}$$

---

[1] Simeon-Denis Poisson (1781–1840), French mathematical physicist. Named after him are the Poisson integral, Poisson's equation and the Poisson constant.

from which follows the gauge factor:

$$K = 1 + 2\mu + \frac{d\rho/\rho}{dl/l} \qquad (8.11)$$

For most metals the strain dependency of the resistivity can be neglected, hence the gauge factor of metals equals $K = 1 + 2\mu$. To find an approximate value for the Poisson ratio, we assume the volume $V = A \cdot l$ of the wire unaffected by stress:

$$\frac{dV}{V} = \frac{dA}{A} + \frac{dl}{l} = 0 \qquad (8.12)$$

resulting in

$$\frac{dl}{l} = -\frac{dA}{A} = -2\frac{dr}{r} \qquad (8.13)$$

The Poisson ratio in this ideal case is, therefore, equal to $\mu = 0.5$ so the gauge factor of a metal strain gauge is $K = 2$. In other words: *the relative resistance change equals twice the strain*. This rule of thumb holds only for metal strain gauges and is an approximation. Actual values of the Poisson ratio range from 0.25 to 0.35.

The maximal strain of a strain gauge is not large: about $10^{-3}$. For this reason, strain is often expressed in terms of *microstrain* ($\mu$strain): 1 $\mu$strain corresponds to a relative change in length of $10^{-6}$. Consequently, the resistance change is small too: a strain of 1 microstrain results in a resistance change of only $2 \cdot 10^{-6}$.

The resistance of a strain gauge changes not only with stress but also with temperature. Two parameters are important: the temperature coefficient of the resistivity and the thermal expansion coefficient. Both effects must be compensated for, because resistance variations due to stress are possibly much smaller than changes provoked by temperature variations. The effect on resistivity is minimised by a proper material choice, for instance constantan, an alloy of copper and nickel with a low temperature coefficient.

The small resistance change of strain gauges is measured invariably in a bridge (see Section 7.1.4). The bridge may contain just one, but more often two or four active strain gauges, resulting in a "half bridge" and a "full bridge" configuration, respectively (Figure 8.3). The advantages of a half bridge and a full bridge are an effective temperature compensation, a higher sensitivity and a better linearity of the sensitivity.

The general expression for the bridge circuit in Figure 8.3 is given by:

$$V_o = \left( \frac{R_2}{R_1 + R_2} - \frac{R_4}{R_3 + R_4} \right) V_i \qquad (8.14)$$

When in equilibrium, the bridge sensitivity is maximal if all four resistances are equal. We consider first the case of three fixed resistors and one strain gauge, for

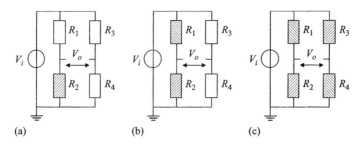

**Figure 8.3** *Bridge configurations: (a) single element bridge; (b) half bridge; (c) full bridge*

instance $R_2$ in Figure 8.3(a). Assuming $R_1 = R_3 = R_4 = R$ and $R_2 = R + \Delta R$ (which means the strain gauge has resistance $R$ at zero strain), the bridge output voltage is:

$$V_o = \frac{\Delta R}{2(2R + \Delta R)} \cdot V_i = \frac{1}{4} \cdot \frac{\Delta R}{R} \cdot \frac{1}{1 + (\Delta R/2R)} \cdot V_i \qquad (8.15)$$

The output is zero at zero strain; the bridge transfer is approximately:

$$V_o \approx \frac{1}{4} \frac{\Delta R}{R} V_i \qquad (8.16)$$

Better bridge behaviour is achieved when both $R_1$ and $R_2$ are replaced by strain gauges, in such a way that, upon loading, one gauge experiences tensile stress and the other compressive stress (Figure 8.3(b)). In practice this can be realized for instance in a test piece that bends when loaded: the gauges are fixed on either side of the bending beam, such that $R_1 = R - \Delta R$ and $R_2 = R + \Delta R$. Another possibility is to mount the two strain gauges under different angles, for instance perpendicular to each other. The output of the half bridge becomes:

$$V_o = \frac{1}{2} \frac{\Delta R}{R} \cdot V_i \qquad (8.17)$$

The transfer is linear, and twice as much compared with the bridge with only one gauge. It is easy to show that the transfer of a four-gauge or full bridge is doubled again.

When the gauges have equal gauge factors and equal temperature coefficients, the temperature sensitivity of the half and full bridge is substantially reduced compared to a bridge with just a single active element. Assume in a half bridge the two active resistances vary according to $R_1 = R - \Delta R_S + \Delta R_T$ and $R_2 = R + \Delta R_S + \Delta R_T$, where $\Delta R_S$ is the change due to strain and $\Delta R_T$ the change due to temperature. The other two resistance values are $R_3 = R_4 = R$. Substitution of these values in equation (8.14) results in

$$V_o = \frac{1}{2} \frac{\Delta R_S}{R + \Delta R_T} V_i \qquad (8.18)$$

In equilibrium ($\Delta R_S = 0$), the output voltage (the offset) is independent of $\Delta R_T$. A similar expression applies for the full bridge:

$$V_o = \frac{\Delta R_S}{R + 2\Delta R_T} V_i \qquad (8.19)$$

Note that only the temperature coefficient of the offset is eliminated, not that of the sensitivity.

**Example 8.3 Temperature sensitivity of a full strain gauge bridge**
A full bridge consists of strain gauges whose temperature coefficient (tc) is specified as $10^{-4}$/K $\pm 5$%. This tc causes a temperature dependent sensitivity, and their unequal values a temperature coefficient in the bridge offset. Find these bridge parameters.

Start with equation (8.14) and substitute $R_i = R + \Delta R_i (i = 1, \ldots, 4)$. This results in the general expression:

$$\frac{V_o}{V_i} = \frac{1}{2} \frac{\Delta R_3 - \Delta R_1 + \Delta R_2 - \Delta R_4}{2R + \Delta R_3 + \Delta R_1 + \Delta R_2 + \Delta R_4}$$

where it is assumed that $\Delta R_i / R_i \ll 1$. In a full bridge, $\Delta R_1 = \Delta R_4 = \Delta R_S$ and $\Delta R_2 = \Delta R_3 = -\Delta R_S$. This results in a strain sensitivity $V_o/V_i = \Delta R_S/R$. Temperature changes result in resistance changes $\Delta R_T$; when equal, the numerator remains zero, but the denominator becomes $2R + 4\Delta R_T$. This changes the sensitivity by a relative amount of $2\Delta R_T/R$, or $2 \cdot 10^{-4}$/K. When unequal, the numerator is, worst case, 4 times the maximum difference, that is 4 times 5% of $\Delta R_T$. So the voltage transfer becomes $5 \cdot 10^{-2} \cdot \Delta R_T/R$ per K, which means an offset of $5 \cdot 10^{-6} V_i$ per K.

Since the strain sensitivity is just $\Delta R_S/R$, a temperature change of 1 K gives as much output as does 5 $\mu$strain. This example demonstrates the need for accurate temperature matching of the strain gauges.

To facilitate mounting, manufacturers provide combined strain gauges, in various configuration, on a single carrier. Figure 8.4 shows some examples.

Strain gauges are used for the measurement of force, torque and pressure. All these sensors have in common that the load is applied to a metal spring element on which

**Figure 8.4** *Various combinations of strain gauges on a single carrier*

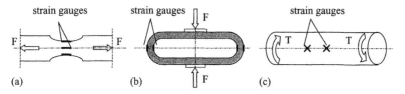

**Figure 8.5** *Load cells: different spring element designs: (a) column type; (b) yoke type; (c) torsion bar*

strain gauges are cemented. The construction of the spring element and the strain gauge arrangement determine the major properties of the device.

Figure 8.5 shows several designs of spring elements for the measurement of force, for various ranges. Figure 8.5(a) displays a typical construction for large loads (up to 5000 metric tons). The central part of the transducer is a metal bar with reduced cross sectional area where gauges are mounted. The sensing part is encapsulated in a protective housing with sealing elements to prevent damage. Load cells for smaller forces have differently shaped spring elements (Figure 8.5(b)). The position of the four strain gauges on the spring element is chosen such that one pair of gauges is loaded with compressive stress and the other pair with tensile stress, to optimise bridge sensitivity and temperature compensation. Figure 8.5(c) shows how torsion can be measured with strain gauges. A torsion in the direction indicated will elongate the gauges mounted from top left to bottom right, and shorten the other two. In all these applications the strain gauges are configured in a bridge for optimal performance.

Proper mounting and configuring of the gauges allow the measurement of very small strain values, down to 0.1 μstrain. Actually, the problem has been shifted towards the bridge amplifier. Its offset, drift and low-frequency noise obscure the measurement signal. The way out is *modulation*: the bridge circuit supply voltage is not a DC but an AC voltage with fixed amplitude and frequency, followed by synchronous detection (Section 4.3). Using this method, strains down to 0.01 μstrain can be measured easily.

## 8.3. Capacitive displacement sensors

The capacitance (or capacity) $C$ of an isolated conducting body is defined by $Q = C \cdot V$, where $Q$ is the charge on the conductor and $V$ the potential (relative to "infinity", where the potential is zero by definition). To put it differently: when a charge from infinite distance is transferred to the conductor, its potential becomes $V = Q/C$.

In practice we have a set of conductors instead of just one conductor. When a charge $Q$ is transferred from one conductor to another, the result is a voltage difference $V$ equal to $Q/C$; the conductors are oppositely charged with $Q$ and $-Q$, respectively.

Again, for this pair of conductors, $Q = C \cdot V$, where $V$ is the voltage difference between these conductors.

A well-known configuration is a set of two parallel plates, the flat-plate capacitor, with capacitance

$$C = \varepsilon \cdot \frac{A}{d} \tag{8.20}$$

where $A$ is the surface area of the plates and $d$ the distance between the plates. This is only an approximation: at the plate's edges the electric field extends outside the space between the plates, resulting in an inhomogeneous field (*stray field* or *fringe field*).

Many capacitive displacement sensors are based on flat-plate capacitors. Usually the parameters $A$ or $d$ are used for this purpose. When using the approximated expression (8.20) to evaluate the displacement from the measurement of $C$, an error is made due to the stray fields at the edges of the plates. They cause a non-linear relationship between displacement and capacitance. The non-linearity can be reduced by the application of *guarding*, a technique that is already discussed in Section 7.1.3, where it is applied to eliminate leakage currents. Similarly, to eliminate stray fields, the active plate of the capacitor (the plate that is connected to a voltage) is completely surrounded by an additional conducting plate in the same plane, and isolated from the active electrode. This is the *guard electrode*. The potential of this guard electrode is made equal to that of the active electrode (*active guarding*), using an amplifier. The result is that the electric field is homogeneous over the total area of the active electrode.

Figure 8.6 shows, schematically, two basic configurations for linear displacement measurements. In these examples the parameter $A$ (surface area of the active part) is the variable parameter. The object whose position or displacement has to be measured is mechanically fixed to the upper plate (white rectangle). In Figure 8.6(a), a linear displacement of this plate in the indicated direction introduces a capacitance change which is, ideally, $\Delta C = \varepsilon \cdot \Delta x \cdot b / d$, or a relative change $\Delta C / C = \Delta x / x$. All other parameters need be constant during movement of the plate.

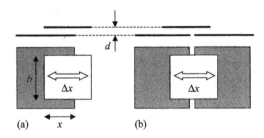

**Figure 8.6** *Basic capacitive displacement sensors with variable surface area: (a) single; (b) differential*

Figure 8.6(b) is a differential transducer (see Section 3.6.3): a displacement causes two capacitances to change simultaneously but oppositely: $\Delta C_1 = -\Delta C_2 = \varepsilon \cdot \Delta x \cdot b/d$. In neutral position ($\Delta x = 0$) a change in $d$ or $b$ (due, for instance, to backlash or a temperature change) does not cause a zero error.

Another advantage of the differential configuration is the extended dynamic range: in the reference or initial position the capacitances of the two capacitors are equal. As only the difference is processed, a small displacement results in a small signal that can electronically be amplified without overload problems. A single capacitance in neutral position may produce a considerable output signal, which could be compensated by additional electronic circuitry. However, such an electronic compensation must be very stable, making this solution less attractive.

A disadvantage of the configurations shown in Figure 8.6 is the need for an electrical connection to the moving plate, to supply and read the measurement signals. Figure 8.7(a) shows a configuration without the need for such a connection.

The moving plate acts here as a coupling electrode between two equal, flat electrodes and a read-out electrode. All fixed electrodes are in the same plane. In the reference position the coupling electrode is just midway between the two active plates, which carry periodic signals with opposite polarity. In the middle position, both (balanced) signals are coupled to the read-out electrode, and compensate each other resulting in a zero output signal. When the moving electrode shifts to an off-centre position, the coupling becomes asymmetrical, resulting in an output signal proportional to the displacement $\Delta x$ and a polarity (here phase 0 or $\pi$) indicating the direction of the movement. With the simplified model of Figure 8.7(b) it is easy to verify that the current through the readout electrode satisfies:

$$I_o = \frac{j\omega C_1 V_1 + j\omega C_2 V_2}{1 + C_1/C_3 + C_2/C_3} \quad (8.21)$$

which means, assuming $V_1 = -V_2 = V_i$, the output current $I_o$ is directly proportional to the capacitance difference $C_1 - C_2$. In a differential measurement

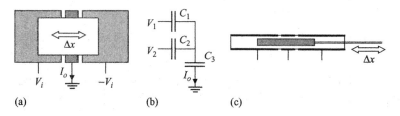

**Figure 8.7** *Differential capacitive displacement sensor with current read-out: (a) top view of a flat plate capacitor; (b) electronic model; (c) cross section of a cylindrical capacitor*

configuration, $C_1 = C_0 + \Delta C$ respectively $C_2 = C_0 - \Delta C$, hence

$$I_o = j\omega \cdot \frac{2\Delta C}{1 + 2C_0/C_3} \cdot V_i \qquad (8.22)$$

The output current appears to be proportional to $\Delta C$ and consequently to the displacement $\Delta x$. Capacitive displacement sensors can also be configured in a cylindrical configuration (Figure 8.7(c)). The principle is the same, but a cylindrical set-up is more compact, has fewer stray capacitances and therefore a better linearity. This type of capacitive sensor is called a *linear variable differential capacitor* or LVDC. The sensor exhibits extremely good linearity and a low temperature sensitivity (down to 10 ppm/K).

Capacitance changes can be measured using various interface circuits. We discuss three major methods (Figure 8.8):

- impedance measurement (preferably in a bridge);
- current-voltage measurement (with an operational amplifier);
- time measurement (charging and discharging $C$ with constant current).

Each of these methods will be shortly analysed. For a half bridge configuration, the transfer (Figure 8.8(a)) is:

$$\frac{V_o}{V_i} = \frac{1}{2} - \frac{C_1}{C_1 + C_2} = \frac{1}{2} \cdot \frac{C_2 - C_1}{C_1 + C_2} \qquad (8.23)$$

In a single sensor configuration, $C_1 = C + \Delta C$ and $C_2 = C$, hence:

$$\frac{V_o}{V_i} \approx -\left(\frac{\Delta C}{4C}\right)\left(1 - \frac{\Delta C}{2C}\right) \qquad (8.24)$$

In a differential or balanced configuration $C_1 = C + \Delta C$ and $C_2 = C - \Delta C$, thus:

$$\frac{V_o}{V_i} = -\frac{\Delta C}{2C} \qquad (8.25)$$

Clearly, the differential method yields better linearity and a wider dynamic range, similar to a differential strain gauge bridge. The transfer for the current-voltage systems of Figure 8.8(b) and (c) is:

$$\frac{V_o}{V_i} = \frac{C + \Delta C}{C_3}; \quad \frac{V_o}{V_i} = 2\frac{\Delta C}{C_3} \qquad (8.26)$$

respectively for a single and a differential configuration. Again, the differential mode has clear advantages over the single mode.

An interface circuit that combines the advantages of a time signal output and high linearity is given in Figure 8.8(d). The unknown capacitance is periodically charged

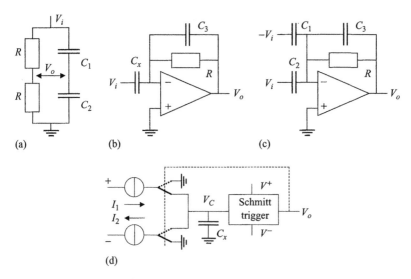

**Figure 8.8** *Interface circuits for capacitive sensors: (a) half-bridge; (b) current-voltage measurement (single); (c) current-voltage measurement (differential); (d) time measurement*

and discharged with currents $I_1$ and $I_2$. During charging, $V_c$ increases linearly with time until the upper hysteresis level $V_s$ of the Schmitt trigger is reached. At this moment the switches turn over and the capacitance is discharged, till the lower hysteresis level $-V_s$ is reached. The result is a periodic triangular signal over the capacitance, with fixed amplitude between the levels $\pm V_s$ of the Schmitt trigger and a frequency related to the capacitance according to:

$$f = \frac{I}{2C_x V_s} \tag{8.27}$$

where charge and discharge currents are supposed to be equal: $I = I_1 = I_2$. The output of the Schmitt trigger is a symmetric rectangular signal with the same frequency. The method yields accurate results, since the frequency only depends on these fixed levels and the charging and discharging currents.

### Example 8.4
In the circuit of Figure 8.8(d) the parameters have the values: $I = 1$ mA; $C_x(x = 0) = C_0 = 100$ pF; $V_s = 5$ V. Find the frequency and the sensitivity to capacitance changes.

According to equation (8.27) the frequency of the triangular signal is 1 MHz. The sensitivity of the interface circuit follows from the derivative of the frequency to the capacitance; around the value $C_x = C_0$ this is $\Delta f/C_x = -f_0/C_0$, or $-10$ kHz/pF.

A further method to interface a capacitive sensor is based on oscillation. The unknown capacitance is part of an oscillator circuit, which generates a periodical signal with a frequency that depends on the value of the capacitance. Obviously, the other circuit elements must be constant and the oscillation process must be guaranteed over the entire capacitance range. Such circuits may therefore become critical and complex.

## 8.4. Magnetic and inductive displacement sensors

All sensors for measuring linear and angular displacement discussed in this section operate on the basis of a magnetic field. Either the magnetic field strength itself is a parameter related to the measurand, or the field creates an induction voltage that depends on a geometric parameter. Magnetic types use (permanent) magnets and magnetic field sensors to convert positional information into an electrical signal. In inductive sensors, a voltage or current is induced in proportion to the distance or angle that has to be measured. Further, component parameters such as self-inductance or mutual inductance can be modulated by a geometrical measurand.

### 8.4.1. *Magnetic proximity and displacement sensors*

The sensors in this section are based on a position-dependent magnetic field strength $H(x)$. The strength of a magnetic field, originating from a permanent magnet or a coil depends on the distance from the magnetic source. The field strength decreases with increasing distance in a way that depends on the pattern of magnetic field lines. Either the magnetic source or the sensor is connected to the moving part of the construction. The field strength can be measured by any magnetic sensor, for instance a Hall sensor (see Section 7.3). This type of displacement sensor in general has a limited range, a strong non-linear sensitivity and is sensitive to the orientation of the source relative to the sensor. Ferromagnetic construction parts in the neighbourhood of the sensing system may disturb the field pattern and hence the transfer characteristic of the sensor system. In general, magnetic displacement sensors of this type are only suitable for low-accuracy applications, for instance presence detection of an object (on-off mode).

The range can easily be extended by applying an array of sensors on the fixed part and one magnet on the moving part of a construction (see Figure 8.9). Each sensor $S_i$ responds to the passing magnet in a similar manner, over a restricted range, but the combination of outputs provide unambiguous information on the position of the magnet over a range covered by the sensor array.

**Example 8.5 Displacement measurement using an array**
A linear array system according to Figure 8.9 consists of 10 sensors positioned 10 cm apart, the first at $x = 0$. The response of each sensor equals

$$V_i(x) = e^{-(x-10i)^2/50}, \quad i = 0, \ldots, 9$$

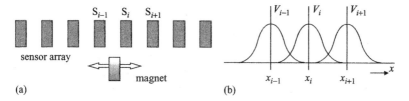

**Figure 8.9** *Sensor array to extend the range: (a) magnet array; (b) subsequent sensitivity curves*

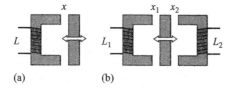

**Figure 8.10** *Displacement sensors based on variable self-inductance: (a) single; (b) differential*

with $x$ the position of the magnet in cm. Determine the output signals $V_0(0)$, $V_0(10)$ and $V_2(17)$. Derive the position of the magnet when the outputs $V_0 = 0.7$ and $V_1 = 0.5$, and all other outputs almost 0.

$V_0(0) = e^0 = 1$ (the maximum output); $V_0(10) = e^{-2} = 0.135$ (magnet in front of sensor 1, signal at adjacent sensor); $V_2(17) = e^{-9/50}) = 0.835$.

From $V_0 = 0.7$ it follows $x = \pm 4.2$ cm, and from $V_1 = 0.5$ it follows $x = 10 \pm 5.9$ cm. The combination gives $x = +4.15 \pm 0.5$ cm.

### 8.4.2. Inductive proximity and displacement sensors

The impedance of a magnetic circuit is determined by its geometry and material parameters. A change in geometry, for instance due to a displacement of part of the circuit, will change the impedance. The general principle of such a displacement sensor is depicted in Figure 8.10. The self-inductance of the configuration is, approximately, equal to $L = n^2/R_m$, with $R_m$ the reluctance (see Section 2.3.2). The latter depends on the length and the cross section area of the flux through the structure as well as the material. Using expressions from Table 2.5 we find for the reluctance:

$$R_m = \frac{1}{\mu_0 \mu_r} \frac{l_{core}}{A} + \frac{1}{\mu_0} \frac{2x}{A} \tag{8.28}$$

where $l_{core}$ is the length of the flux path through the core (with relative permeability $\mu_r$), $2x$ the length of the flux path through the air gap and $A$ the cross section area of the core. So the self-inductance is (approximated):

$$L = \frac{\mu_0 n^2 A}{l_{core}/\mu_r + 2x} \qquad (8.29)$$

When displacements are to be measured relative to a reference or initial position $x_o$ the sensitivity can be expressed as:

$$\frac{\Delta L}{L_o} = \frac{-2\Delta x}{l_o + 2\Delta x} \approx \frac{-2\Delta x}{l_o} \qquad (8.30)$$

where $l_o = l_{core}/\mu_r + 2x_o$ is the effective length of the flux path in the initial position $x = x_o$. The sensor has a restricted range, because the sensitivity strongly decreases with increasing $x$ as can be seen from equation (8.29). A differential configuration gives some improvement (Figure 8.10(b)). The output is the difference between $L_1$ and $L_2$, so the sensitivity is doubled. Moreover, the non-linearity is reduced, as can be seen from (8.31), which shows the sensitivity of this differential configuration, where again $L_o$ and $l_o$ are the self-inductance and the effective flux path in the initial position.

$$\frac{L_2 - L_1}{L_o} = \frac{-4\Delta x}{l_o - 4(\Delta x)^2/l_o} \approx \frac{-4\Delta x}{l_o} \qquad (8.31)$$

**Example 8.6 Non-linearity of inductive displacement sensors**
The flux path of the sensors in Figure 8.10 amounts to 4 cm. Calculate the non-linearity error at a displacement of 0.4 mm, in both configurations.

The non-linearity error of the single sensor is about $-2\Delta x/l_0 = 2\%$, that of the differential sensor about $4(\Delta x/l_0)^2 = 0.04\%$.

Another sensor configuration with coils is shown in Figure 8.11(a): a coil with movable core. The range is substantially wider but again, the self-inductance varies in a non-linear manner with displacement. Also in this case, the differential configuration in Figure 8.11(b) has a sensitivity that is twice as high and also it has a better linearity. The two self-inductances are measured in a bridge configuration (Figure 8.11(c)), to minimize temperature sensitivity and non-linearity. This type

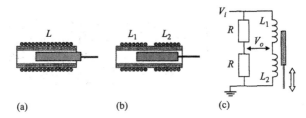

**Figure 8.11** *Displacement sensors with movable core: (a) single; (b)differential; (c) bridge circuit*

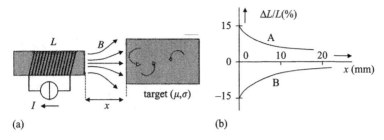

**Figure 8.12** *Eddy current proximity sensor: (a) illustration of the principle; (b) typical sensitivity curves: A ferromagnetic material and B conductive material*

of inductive displacement sensor should not be confused with the LVDT, which will be introduced in Section 8.4.4.

### 8.4.3. Eddy current displacement sensors

Eddy currents originate from induction: free charge carriers (electrons in a metal) experience Lorentz forces in an alternating magnetic field, and cause currents to flow in that material. In an eddy current sensor this effect is used to measure displacement (Figure 8.12(a)).

An eddy current proximity sensor consists of a coil with core. When the coil is activated by an AC current, an induction field $B(x, y, z)$ is present around the coil. The shape of that field depends on the shape of the coil and core, and on conductive and ferromagnetic objects in its vicinity. The strength of the field can be expressed in terms of magnetic flux $\Phi$, and since $\Phi = L \cdot I$ the flux is proportional to the current (Table 2.4). Charge carriers in a conductive object in the vicinity of the coil experience Lorentz forces, resulting in currents flowing more or less randomly through the structure. These currents produce a magnetic field that counteracts the original field from the coil. Hence, the total flux from the coil reduces and, since $L = \Phi/I$, the self-inductance decreases due to the presence of the conductive object. This effect is stronger in materials with a low resistivity, because of the higher currents. In conclusion: the sensor impedance *decreases* upon the approach of a *conducting, non-ferromagnetic material*.

The situation is different when the target is ferromagnetic. The reluctance (Section 2.3.2) of a ferromagnetic material is, approximately, $R_m = l/\mu_0 \mu_r A$, where $l$ is the length of the field line under consideration in the magnetic structure and $A$ the cross section. Further, the self-inductance equals $L = n^2/R_m$. Hence, when a ferromagnetic material approaches a coil, the impedance of the coil *increases*. This effect appears in materials having a permeability larger than 1.

Obviously, the sensitivity of an eddy current sensor depends strongly on the material of the target (Figure 8.12(b)). For highly conductive targets or targets with a high permeability, the maximum change in self-inductance is about 20% when the object approaches the sensor from infinite to zero distance. The sensor impedance is a rather complicated function of the distance: the frequency, the nature of the material, and the orientation of the object have substantial influence on the sensor sensitivity. Other metal or ferromagnetic objects in the neighbourhood of the sensor may reduce the measurement accuracy. However, eddy current sensors have a high resolution, due to the intrinsically analogue operation.

The measurement frequency of an eddy current sensor is in the order of 1 MHz. Some types contain a second or compensation coil, arranged in a bridge or a transformer configuration. Eddy current sensors integrated with electronic circuits on silicon is a subject of current research [2, 3], to extend their applicability.

Eddy current sensors are used in a variety of applications. They are suitable for testing the surface quality of metal objects (cracks increase the resistance), object classification based on material properties (the sensor impedance depends on the resistivity); distance measurement; and for event detection (holes, edges, profiles etc. in metal objects) [4, 5]. Non-ferro and isolating objects can be detected as well, by applying a thin conducting layer on the target object, for instance an aluminium foil.

### 8.4.4. Transformer-type sensors

Most of the displacement sensors discussed thus far show a rather non-linear transfer characteristic. Even the balanced coil with movable core (Figure 8.11) has a non-linear characteristic. A much better linearity is obtained with a transformer with variable core, the *linear variable differential transformer* (LVDT). This device consists of one primary coil and two secondary coils, positioned symmetrically with respect to the primary coil (Figure 8.13(a)), and wound in an opposing direction.

When the core is in its centre position the voltages over the secondary coils are equal but with opposite polarity (due to the winding directions). The total output voltage of the two coils in series is zero. When the core shifts from its neutral position,

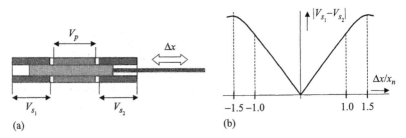

**Figure 8.13** *Principle of operation and sensitivity characteristic of an LVDT: (a) construction; (b) transfer characteristic*

the output voltages are not equal anymore, due to the resulting asymmetry. The output amplitude varies linearly with the displacement of the core; the phase with respect to the primary voltage is either 0 or $\pi$, depending on the direction of the displacement. The coils are wound around the cylindrical tube in a special way, to obtain a linear characteristic over a large part of the mechanical travel of the device (Figure 8.13(b)).

The LVDT is operated with AC voltages; their amplitude and frequency can be chosen by the user within a wide range (typically 1 to 10 V, 50 Hz to 10 kHz). Its sensitivity is expressed in mV (output) per volt (input) per m (displacement), so in $mV \cdot V^{-1} \cdot m^{-1}$.

The measurement range of an LVDT is set by its size and construction. The smallest types have a $\pm 1$ mm range, the largest up to $\pm 50$ cm (or a stroke of 1 m).

Users who like to avoid the handling of AC signals (amplitude and phase measurement) can choose the (more expensive) DC-DC LVDT, with built-in oscillator and phase sensitive detector: both input and output signals are DC voltages.

Inductive rotational sensors for the full range of $2\pi$ are the *resolver* and the *synchro*. These types are also based on a variable transformer, with fixed and revolving coils. The coupling between primary and secondary coils depends on the angle between the coils. The resolver (Figure 8.14) consists of two fixed coils (stator) whose axes make an angle of $\pi/2$, and one rotating coil, the rotor. The shaft of this rotor is connected to the rotating object whose angle should be measured.

One possible mode is to supply voltages to the stator coils, with a phase difference $\pi/2$ rad: $V_1 = V \cos \omega t$ and $V_2 = V \sin \omega t$, respectively. Both coils induce voltages in the rotor coil, where they are added together. Assuming equal amplitudes of the stator voltages, the rotor voltage is:

$$V_3 = a \cdot V \cdot \cos \omega t \cos \alpha + a \cdot V \cdot \sin \omega t \sin \alpha = a \cdot V \cdot \cos(\omega t + \alpha) \quad (8.32)$$

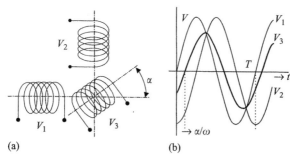

**Figure 8.14** *Operating principle of a resolver: (a) rotor and stator coils; (b) coil voltages*

where the factor $a$ accounts for the transfer from stator to rotor. Apparently, the phase of $V_3$ relative to $V_1$ is equal to the geometric angle of the rotor (Figure 8.14(b)). Transmission of the rotor voltage to the terminals is performed by either sliprings and brushes or an additional coil, acting as the secondary of a transformer with fixed transfer (the brushless resolver).

In another mode the input voltage is supplied to the rotor, and the sensor output consists of two stator voltages equal to:

$$V_{s1} = a \cdot V \cdot \cos \omega t \cos \alpha$$
$$V_{s2} = a \cdot V \cdot \sin \omega t \sin \alpha$$
(8.33)

from which the shaft angle can be derived by proper signal processing.

A configuration with three stator coils is called a synchro. The three coils make geometric angles of 120° and are supplied with three sinusoidal voltages with phase differences of 120° too. The principle is the same as for the resolver.

Resolvers and synchros have a measurement range of $2\pi$ rad. The performance is limited by the read-out electronics, in particular the phase measurement circuitry. The accuracy can be better than 0.001 rad whereas a resolution down to 0.0001 rad can be achieved. Resolvers operate at frequencies between 1 kHz and 10 kHz, synchros traditionally at 400 Hz.

A variation on the resolver is the inductosyn®. We discuss only the linear inductosyn here: the rotational version operates in the same way. The linear inductosyn (Figure 8.15) consists of a fixed part, the *ruler*, and a movable part, the *slider*.

Both ruler and slider have meander shaped conducting strips on an isolating carrier (like printed circuit boards), and are isolated from each other by a thin insulating layer. The ruler acts as one elongated coil; its length determines the measurement range. The slider consists of two short coils, with the same periodicity $p$ as that of the ruler. The geometric distance between the two parts of the slider is $\frac{1}{4}$ period (actually $p/4$ plus a number of full periods). When the slider moves along the ruler, the inductive coupling between these parts varies with the same periodicity as the geometry of the structure. Due to the $\frac{1}{4}$ period distance shift, the induced voltages

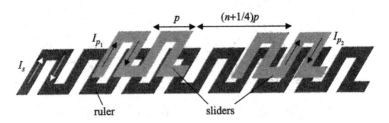

**Figure 8.15** *Lay-out of an inductosyn®*

show a phase shift of $\pi/2$ rad, which can be used to determine the direction of the displacement (for a full explanation of this principle, see Section 8.5.4 on optical encoders).

The inductosyn is activated in a similar way to the resolver. Here too, the input voltage can be applied to either the ruler or the slider. In the last case, the ruler receives the sum of two induced voltages, according to:

$$V_o = a \cdot V \cdot \cos\left(\frac{2\pi x}{p}\right) \cos \omega t + a \cdot V \cdot \sin\left(\frac{2\pi x}{p}\right) \sin \omega t$$

$$= a \cdot V \cdot \cos\left(\omega t + \frac{2\pi x}{p}\right) \tag{8.34}$$

where $x$ is the displacement and $p$ the periodicity of the structure (also called the cyclic length). Despite a very short distance between ruler and slider (typically a few μm) the coupling between these parts is rather weak, resulting in a small output signal. Nevertheless, the amplitude is almost independent of the slider position and can be amplified up to a value where the phase can be measured with high accuracy.

To measure absolute displacement over the total extent of the ruler, the system must keep track of the number of electrical periods that have passed. In this way, an almost infinite range can be achieved (available up to 35 m), combined with a high resolution (1 μm or better). The rotational version of the inductosyn has a resolution better than 0.5 arc seconds.

## 8.5. Optical displacement sensors

In this section the measurand (a linear or angular displacement) modulates a particular parameter of an optical carrier (light beam), for instance the intensity, the deflection or the length of the travelled path.

### 8.5.1. *Optical distance sensing by intensity*

A simple though rather inaccurate method to measure displacement is based on the optic inverse square law: light intensity from a point source decreases with the square of the distance (see Figure 2.3). Figure 8.16 shows the general set-up.

In the direct mode (Figure 8.16(a)) a source (LED) is mounted on the fixed part of the construction, whereas the detector is positioned on the moving part (or vice versa). The detector's output varies with distance between the two parts. The output not only depends on distance but also on the source intensity and the sensitivity of the detector. These parameters should have sufficient stability to be able to distinguish them from displacement-induced intensity changes. The stability requirements can be weakened by applying a second detector, with fixed position relative to the

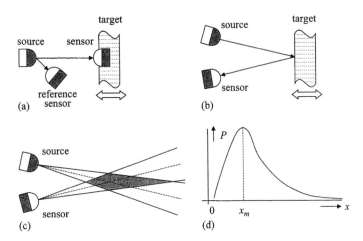

**Figure 8.16** *Optical proximity sensors: (a) direct mode; (b) reflection mode; (c) active range; (d) typical transfer characteristic*

source, as also shown in Figure 8.16(a). The output quantity is the difference or (better but more complicated to implement) the ratio of both detector signals. If both detectors have stable, identical sensitivity, source instability can be completely eliminated. The final accuracy is limited by changes in the difference between the parameters of the detectors.

In the reflection mode (Figure 8.16(b)), both source and detector are mounted on the fixed part of the construction. The emitter casts a beam on the object which scatters this light in all directions (diffuse reflection; we disregard here specular reflection for simplicity). Part of the reflected light arrives at the detector: its intensity depends on the distance from the source-detector system to the moving object and, unfortunately, on the reflection properties of the object as well. Actually, this principle allows non-contact distance sensing (in a mechanical sense). When configured for small distances, the sensor system is referred to as an *optical proximity sensor*. Source and detector are mounted in a single housing. Their optical axes make a specific angle; this angle is an important parameter for the sensitivity characteristic, in particular the useful range of the sensing system (Figure 8.16(c)). Figure 8.16(d) shows a typical transfer characteristic of such a proximity sensor. Both the lower and higher limit of the detection range is set by the configuration geometry and by the directivity properties of the source and detector: when the object is very close to the detector the scattered light cannot reach the detector; at large distances the sensitivity decreases due to the inverse relation between intensity and position.

**Example 8.7 Transfer characteristic of an intensity displacement sensor**
Suppose in Figure 8.16(b) that all the light emitted by the source falls on the reflective part of the target, and the whole light spot on the target is "seen" by the detector. How does the detector output depend on the distance $x$ between the source and the target? And how is that

relation affected if the target has small dimensions, such that the reflective part of the target falls completely within the light cone of the source?

If the light beam falls completely on the target for all distances that are considered, the total available optical power arrives at the target, independent of the distance. The light spot on the target acts as a (secondary) light source, and part of the light is reflected into the direction of the detector. The detector output is proportional to the radiance, so inversely proportional to the square of the distance: $P \propto 1/x^2$.

In the second case, the whole target remains illuminated, for all values of the distance. So, the irradiance drops with the square of the distance to the source. Combined with the inverse square relation between the detector output and the illuminated target, this results in an output that drops with the 4th power of the distance: $P \propto 1/x^4$.

### 8.5.2. Optical distance sensing by triangulation

A triangle is completely determined by three parameters (sides and angles), a property that is used to find an unknown distance. Assume $x$ in Figure 8.17(a) is an unknown distance. When $a$ and the angles $\alpha$ and $\beta$ are known, $x$ follows from the equation

$$x = a \frac{\sin \beta}{\sin(\alpha + \beta)} \tag{8.35}$$

This property is used to find an unknown distance, by projecting this distance onto a triangle with sufficiently known parameters. We discuss two modes: the direct mode, where the light source is mounted on the moving part (as for instance on top of a vehicle whose position has to be measured), and the indirect or reflection mode, where the displacement of the moving part is measured just using light that is reflected from the target. For simplicity we consider only the case of a one-dimensional displacement measurement.

The most simple situation is depicted in Figure 8.17(b). Here, the position $x$ of a moving target is measured using a light source fixed to the target. The light is projected through a lens with focal length $f$ onto a position-sensitive device. This

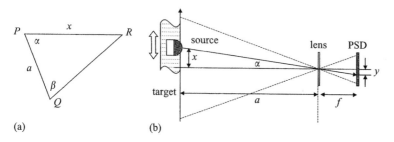

**Figure 8.17** *Position measurement using triangulation: (a) principle of triangulation; (b) application with a PSD, for a one-dimensional case and direct mode*

device can be either a PSD (see Section 7.5.3) or a linear diode array. We consider a configuration with a PSD. In this triangulation system, the known parameters in the left triangle are the fixed distance $a$ and the right angle between the optical axis and the straight target trajectory. The angle $\alpha$ is derived from the output of the PSD. From Figure 8.17(b) it follows for the relation between the object position $x$ and the light spot position $y$ on the PSD:

$$\frac{x}{a} = -\frac{y}{f} \tag{8.36}$$

The measurement range is limited by the dimension of the PSD and the distance $a$.

In situations where it is unpractical to have a light source on the moving target, the reflection mode is preferred. The geometric configuration is more complicated: the source and detector should be positioned in such a way that over the whole measurement range the reflected light falls on the PSD. Moreover, the whole range of the PSD should be used for an optimal performance of the system.

Figure 8.18 shows a possible configuration of a one-dimensional displacement sensing system using the reflection mode. The distance to be measured is $x$, running from the light source to the point where the beam hits the diffusively reflecting target. The incident light is scattered, and part of it is focussed on the PSD. The detector is positioned at the focal point of the optical system. A displacement of the target along the indicated direction causes a displacement of the projected light spot on the sensor. The relation between the distance $x$ to be measured and the position $y$ on the detector is found using the goniometric relation:

$$\tan \alpha = \tan(\beta + \gamma) = \frac{\tan \beta + \tan \gamma}{1 - \tan \beta \tan \gamma} \tag{8.37}$$

where $\alpha$ is the angle between the optical axes of the source and the detector, respectively. With $y/f = \tan \beta$ and $a/x = \tan \gamma$ we arrive at

$$x = a \cdot \frac{y \tan \alpha + f}{f \tan \alpha - y} \tag{8.38}$$

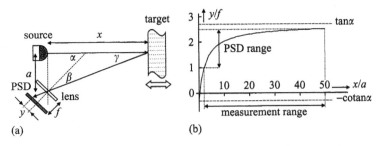

**Figure 8.18** *Example of a triangulation system in reflection mode: (a) configuration; (b) sensitivity curve for $\alpha = 70°$*

A plot of this relation between PSD position and target position is depicted in Figure 8.18(b), for the case $\alpha = 70°$. The value of $y/f$ ranges from $-\cotan \alpha$ to $\tan \alpha$. However, the measurement range depends on the length of the PSD as shown in this figure. This range can further be optimised by a proper choice of the parameters $a$, $f$ and $\alpha$. At a fixed length of the PSD, the measurement range increases for increasing $a$ and decreasing $f$.

Although the measurement range can be made rather large, as is shown by Figure 8.18(b), it is not recommended to use the rightmost part of the sensitivity curve. Here, the output (position on the PSD) changes only slightly with the target displacement, making the system susceptible to errors. For instance, a small error in the current-to-voltage converter of the PSD is equivalent to a relatively large displacement of the target.

Figure 8.18 also shows the effect of the parameter $\alpha$ on the range and the sensitivity. For the special case $\alpha = \pi/2$ (the optical axes of the source and the detector are perpendicular) the transfer is linear: $y/f = x/a$, a situation that is similar to the one in Figure 8.17; the range however is rather small. For $\alpha = 0$ (parallel optical axes) the transfer is simply the inverse: $y/f = -a/x$. This is a suitable choice for a wider range application.

**Example 8.8 Measurement range of an optical triangulation system with PSD**
An optical displacement sensor according to Figure 8.18 should have a measurement range from 20 cm to 200 cm. Discuss a design with $\alpha = 0°$, that meets this requirement. Determine the minimum active length of the PSD and its position with respect to the lens.

For this particular case $y/f = -a/x$. A proper value for the distance $a$ is 4 cm. The extreme positions of the light spot on the PSD are found to be $\frac{-4}{20} = -0.2$ and $\frac{-4}{200} = 0.02$. With a lens with focal length 10 cm the extreme positions are $-2$ cm and $-0.2$ cm. So the minimum length of the PSD should be about 2 cm.

The size of the light spot image on the sensor depends on the beam diameter and divergence, and may be substantially larger than the required resolution of the detector. For a PSD this does not matter because this device responds to the *optical centre of intensity*, which lies on the main axis of the light beam. Obviously the spot should not be wider than the width of the PSD.

When using a diode array, the light spot may simultaneously activate several elements of the array. Therefore, the element in the centre of all activated elements should be selected as being the best position of the spot.

### 8.5.3. *Optical tachometers*

Figure 8.19 presents two simple optical *tachometers*. One or more segments of a reflective material, such as aluminium foil, are put along the circumference of the rotating shaft or on the head of the shaft.

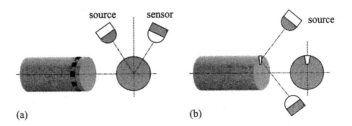

**Figure 8.19** *Optical tachometers: (a) reflection from the circumference; (b) reflection from the head*

In the case of a reflective shaft one should take an absorbing material, like black paint or black paper. The output signal from the photodetector is a pulse shaped signal, with pulse frequency $f = n \cdot p/2$ Hz, with $n$ the number of revolutions per second and $p$ the number of alternations between reflective and non-reflective material along the circumference. The measured angular speed equals $2\pi n = 4\pi f/p$ rad/s.

At high speed, only one pair of black-and-white areas suffices ($p = 2$). Accurate measurement of a lower speed requires a larger number of intensity changes per revolution. Optical tachometers of this type are available in various sizes and ranges, up to 12 000 rpm.

### 8.5.4. Optical encoders

The optical sensors described in this section are digital in nature. They convert, through an optical intermediate, the quantity that has to be measured into a binary signal, representing a digitally coded measurement value. Sensor types that belong to this category are optical encoders, designed for measuring linear and angular displacement, optical tachometers (measuring angular speed or the number of revolutions per unit of time, as discussed in 8.5.3) and optical bar code systems for identification purposes.

Optical encoders are composed of a light source, a light detector and a coding device. The latter consists of a flat strip or disk, containing a pattern of alternating opaque and transparent segments (transmission mode) or alternating reflective and absorbing segments (reflection mode). Both possibilities are illustrated in Figure 8.20.

The coding device can move relatively to the assembly of source and detector, causing the radiant transfer between them switching between a high and a low value. For use in the transmission mode, the encoder consists either of a translucent material (glass, plastic, mylar), covered with a pattern of an opaque material, or just the reverse, for instance a metal plate with slots.

Measurement of Mechanical Quantities    223

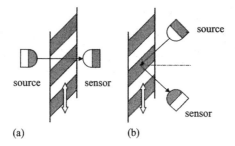

**Figure 8.20** *Optical encoders in: (a) transmission mode; (b) reflection mode*

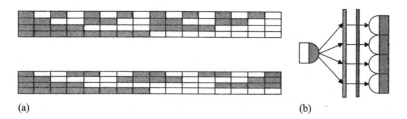

**Figure 8.21** *Linear absolute encoders: (a) rulers (top: natural code; bottom: Gray code); (b) optical system: source – diffuser – ruler – sensors*

Two basic encoder types are distinguished: absolute and incremental encoders. An absolute encoder gives instantaneous information about the absolute displacement or the angular position. Figure 8.21(a) gives two examples of linear absolute encoders.

Each (discrete) position corresponds to a unique code, which is obtained by the optical read-out system. The acquisition of absolute position with a resolution of $n$ bits requires at least $n$ optical tracks on the encoder and $n$ separate detectors (Figure 8.21(b)). Obviously, the maximum resolution of an absolute encoder is set by that of the LSB (least significant bit) track, which evidently is the outer track of an angular encoder, and the chosen size of the detectors.

The natural binary code is polystrophic, which means that two or more code bits may change for a 1 LSB displacement (Figure 8.21(a) top). Mechanical tolerances will introduce error codes between two adjacent binary words (compare glitches in DA converters, Section 6.1). For this reason, most encoders use the Gray code, which requires also the minimum number of tracks at a given resolution, but is monostrophic: at each LSB displacement only one bit will change at a time (Figure 8.21(a), bottom).

Serious disadvantages of an absolute encoder are its limited resolution, the large number of output sensors and the consequently high price. A much better resolution is achieved by the *incremental encoder*, of which Figure 8.22 shows a typical example for linear displacement. It has only one track, consisting of a large number

**Figure 8.22** *Linear incremental encoder with coded markers*

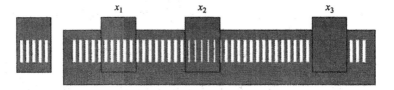

**Figure 8.23** *Linear incremental encoder with fixed scale and moving mask*

of transparent slots in the scale. In the angular type these slots are arranged radially near the circumference of the disc.

The output of the sensor changes alternately from low to high when the encoder is displaced relative to the source-sensor system, generating a number of pulses equal to the number of transitions from light to dark (or vice versa). Although the actual output signal does not contain information about the absolute position of the encoder, this position can nevertheless be found by counting the number of output transitions or pulses, starting from a fixed reference position. Such a reference position is marked by an extra slot in the disc or scale. Long scales have multiple markers, which are coded to make them distinguishable from each other. In Figure 8.22 the markers are identical but positioned on different distances. The number of lines between two adjacent markers is a unique code for the marker position.

If the incremental encoder is used in the same way as the absolute encoder (one detector for each line) its resolution would not be much better: the width of the slot should not be less than the size of the detector. However, when the sensor head (with the source and sensor) is combined with a mask a much higher resolution can be achieved, as is illustrated in Figure 8.23. The mask has an extension of several line widths, with the same pitch as that of the scale, and fully covers the light beam between source and sensor.

The light transfer changes periodically from a minimum to a maximum value and back for each displacement over the pitch of the optical pattern, as shown in Figure 8.23. With the mask at position $x_1$ the transmission is maximum; at $x_3$ minimum. In position $x_2$ light is partly transmitted.

If scale and mask are perfectly matched and aligned, the minimum light transfer is zero over the whole area of the mask, and the maximum transfer is 50%, irrespective of the slot width. So the size of the detector does not limit the resolution. However, due to mechanical tolerances and not perfectly collimated light the minimum

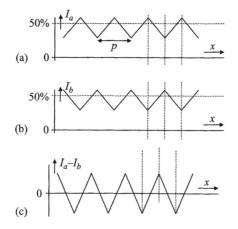

**Figure 8.24** *Elimination of the shifted base line: (a) transfer characteristic of an incremental encoder; (b) as (a) but inversed phase; (c) differential output, with zero base line*

transfer is not completely zero. These effects become more significant at decreasing line width: the base line of the signal will shift upwards and the difference between "light" and "dark" becomes smaller (Figure 8.24(a)).

To improve the discernibility of the light-dark transition, a second source-sensor pair (and mask) is added, positioned half a period from the first (actually half a period plus a number of full periods). This results in two anti-phase outputs, the difference of which has a zero base line and twice as large an amplitude (Figure 8.24(c)).

The incremental encoders discussed so far do not provide information about the direction of displacement or rotation. Figure 8.25 clarifies how to solve this problem. Next to the two anti-phase signals two *quadrature* signals are generated. These are signals with a phase difference of $\pi/2$, created by another set of source-sensor pairs located at a distance a quarter of the decoder period from the other sets. Again, the two outputs $I_c$ and $I_d$ of this additional pair are subtracted to eliminate the shift of the base lines. A displacement results in two differential signals $I_a - I_b$ and $I_c - I_d$ having a phase difference equal to $\pi/4$. Since only transitions have to be detected, these signals are converted to a binary format, still with $\pi/4$ phase difference (the signals $y_1$ and $y_2$ in Figure 8.25). At any time they can be represented by a two-bit word. When moving in the positive $x$-direction, the order of the codes is 00 – 01 – 11 – 10 – 00 etc. However, a movement in the reverse direction will reverse this order too; from the sequence the direction of the movement can be reconstructed.

The measurement head of commercial incremental encoders contains five gratings (on a single plate), to obtain direction sensitivity, zero crossing accuracy and a reference position all in one.

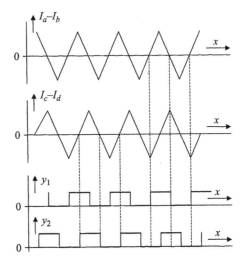

**Figure 8.25** *Determination of displacement direction, using quadrature signals*

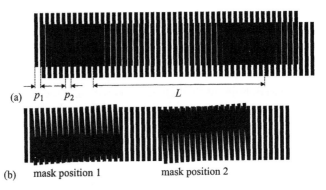

**Figure 8.26** *Incremental encoder employing: (a) the vernier effect; (b) the Moiré effect*

The incremental encoder offers the possibility to interpolate between successive slots using the triangular relationship between detector output and displacement. In some commercial types the triangular shape is first converted into a sinusoidal shape, which facilitates the interpolation process.

The intrinsic resolution of an incremental encoder is set by the number of slots per unit distance or rotation. Apart from interpolation, the resolution can further be increased over the intrinsic resolution by employing the *vernier effect* or the *Moiré effect*[2]. Two identical scales with patterns having a slightly different pitch or angle cause alternately light and dark zones along the encoder (Figure 8.26), the *vernier fringes* and *Moiré fringes*, respectively. At a relative displacement of one period of

---

[2] Ernst Moiré (1857–1929), Swiss photographer.

a single pattern, the vernier fringe pattern shifts over a full period $L$, in the same (horizontal) direction. In a Moiré configuration the fringe zones move in a direction perpendicular to the relative scale movement.

The vernier fringe period equals $L = |p_1 p_2/(p_1 - p_2)|$: the configuration behaves as an optical displacement amplifier. Its gain factor is inversely proportional to the pitch difference and can be very large for pitches having slightly different values. The fringe zones have an extension that allows the use of normally sized sensors.

Direction sensitivity is obtained in the same way as is illustrated in Figure 8.25, using two sets of source-detector pairs. Here too, the system allows interpolation between two adjacent positions with maximum and minimum output, similar to the normal encoder.

### 8.5.5. *Interferometric displacement sensing*

Interferometry is based on the phenomenon of (optical) interference, occurring when two waves of equal wavelength coincide. The resulting intensity varies with the phase difference between the waves. In an interferometric sensor such a phase difference occurs due to a path length difference between two waves, one of which usually follows a reference path, the other a path whose length is modulated by a movable part of the construction.

At equal amplitudes of the individual waves, the total intensity doubles when the waves are in phase (*constructive* interference), and drops to zero when in anti-phase (*destructive* interference). If the wave amplitudes are not the same, the minimum intensity is not zero, and the interference effect becomes less pronounced.

The wave form of monochromatic light is described by $A \cdot \cos(kx - \omega t)$, where $A$ is the wave amplitude (for both the electrical field and magnetic field components), $k = 2\pi n/\lambda$ is the propagation constant and $x$ the coordinate in the direction of propagation. When two waves travel distances $x_1$ and $x_2$, respectively, their amplitudes are described as $A_1 \cos(kx_1 - \omega t)$ and $A_2 \cos(kx_2 - \omega t)$. Both waves fall on the same detector, where the wave functions are added:

$$A_{tot} = A_1 \cos(kx_1 - \omega t) + A_2 \cos(kx_2 - \omega t) = A(x) \cos\{\varphi(x) - \omega t\} \quad (8.39)$$

with

$$A(x) = \sqrt{A_1^2 + A_2^2 + 2A_1 \cos k(x_1 - x_2)} \quad (8.40)$$

and

$$\tan \varphi(x) = \frac{A_2 \sin k(x_1 - x_2)}{A_1 + A_2 \cos k(x_1 - x_2)} \quad (8.41)$$

Equation (8.39) describes a periodic signal with frequency $\omega$ and which amplitude and phase both vary with $k(x_1 - x_2)$, hence with the difference between the travelled distances.

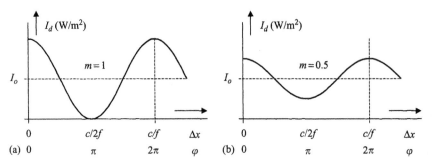

**Figure 8.27** *Detector output as a function of the relative displacement: (a) $m = 1$; (b) $m = 0.5$*

Since the photodetector responds to the light intensity, that is the average wave energy, its output is also proportional to the square of the wave function amplitude $A(x)$ in (8.40). So the detector output is written as:

$$I_d = |A(x)|^2 = A_1^2 + A_2^2 + 2A_1 \cos k(x_1 - x_2) = I_0(1 + m \cos k \Delta x) \quad (8.42)$$

Apparently, the detector output changes sinusoidally with the path difference $\Delta x$. The differential sensitivity is at maximum for odd multiples of $\pi/2$ (the so-called quadrature point), and minimal at multiples of $\pi$. Only when both light wave amplitudes are equal, $m = 1$, resulting in an optimal interference effect. This is why the factor $m$ is called the *interference visibility* (Figure 8.27).

In an interferometer the two waves travel along separate paths, one with a fixed length (the reference) and the other having a length that is set by a movable object whose displacement has to be measured. The interferometer adds up the two waves, and determines their resulting amplitude that varies with the optical path difference, hence with the displacement of the object.

**Example 8.9 Range and resolution of an interferometer**
The light source of a particular interferometer is a laser with visible light, wavelength 600 nm. Discuss the range and resolution of this instrument.

The unambiguous range is half the wavelength, so 300 nm. Suppose the resolution of the intensity measurement is 1%. The maximum sensitivity is around $\pi/2$ or $3\pi/2$ of one fringe period, and amounts to $2mI_0/\lambda$. For $m = 1$ this is 1% of 300 nm, so 3 nm. The resolution is substantially worse near 0 and $\pi$ of the fringe period. Practical interferometers use quadrature signals to increase the overall resolution.

Interference is only visible with two monochromatic, coherent light beams. The creation of such a pair of coherent beams starts with a single, monochrome light source (for instance a laser diode). The emitted light beam is split up into two

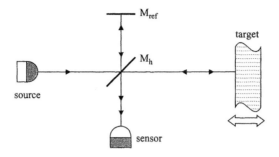

**Figure 8.28** *Classical interferometer configuration*

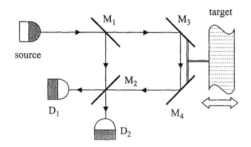

**Figure 8.29** *Mach-Zehnder interferometer configuration*

separate beams, for instance using a semitransparent mirror. Figure 8.28 shows such a system, known as the classical interferometer or Michelson interferometer[3].

The beam splitter $M_h$ is a semitransparent mirror: its transmission and reflection coefficients are (about) 50%. The two light beams arrive at the detector via different paths: one runs from the source, via $M_h$ and $M_{ref}$ and back through $M_h$ to the detector; it has a fixed length. A second beam travels via the (moving) target and ends at the detector as well: its total path length varies with the position of the target. Evidently, all optical components should be accurately aligned to achieve a proper operation.

A major disadvantage of the interferometer configuration of Figure 8.28 is the optical feedback towards the source, which can introduce instability of the output frequency. In the configuration of Figure 8.29 such feedback is minimized by using an extra beam splitter. Both detectors receive two beams: one via $M_1$ and through $M_2$ that has travelled over a fixed distance, another that has reflected at $M_3$ and $M_4$ connected to the target, and hence has travelled a distance set by the position of the target. This type of interferometer is referred to as the *Mach-Zehnder interferometer*.

---

[3] Albert Abraham Michelson (1852–1931). German-American physicist, Nobel Prize in Physics, 1907.

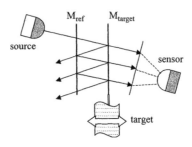

**Figure 8.30** *Fabry-Perot interferometer configuration*

A popular type of interferometer is the Fabry-Perot configuration (Figure 8.30). It is essentially an optical cavity made up from two semitransparent mirrors in parallel. The moving object is fixed to one of these mirrors. The detector receives light that has travelled a number of path lengths in the cavity. According to the basic principle of interferometry, the output varies sinusoidally with the path length, which is set by the distance between the mirrors. It has a minimum for a half wave length difference and a maximum when the difference is a complete wave length.

A special arrangement allows the measurement of angular speed, utilizing the *Sagnac effect*. Light is forced to travel a circular path, by means of an optic fibre loop. Actually there are two light waves, one travelling clockwise through the loop, the other anticlockwise. Both waves arrive at the same detector, where they interfere. The intensity is maximal when the travel times are equal, which is the case in an inertial system at rest. When the loop turns around its axis, the travel times of the two oppositely travelling waves differ somewhat, resulting in a phase shift at the detector and hence a change in the interference pattern. In this configuration the Sagnac interferometer acts as a *gyroscope*.

Since an interferometer has a very high sensitivity to displacement parameters it requires careful adjustment of the optical components. Furthermore, a frequency shift $\Delta f$ induces an output change equivalent to a phase shift of $4\pi \Delta f/c$, where $c$ is the speed of light. An interferometer, therefore, needs a very stable source, to minimize phase noise in the output signal. Altogether, an interferometer is quite an expensive instrument.

The accuracy of a laser interferometer depends largely on the accuracy and stability of the wavelength. The wave length in air depends (non-linearly) on the index of refraction, and hence varies with *temperature*, pressure and air humidity. The wavelength of a helium-neon laser in vacuum is 0.632 991 4 μm; in air with temperature $t = 20°C$, pressure $p = 1013$ hPa and 50% humidity this is 0.632 820 μm.

The accuracy of a non-compensated interferometer in a conditioned room is not much better than 10 μm per meter (so $10^{-5}$), mainly due to pressure dependency. When compensated for environmental parameters (which is never perfect), an accuracy of 3 μm per meter can be obtained, within a temperature range of 0 to 40°C. The range of an interferometer runs from several cm up to several tens of meters. The resolution is set by that of the interference pattern measurement.

For particular applications special interferometric instruments have been constructed. For example, wear of metal or other surfaces can be measured with a resolution better than 50 nm, using interferometry. Obviously, the absolute accuracy is not important in this application: only differences in height over the surface are to be measured with high resolution. On the other side of the range spectrum interferometers with a length of several km have been constructed to determine the speed of gravitation waves.

## 8.6. Piezoelectric force sensors and accelerometers

The piezoelectric effect couples the mechanical domain to the electrical domain. It is the basis for piezoelectric sensors, suitable devices for the measurement of force, pressure, torque and related force quantities in a variety of applications. When combined with a calibrated seismic mass, acceleration also can be measured.

Due to the reversibility of the piezoelectric effect, this property can also be explored for the construction of piezoelectric actuators, in particular for small displacements (Braille cells, inkjet printers, shutter drives of cameras, to mention a few applications).

First a short review of the piezoelectric effect and piezoelectric materials is given. Next we discuss particular properties of piezoelectric sensors, and finally the interfacing of such sensors is highlighted.

### 8.6.1. *Piezoelectricity*

Piezoelectricity is encountered in crystalline materials with asymmetric structure. Deformation of such materials results in electric polarisation: positive and negative charges are displaced relatively to each other, resulting in surface charges at opposite faces of the material. The charge per unit area is proportional to the applied stress. Generally, the material is shaped as small rectangular blocks, cylinders, plates or even sheets, with two parallel faces provided with a conducting layer.

In such regular structures the charge is proportional to the applied force. The construction behaves as a flat-plate capacitor, for which $Q = C \cdot V$; so the output signal is available as a voltage. The sensitivity of such piezoelectric sensors is characterized by the charge sensitivity $S_q = Q/F[C/N]$ or voltage sensitivity $S_u = V/F = S_q/C[V/N]$.

The piezoelectric effect is reversible. A voltage applied to the piezoelectric material results in a (small) deformation; this property is used in piezoelectric actuators.

Piezoelectricity is found in three groups of materials:

- natural PE materials: a well-known example is quartz (crystalline $SiO_2$);
- ceramic materials (polycrystalline), for instance barium titanate ($BaTiO_3$);
- polymers, for instance PVDF (or $PVF_2$).

Materials of both last groups are not piezoelectric by themselves, but are made so by poling. Such materials consist of very small piezoelectric particles (as in ceramic materials) or molecules having a dipole moment (as in some polymers). These microdipoles are randomly oriented; the observed net polarization is therefore almost zero. When the material is heated up to the so-called *Curie temperature*, the dipoles gain high mobility. In this state, a high electric field is applied; the dipoles are oriented in the direction of the field. After slowly cooling down, the dipoles retain their orientation, giving the material its macro-piezoelectric properties.

Quartz has a rather low but stable piezoelectricity, with a sensitivity of about 2 pC/N. The piezoelectricity of poled ceramic materials is much higher, ranging from 100 up to over 1000 pC/N. Poled polymers have a sensitivity of around 25 pC/N. From all known polymers PVDF has the highest piezoelectricity. This material is available in sheets with various thickness (6 μm to 100 μm). Poling is performed during stretching (in one or both directions), to obtain reasonable piezoelectricity [6].

Table 8.1 shows major piezoelectric properties (piezoelectric charge constant, coupling constant and Curie temperature) for a representative from each of the three groups. The double subscripts denote the direction of the electric field and mechanical stress, respectively, where the 3-direction coincides with the direction of the polarization. For the definition of the various material properties, see Section 2.3.

**Example 8.10**
Find the charge sensitivity $S_q$ and the voltage sensitivity $S_v$ of a piezoelectric force sensor, fabricated as a cube of the piezoelectric material PZT with size 1 cm. The dielectric constant of PZT is $\varepsilon_{33} = 1800$.

The charge sensitivity equals the piezoelectric charge constant. In the 3-3 direction this is $d_{33} = 3.84 \cdot 10^{-10}$ C/N. The capacitance of the cube is $C = \varepsilon_0 \varepsilon_{33} A/d = 8.8 \cdot 10^{-12} \cdot 1800 \cdot 10^{-4}/10^{-2} = 158$ pF, so $S_v = S_q/C = 2.4$ V/N.

Table 8.1 *Various properties of piezoelectric materials*

| Parameter | Symbol | PZT | Quartz | PVDF | Unit |
|---|---|---|---|---|---|
| Curie temperature | $\Theta_C$ | 285 | 553 | 100–150 | °C |
| PE charge constant | $d_{33}$ | 384 | −2 | −18 | pC/N |
|  | $d_{31}$ | −169 | −2 |  | pC/N |
| Coupling constant | $k_{33}$ | 0.7 | 0.09 |  | − |
|  | $k_{11}$ | 0.33 | 0.09 |  | − |

### 8.6.2. Piezoelectric sensors

Piezoelectric sensors are pre-eminently suited to dynamic measurement of force and acceleration. The market offers an enormous number of piezoelectric accelerometers. In most commercial accelerometers the sensitive element consists of piezoelectric ceramics, because of the robustness and high temperature range. More recently it has been shown that piezoelectric PVDF can also be applied for the measurement of linear and angular acceleration [7, 8].

Piezoelectric sensors also have disadvantages. The surface charge produced by an applied force might be neutralized easily by charges from the environment, for instance along the sides of the crystal or another leakage path. This makes the sensor act as a high-pass filter for input signals and impedes a fully static measurement. Further, crystal deformation may also be induced by temperature changes (all piezoelectric materials are also pyroelectric). As long as these changes are slow, they would not limit the applicability because of the intrinsic high-pass character.

An accelerometer behaves as a second-order spring-mass system, due to the seismic mass that moves relative to the case, and the spring between these parts. These sensors have, therefore, a sensitivity characteristic peaking at the resonance frequency, which may range from 1 kHz to 250 kHz, depending on the size. The spring is rather stiff, hence the damping factor of the mass-spring system is small, resulting in a rather sharp peaking. Care should be taken not to use the sensor close to this resonance frequency.

Piezoelectric sensors have a direction-dependent sensitivity. The transversal sensitivity amounts to several % of that along the main axis. For simultaneous measurement of acceleration in three directions, three-axes accelerometers are available; usually they consist of three separate sensors, mounted orthogonally in a single housing (Figure 8.31). The smallest dimensions of such sensors are comparable with a die, and have a weight of down to 20 grams.

In Figure 8.32 two simple models of a piezoelectric sensor are shown: a current source model and a voltage source model. The primary signal of a piezoelectric sensor is charge, so the current source in the model has the value $\dot{Q} = dQ/dt$. The source impedance behaves as an electric capacitance $C_e$, corresponding to that of a flat-plate capacitor with a (piezoelectric) dielectric. According to the relation $Q = C \cdot V$, the equivalent value of the voltage source is $Q/C_e$.

**Figure 8.31** *Example of a three axis piezoelectric accelerometer [Ziegler]*

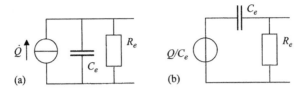

**Figure 8.32** *Two equivalent models of a piezoelectric sensor: (a) current source; (b) voltage source*

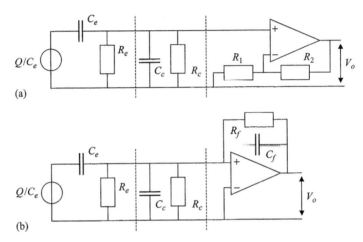

**Figure 8.33** *Interfacing a piezoelectric sensor: (a) voltage amplifier; (b) charge amplifier*

This simple model can be extended by an electrical resistance $R_e$, modelling the finite resistance through which the charge can flow away.

### 8.6.3. Interfacing piezoelectric sensors

The charge signal provided by a piezoelectric sensor can be measured in two different ways. The first method is based on the relation between charge and voltage over a capacitor: $Q = C \cdot V$. Hence, the interface circuit consists of a voltage amplifier. In the second method the charge is forced to flow through a capacitor. The voltage across this capacitor is proportional to the charge. This type of interface is commonly called a *charge amplifier*, but a better name would be *charge-voltage converter*.

Figure 8.33 shows examples of both these interface circuits. Due to the capacitive character of the sensor, the cable capacitance can have a substantial influence on the signal transfer of the system. Therefore, we include also a model of the connecting cable: the parallel connection of the cable capacitance $C_c$ and cable resistance $R_c$.

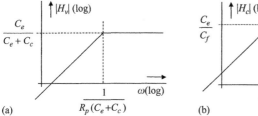 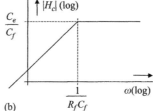

**Figure 8.34** *Transfer characteristics of the interface circuits in Figure 8.33: (a) voltage transfer; (b) charge-to-voltage converter*

First we consider the voltage read-out (Figure 8.33(a)). The sensor voltage is amplified by an operational amplifier in non-inverting mode, to obtain a high input impedance. The transfer of this circuit is given by:

$$H_v = \frac{V_o}{Q/C_e} = A \cdot \frac{j\omega R_p C_e}{1 + j\omega R_p (C_e + C_c)}$$

$$= A \cdot \frac{C_e}{C_e + C_c} \cdot \frac{j\omega R_p (C_e + C_c)}{1 + j\omega R_p (C_e + C_c)} \quad (8.43)$$

where $R_p = R_e \| R_c$ (parallel connection) and $A = 1 + R_2/R_1$. Obviously, the transfer shows a high-pass character; Figure 8.34: for frequencies satisfying $\omega R_p (C_e + C_c) \gg 1$ the transfer has a frequency independent value, equal to $C_e/(C_e + C_c)$. The cable capacitance $C_c$ causes signal attenuation. Hence, the total signal transfer depends on the length of the cable. If for any reason the connection cables are replaced, the system should be recalibrated.

In this respect the interface circuit of Figure 8.33(b) behaves better. The "charge amplifier" consists of an operational amplifier with a feedback capacitance $C_f$ in the inverting mode. Due to the virtual ground of the operational amplifier, the voltage across the sensor and the cable is kept at zero: neither the cable impedance nor the input impedance of the amplifier influences the transfer.

Assuming an ideal operational amplifier, the transfer of the charge amplifier is $-Z_2/Z_1$. In this case, $Z_1 = j\omega C_e$ and $Z_2 = 1/j\omega C_f$, hence the total transfer $H_c$ for the charge amplifier circuit is simply:

$$H_c = \frac{V_o}{Q/C_e} = -\frac{C_e}{C_f} \quad (8.44)$$

which, indeed, does not depend on cable properties. Ideally, the circuit with a capacitor in the feedback path behaves as an integrator: the input current flows through $C_e$, resulting in a voltage equal to $(1/C_f) \int I \, dt$. Here, the input current is generated by the piezoelectric sensor, and equals $dQ/dt$, that is the time derivative of the piezoelectrically induced charge $Q$. Hence, the voltage across $C_f$ is just $Q/C_f$, explaining the name *charge amplifier*.

Unfortunately, the integrator integrates not only the input charge, but also the inevitable offset voltage and bias current of the operational amplifier. To prevent the amplifier from overload, a feedback resistor $R_f$ is applied, in parallel to $C_f$. This results in a circuit transfer given by

$$H_c = \frac{V_o}{Q/C_e} = -\frac{C_e}{C_f} \cdot \frac{j\omega R_f C_f}{1 + j\omega R_f C_f} \qquad (8.45)$$

Just as in equation (8.43), this transfer has a high-pass characteristic: the cut-off frequency is set by the components of the amplifier only: for $\omega R_f C_f \gg 1$, the output voltage is $-(C_e/C_f)(Q/C_e)$. When using a high quality operational amplifier (low offset, low bias current), the value of the cutoff frequency can be chosen down to 0.01 Hz. However, a true static measurement of acceleration or force is not possible.

### Example 8.11 Piezoelectric accelerometer with voltage read-out
A piezoelectric accelerometer with voltage sensitivity 2 V/(ms$^{-2}$) and capacitance $C_e = 1$ nF is connected to a voltage amplifier (gain 10) through a 4 m cable with cable capacitance 80 pF/m and cable resistance 100 M$\Omega$/m. Determine the sensitivity and the cut-off frequency of the total system.

The cable capacitance is $4 \cdot 80 = 320$ pF. Due to this cable capacitance the signal is attenuated with a factor $1000/1320 = 0.76$. The total sensitivity becomes $2 \cdot 0.76 \cdot 10 = 15.2$ V/(ms$^{-2}$). The cable resistance amounts $100/4 = 25$ M$\Omega$, so the cut-off frequency is $f = 1/\{2\pi R_c(C_e + C_c)\} = 4.8$ Hz.

### Example 8.12 Piezoelectric accelerometer with charge read-out
The accelerometer from Example 8.11 is connected to a charge amplifier. The feedback components have the values $C_f = 1000$ pF and $R_f = 100$ M$\Omega$. Determine the sensitivity and the cut-off frequency of the total system.

The sensitivity of the sensor is multiplied by the transfer of the charge amplifier; this results in a total sensitivity of $2 \cdot (10^{-9}/10^{-10}) = 20$ V/(ms$^{-2}$). The cut-off frequency is $f = 1/(2\pi R_f C_f) = 1.6$ Hz.

## 8.7. Acoustic distance measurement

In this section we discuss methods of measuring distance or displacement using (ultra)sound. Similar to optical distance measurements, using an optical source and detector, ultrasound measurements require the use of an acoustic source (or transmitter) and detector (or receiver). First we will introduce some acoustic parameters. Next some principles and properties of acoustic transducers are detailed. Finally, acoustic measurement strategies are presented.

### 8.7.1. *Acoustic parameters*

In order to understand and properly evaluate measurements using ultrasound, it is necessary to have insight into the various acoustic parameters of both the transducers and the medium through which the sound propagates. We give a review of the major acoustic parameters and some definitions.

**Sound intensity**
The rate at which acoustic energy is radiated is called the acoustical or sound *power* (W). Sound intensity is the rate of energy flow through a unit surface area, hence *intensity* is expressed as W/m² (compare the quantity *optical flux*). A sound wave is characterized by two parameters: the sound pressure (a scalar, the local pressure changes with respect to ambient) and the particle velocity (a vector). The intensity is the time-averaged product of these two parameters. It may vary from zero (when the two signals are 90° out of phase) to a maximum (at in-phase signals). Most sensors do not measure sound intensity, but sound pressure only.

**Sound propagation speed**
The speed of sound is not the same as the particle velocity introduced above. In general, the propagation speed of sound in a material is given by the equation

$$v_a = \sqrt{\frac{c}{\rho}} \tag{8.46}$$

where $c$ is the stiffness and $\rho$ the specific mass of that material. For ideal gases, the expression for the speed of sound is:

$$v_a = \sqrt{\frac{c_p}{c_v} \cdot \frac{p}{\rho}} = \sqrt{\frac{c_p}{c_v} \frac{R}{M} \Theta} \tag{8.47}$$

where $\Theta$ is the absolute temperature. Substitution of numerical values for air yields:

$$v_a = 331.4(1 + 1.83 \cdot 10^{-3}\vartheta) \tag{8.48}$$

with $\vartheta$ the temperature in °C. Hence, as a rule of thumb, the temperature coefficient of $v_a$ is about 2% per 10 K. The speed of sound is roughly $10^6$ times lower than the speed of light. Hence, acoustic time-of-flight is easier to measure than optical time-of-flight. On the other hand, the lower speed of sound results in a relatively long measurement time. Moreover, the much larger wave length of sound waves compared to light waves makes it more difficult to manipulate with sound beams (for instance focussing or the creation of narrow beams).

**Acoustic impedance**
The acoustic impedance is defined as the ratio between the (sinusoidal) acoustic pressure wave $p$ and the particle velocity $u$ in that wave. For a sound wave that

**Table 8.2** *Acoustic properties of some materials*

| Medium | Density (kg/m$^3$) | Velocity (m/s) | Impedance (kg/m$^2$s) |
|---|---|---|---|
| ceramics | $7.5 \cdot 10^3$ | $3.2 \cdot 10^3$ | $3 \cdot 10^7$ |
| quartz | $2.6 \cdot 10^3$ | $5.7 \cdot 10^3$ | $1.5 \cdot 10^7$ |
| air | 1.3 | 330 | $0.4 \cdot 10^3$ |
| water | $10^3$ | $1.5 \cdot 10^3$ | $1.5 \cdot 10^6$ |

propagates only in one direction, the acoustic impedance $Z$ is found to be

$$Z = \frac{p}{u} = \sqrt{\rho \cdot c} = v_a \cdot \rho \tag{8.49}$$

The acoustic impedance is an important parameter with respect to the transfer of acoustic energy between two media. Similar to electromagnetic waves, which exhibit reflection and refraction on the interface of two media with different optical properties (refractive index; dielectric constant), sound waves are also reflected at the interface of two media with different acoustic properties. When a sound wave arrives at such a boundary, part of the acoustic energy is reflected back and the remaining part will enter the other medium. The amount of reflected or transmitted energy can be expressed in terms of the acoustic impedances. For sound waves, Snell's law[4] applies to calculate the direction and amplitude of the refracted sound. In the special case of a wave perpendicular to the boundary plane, the ratio between reflected and incident power (the reflection ratio) is given by:

$$R = \left(\frac{Z_1 - Z_2}{Z_1 + Z_2}\right)^2 \tag{8.50}$$

$Z_1$ is the acoustic impedance of the medium of the incident wave, $Z_2$ that of the medium behind the boundary plane. Obviously, there is no reflection (hence total transmission) if both media have equal acoustic impedances. On the other hand, when the impedances show a strong difference, the power transmission is small. To obtain an impression of the range of acoustic impedances, Table 8.2 displays the values for various materials.

Transmission from an acoustic "soft" material to an acoustic "hard" material is poor. This is an important issue in acoustic sensors for application in air. Considerable effort is put into matching of the acoustic properties to improve the energy transfer. Especially in air, this is not easy because of the wide gap between impedances of air and most sensor materials. Since many systems consist of a source and a detector, the power loss occurs twice, resulting in a low signal-to-noise ratio in acoustic sensing systems.

---

[4] Willebrord Snel van Royen (Snellius) (1580–1626), Dutch mathematician and physicist.

## 8.7.2. General properties of acoustic transducers

Commonly used acoustic transducers belong to one of the following types:

- piezoelectric;
- electrostatic;
- electromagnetic; and
- magnetostrictive.

Only the first two types are suitable for ultrasonic applications (frequency above 20 kHz). All transduction effects are reversible, which means that an acoustic transducer might be used as a transmitter as well as a receiver. Time-of-flight (ToF) systems may have both a transmitter and receiver, or only one transducer, alternately in use as a transmitter and a receiver.

It is important to understand the properties of ultrasonic transducers, to be able to evaluate results from acoustic measurements. First we will study the wave pattern of an acoustic transmitter. For the calculation of the wave pattern we use the simple piston model: a circular plate with diameter $D$, of which all the points oscillate with the same amplitude and phase (Figure 8.35). Using this model, the acoustic pressure in all points of the hemisphere can be calculated. The pressure as a function of distance $r$ and angle $\varphi$ is given by [9]

$$p(r,\varphi) = 2p(r)J_1\left(\frac{\pi D}{\lambda}\sin\varphi\right)\left(\frac{\pi D}{\lambda}\sin\varphi\right)^{-1} \quad (8.51)$$

where $\lambda$ the wavelength of the sound signal, $D$ the plate diameter and $J_1(\pi D \sin\varphi/\lambda)$ is an order-1 Bessel function. The acoustic pressure appears to have peaks and lows in particular directions and at particular distances from the transmitter, due to interference. After all, each point of the plate generates a sound wave; the total sound field is the sum of all these elementary waves. At places where the sound waves are (partly) out of phase, the sound intensity is low; at places where the waves are in phase, the sound intensity is high.

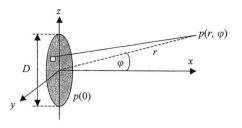

**Figure 8.35** *Calculation of the wave pattern using the piston model*

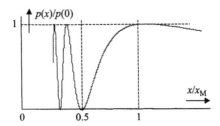

**Figure 8.36** *Sound pressure along the main axis of an acoustic transducer*

When we consider only the acoustic pressure along the main or acoustic axis, the pressure is found to be

$$p(x) = 2p(0) \sin\left\{\frac{\pi}{\lambda}\left[\sqrt{\frac{D^2}{4} + x^2} - x\right]\right\} \quad (8.52)$$

where $p(0)$ is the amplitude of the acoustic pressure on the plate surface ($x = 0$) and $x$ the distance from the plate on the main axis. A picture of the intensity along the acoustic axis is given in Figure 8.36. The place of the last maximum, at distance $x_M$ from the plate, is taken as the boundary between the *near field* (or Fresnel zone[5]) and the *far field* (or Fraunhofer zone[6]) of the transducer.

From equation (8.52) $x_M$ is found to be:

$$x_M = \frac{D^2 - \lambda^2}{4\lambda} \quad (8.53)$$

**Example 8.13 Calculation of the boundary between near and far fields**
An acoustic transducer operating at $f = 4$ MHz in water has a diameter of 8 mm. From where starts the far field?

From Table 8.2 we have $v = 1500$ m/s. The wavelength is $\lambda = v/f = 1500/4 \cdot 10^6 = 15/40$ mm. Substitution of the numerical values into equation (8.53) gives $x_M \approx 40$ mm.

Apparently, the near field covers just a short distance from the transmitter. To obtain an impression of the wave pattern all over the hemisphere, equation (8.51) should be solved. This results in the directivity diagram or radiation diagram of the transducer. For a circular piston, this directivity diagram is rotationally symmetric. Figure 8.37 gives an example for two values of the parameter $D^2/4\lambda$, only valid in the far field. In general, the diagram consists of a main lobe and side lobes. The width of the main lobe (the angle of divergence or the cone angle) follows from

---

[5] Augustin Jean Fresnel (1788–1827), French physicist.
[6] Joseph von Fraunhofer (1787–1826), German scientist, inventor and entrepreneur.

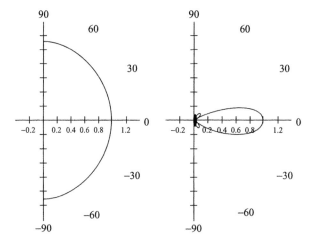

**Figure 8.37** *Directivity plots for $r/\lambda = 0.25$ (left) and $r/\lambda = 1$ (right)*

equation (8.51). An approximated value for the half angle $\varphi$ of the main lobe is given by

$$\sin \varphi \approx 1.22 \cdot \frac{\lambda}{D} \tag{8.54}$$

This approximation only holds for small values of $\lambda/D$. According to this equation the width of the sound beam depends on the ratio of the wavelength and the transducer diameter, hence on the frequency and the medium. Also the number of side lobes and their direction depend on $\lambda/D$.

It is important to note that the smaller the size of the transducer, or the larger the wavelength, the more it behaves as a point source. For a narrow beam (in most cases required to achieve optimal signal-to-noise ratio), the transducer should be large compared to the wavelength. This means, to obtain narrow beams, we have to choose either a large transducer or a higher frequency. The size of the transducer is often limited due to construction restrictions. A higher frequency is not always an option either: sound waves in air experience damping that is proportional to the square of the frequency. So, doubling the frequency reduces the range of the transducer by a factor of four.

### 8.7.3. Electrostatic and piezoelectric transducers

We will now summarize the operation principle and properties of two types of acoustic sensors: the electrostatic and the piezoelectric transducer. Further, the properties of an acoustic array are shortly reviewed.

An *electrostatic transducer* consists of two flat plates, one fixed to the housing, the other movable relative to the fixed plate, together constituting a flat-plate capacitor

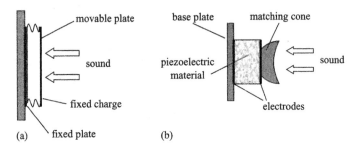

**Figure 8.38** *Ultrasonic transducers: (a) electrostatic; (b) piezoelectric*

(Figure 8.38(a)). For this construction, the equations $V = Q/C$ and $C = \varepsilon_0 \varepsilon_r A/d$ apply.

In the receiver mode, the moving plate is charged with a more or less constant charge, by connecting the plate via a resistor to a rather high DC voltage of some hundred volts. Sound waves (moving air particles) bring the plate into motion. Due to the constant charge and the changing plate distance, the voltage across the plates varies in accordance with the sound wave. For obtaining a high sensitivity, the plate is made very thin. In the transmitter mode, an AC voltage is put on the movable plate. The resulting electrostatic force causes the plate to vibrate in accordance with the applied voltage.

To make the transducer robust, the thin movable plate is sustained not only at its edges, but at a larger number of points distributed over the entire surface, for instance by a circular grooved structure. This explains the large deviations in the directivity diagrams of real electrostatic transducers as compared to those given in Figure 8.37.

The major characteristics of (commercial) electrostatic transducers are:

- wide frequency band (up to several MHz for the smaller types);
- relatively low price; the smaller the dimensions, the higher price;
- high DC voltage required.

A *piezoelectric acoustic transducer* consists of a piece of piezoelectric material. The operating principle is the same as of piezoelectric pressure or force sensors (Section 8.6). The crystal is configured as a flat plate capacitor (Figure 8.38(b)). An AC voltage applied to the crystal causes the material to vibrate, resulting in the generation of acoustic energy. Conversely, when the crystal is deformed by incident sound waves, a piezoelectric voltage is induced. Hence, the transducer might be used in transmitter mode and in receiver mode, similar to the electrostatic transducer.

Since a ceramic material has a high acoustic impedance compared to air, the power transfer from the electrical to the acoustic domain and vice versa is poor. The transfer can be improved significantly by applying a matching layer of a particular material between the ceramic and the air. The power transfer can further be improved by a horn-shaped piece of an acoustically soft material glued on the ceramic, as shown in Figure 8.38(b). Such a measure also improves the directivity of the transducer. Finally, the power transfer is improved by a proper backing layer, to minimize backward radiation.

Piezoelectric transducers always operate at *resonance*: the high stiffness of ceramic materials results in a narrow bandwidth. The resonance frequency is determined by the dimensions of the crystal, and to a lesser extent by the matching elements. Popular frequencies are 40 kHz and 200 kHz. A typical response of a piezoelectric transducer to a burst signal at the resonance frequency is shown in Figure 8.39. Apparently, the output burst shows strong distortion at the edges. The consequences of this effect on the accuracy of a time-of-flight measurement will be discussed in the next section.

The major characteristics of piezoelectric acoustic transducers are:

- narrow frequency band (operation only in resonance)
- robust
- cheap.

**Acoustic arrays**
As shown before, the beam width of a sound emitter is determined by its lateral dimensions and the frequency of the emitted sound signal. With a properly designed (linear) array of transducers a much narrower beam can be obtained, using spatial interference of the individual acoustic signals. Figure 8.40 illustrates the concept, which is similar to interference in optical systems. This figure shows the central element (on the acoustic axis) and two extra elements of a five-element array (the two elements on the other side of the axis are not shown). For points where the waves of the acoustic elements have travelled distances that differ a multiple of the wavelength, the waves add up (for instance in P, the direction of the first side lobe). At other points, the waves (partly) cancel (for instance in Q). The distance between the elements is an important parameter for the characteristics of the array.

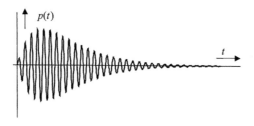

**Figure 8.39** *Typical response of a PE transducer to an input burst signal*

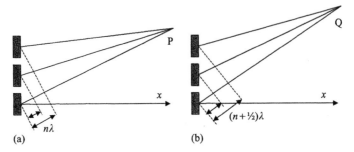

**Figure 8.40** *Beam narrowing and steering by phased arrays: (a) constructive interference in the direction of P; (b) destructive interference in the direction of Q*

An array not only narrows down the beam, but also allows the direction of this beam to be controlled, by varying the phase difference of the applied electrical signals. The control range is rather small, not more than ±30°. Similarly, the sensitivity angle of an acoustic transducer operated in receiver mode can be narrowed using an array of such receivers. At the receiving side, the individual outputs are added after a properly chosen time delay. By controlling the delay times, the main sensitivity axis can be varied over a limited angle. In general, at increasing deflection angle the amplitude of the side lobes (Figure 8.37) increases as well, causing deterioration of the beam quality.

### 8.7.4. *Distance measurements using time-of-flight*

In this section we discuss methods used to measure distance using sound waves. The majority of acoustic distance measurement systems are based on the time-of-flight (ToF) method. In most acoustic ToF systems the time elapsed between the moment of excitation of a sound pulse and the arrival of its echo is measured. The travelled distance $x$ follows directly from the ToF $t$, using the relation $v_a = x/t$ (for direct travel) or $v_a = 2x/t$ for echo systems (where the sound wave travels twice the distance). In the latter case, since the transduction effects are reversible, just a single transducer could be used, by switching from transmitter to receiver mode.

The sound signal can be of any shape. Most popular are the *burst* (a number of periods of a sine wave), a continuous wave with constant frequency (*CW*) and an FM-modulated sine wave (*FM-CW* or "chirp"). We now discuss the major characteristics of these three types.

**Burst**
The transmitter emits a short burst: 5 to 10 periods of a sine wave in the direction of an object. The sound reflects (in accordance with the reflection properties of the object's surface), travels back and is received by the same or a second transducer, which detects the moment of arrival. When a single transducer is used, switching from transmitter to receiver mode is only possible after the complete burst has been

transmitted. This limits the minimum detectable distance: there is a *dead band* or dead zone. Shortening the pulse duration reduces this dead band, but also the sound energy and hence the signal-to-noise ratio is reduced.

Several causes limit the accuracy of the ToF measurement, hence that of the distance measurement. With piezoelectric transducers, the leading and trailing edges of the received burst are less pronounced, due to the second order transfer (with narrow bandwidth) of the transducer (Figure 8.39). The starting point of the echo pulse cannot be determined accurately, as is illustrated in Figure 8.41. Further, noise and spurious reflections mask the arrival time of the echo; this effect becomes important at larger distances.

The most common detection method for the echo burst is a *threshold* applied to the receiver signal. The ToF is set by the first time the echo exceeds the threshold. When noise is present in the received signal, the threshold should be chosen high enough to prevent noise being interpreted as the incoming echo. When the threshold is chosen too high, one or more periods of the received signal can be missed (Figure 8.42). At 40 kHz this results in an error of (a multiple of) about 7 mm, which is the penalty for this simple signal processing scheme.

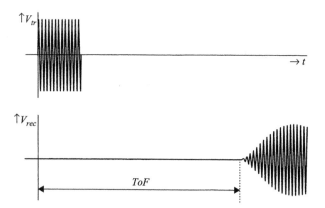

**Figure 8.41** *Distance determination using a burst; $V_{tr}$: burst signal on the input of the transmitter; $V_{rec}$: electrical output signal of the receiver*

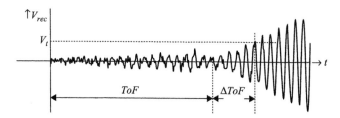

**Figure 8.42** *Noise may introduce large errors in the measured ToF*

A better approach is to use the complete echo signal and not only the starting point, to determine the time of flight, for instance by autocorrelation or cross-correlation. These techniques yield typical errors below 1 mm, depending on the signal-to-noise ratio and the travelled distance, but need longer processing time. Another approach is based on knowledge of the shape of the echo's envelope, which is determined by the transducer's transfer function. Instead of the starting point of the echo signal, where the signal-to-noise ratio is low, the arrival time of a point around the maximum amplitude could be chosen as a measure for the ToF. With this method the typical distance error is less than 0.1 mm.

**Continuous sine wave (CW)**
The average output power of a burst is relatively small, and may result in a poor signal-to-noise ratio, in particular at increased distances. In this respect, the emission of a continuous acoustic signal is better. Distance information is obtained from the phase difference between the transmitted and received waves. This method has two main drawbacks. First, emission and detection cannot share the same transducer, hence two transducers are required. Secondly, the unambiguous measurement range is only half a wavelength. The resolution, however, can be very high: for instance one degree phase resolution and 7 mm wavelength corresponds with a distance resolution of 16 µm. So, for controlling a fixed distance the CW method is preferred over the burst method.

**Frequency modulated continuous waves (FM-CW)**
The unambiguous measurement range of the continuous mode can be increased by frequency modulation of the sound wave. Figure 8.43 shows the transmitted and the received signals for this method.

Assume the frequency varies linearly with time: $f = f_0(1 - k \cdot t)$. At any moment the frequency difference between the transmitted and the received wave equals:

$$\Delta f = f_0(1 - k \cdot t) - f_0\{1 - k \cdot (t - \Delta t)\} = f_0 \cdot k \cdot \Delta t \qquad (8.55)$$

The distance travelled follows from

$$x = \frac{\Delta f \cdot v_a}{2 \cdot k \cdot f_0} \qquad (8.56)$$

The distance is directly proportional to the frequency difference and unambiguous over a time $t_m$ as illustrated in Figure 8.43. The CW-FM method combines the advantages of the first two methods: the transmitted signal is continuous (favourable signal-to-noise ratio), and the range is larger (determined by the frequency sweep). Since the frequency varies, only wide band transducers can be

Measurement of Mechanical Quantities 247

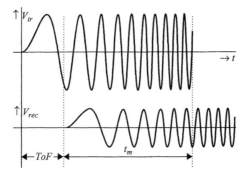

**Figure 8.43** *CW-FM mod; during $t_m$ the frequency difference is proportional to the ToF*

applied (so no piezoelectric transducers). Disadvantages over the burst method are the more complex interfacing and the need for two transducers.

### 8.8. Further reading

We list here a selection from the many textbooks on sensors discussed in this chapter, with a short bibliographic note.

E.E. Herceg, *Handbook of Measurement and Control – An authorative treatise on the theory and application of the LVDT*, Schaevitz Engineering, Pennsauke NJ, 1972, LCCCN 72-88883.
This book gives detailed information on many aspects of LVDTs: evolution and characteristics (both AC- and DC-types), environmental effects and installation, and various applications. In separate chapters the LVDT as sensor for displacement, force, pressure, velocity and acceleration is discussed. In an appendix a short mathematical analysis of the LVDT is given.

G. Gautschi, *Piezoelectric Sensorics*, Springer-Verlag, Berlin etc, cop., 2002, ISBN 3-540-42259-5.
A revised and extended translation of a German text from 1980. A useful, application oriented book on piezoelectric sensors, with details on materials, calibration and interfacing electronics.

R.C. Asher, *Ultrasonic Sensors*, IOP Publ., Bristol, 1997, ISBN 0-7503-0361-1.
The subtitle of this book is "for chemical and process plant"; however, its use is much wider. With a minimum of mathematics it offers a clear picture of the applicability of ultrasound for the measurement of various physical and chemical quantities. Easy to read, with clear illustrations, this book is suitable for undergraduate students and newcomers in this field. Part 1 is on properties of ultrasound, part two on acoustic sensors and in part three (appendices) some topics are discussed in more depth. It is one of the few textbooks written in the first person.

M.G. Silk, *Ultrasonic Transducers for Nondestructive Testing*, Adam Hilger, Bristol, 1984, ISBN 0-85274-436-6.
The book discusses in a concise way many aspects of ultrasonic transducers (and but little on NDT), making the book of interest for a much wider readership. It contains very useful chapters on beam properties and acoustic arrays. A highly instructive work for basic studies on ultrasound measurements.

## 8.9. Exercises

1. The calculated value of the Poisson ratio $\mu$ in 8.2 is 0.5. Prove that real values should be less than 0.5, for materials in the normal elastic region.

2. Resistances $R_1$ and $R_2$ in the measurement bridge have equal values. Resistor $R$ is a wire-wound potentiometer, with total length 4 cm. Calculate the output voltage $V_o$ at a displacement of 4 mm in positive direction relative to the centre position ($x = 0$); the bridge supply voltage is $V_i = 8$ V.

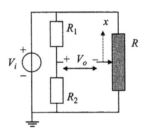

3. The sensitivity of a particular piezoelectric force sensor is specified as 4 pC/N. Its capacitance is 2 nF. The sensor is connected to a voltage amplifier by a cable. The total capacitance of the cable and the amplifier is 80 pF. If the measurement is not corrected for the error due to the cable and input capacitance, what is the relative error in this measurement?

4. Each of the plates of the differential capacitive displacement sensor in the figure have a surface area of 1 cm². For $x = 0$ the plate distance of both capacitors is 1 mm. The dielecric medium is air. Calculate the difference of the capacitances $C_1 - C_2$ when the middle plate moves 1 µm upwards. Fringe fields at the plate edges may be neglected.

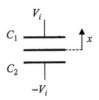

5. The differential capacitance from the preceding exercise is connected to the interface circuit as drawn below. The middle plate is connected to terminal (1). $V_i$ is a sine wave signal with frequency 15 kHz and amplitude 5 V. The feedback capacitance is $C_f = 1$ pF. Calculate the output voltage at a displacement of 1 μm.

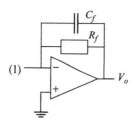

6. The level of a liquid in the rectangular container is measured using the flat-plate capacitor, as shown in the figure. The plates and the container have equal height. The container is filled with a liquid with dielectric constant $\varepsilon_r = 5$. The capacitance appears to be twice as that of the empty tank. Determine the fill factor (in % of the full tank).

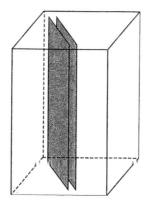

# Chapter 9

# Measurement of Chemical Quantities

In this chapter, the basics of electrochemical sensing and sensors are described. As is the case with physical sensors in the electrical domain, the retrieved information can be represented by a voltage, a current or an impedance, as shown in Figure 9.1.

For chemical sensors based on electrochemical measuring principles, this subdivision turns out to be a fundamental one because different means of (ionic) mass transport are involved. Information represented by a voltage, a current or an impedance is retrieved by potentiometric sensors, by amperometric sensors or by electrolyte conductivity sensors, respectively.

This fixes the outline of this chapter: a section defining some fundamentals is followed by three sections: *potentiometry*, *amperometry* and *electrolyte conductivity*.

## 9.1. Some fundamentals and definitions

In this section, several repeatedly used terms are defined [1].

- An *electrolyte* is a liquid solution through which charge is carried by the movement of ions.
- An *electrode* is a piece of metal (or semi conductor) through which charge is carried by electronic movement.
- An *electrochemical cell* is most generally defined as two electrodes separated by an electrolyte.
- The *cell potential* is the difference in potential across the electrodes of an electrochemical cell.

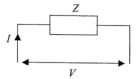

**Figure 9.1** *Any sensor functioning in the electrical domain represents its information as a voltage V, a current I or an impedance Z*

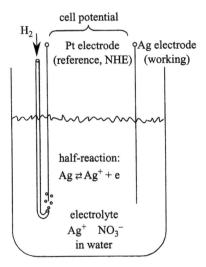

**Figure 9.2** *Visualization of the concept of an electrochemical cell*

In an electrochemical cell, often two independent *half-reactions* take place, each representing the chemical changes at one of two electrolyte/electrode interfaces. Most of the time, we are only interested in one of these two reactions and the electrode at which this occurs is called the *working* electrode. The remaining electrode is called the *reference* electrode and is standardized by keeping the concentrations of the species involved in this half-reaction constant. An internationally accepted reference is the *normal hydrogen electrode* (NHE) where $H_2$ gas at one atmosphere pressure passes a platinum wire via a hollow tube in a solution with pH = 0. All these definitions are visualized in Figure 9.2.

## 9.2. Potentiometry

Since the reference electrode has a constant makeup, its potential is fixed. Therefore, any changes in the electrochemical cell are ascribed to the working electrode. We say that we observe or control the potential of the working electrode with respect to the reference, and that is equivalent to observing or controlling the energy of the electrons within the working electrode [1]. By driving the electrode to more negative potentials the energy of the electrons is raised, and they will eventually reach a level high enough to occupy vacant states on species in the electrolyte. In this case, a flow of electrons from electrode to solution (a *reduction* current) occurs, resulting in a chemical reaction called *reduction*. A few examples of reduction reactions, illustrated in Figure 9.3(a) and (b), are

$$Fe^{3+} + e \rightarrow Fe^{2+} \tag{9.1}$$

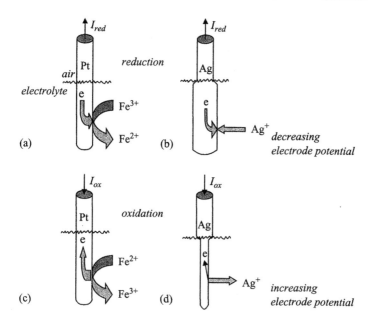

**Figure 9.3** *Illustration of redox reactions at an electrode in an electrolyte: (a) and (b) representing reduction according to (9.1) and (9.2), respectively; (c) and (d) representing oxidation according to (9.3) and (9.4), respectively*

occurring, e.g., at a Pt electrode; both $Fe^{x+}$ ions remain dissolved in the electrolyte, and

$$Ag^+ + e \rightarrow Ag \tag{9.2}$$

occurring at a silver electrode, now getting thicker.

Similarly, the energy of the electrons can be lowered by imposing a more positive potential, and at some point electrons on solutes in the electrolyte will find a more favourable energy on the electrode and will transfer there. Their flow, from solution to electrode, is an *oxidation* current. Some examples of oxidation reactions, illustrated in Figure 9.3c and d, are

$$Fe^{2+} \rightarrow Fe^{3+} + e \tag{9.3}$$

occurring, e.g., at a Pt electrode; both $Fe^{x+}$ ions remain dissolved in the electrolyte, and

$$Ag \rightarrow Ag^+ + e \tag{9.4}$$

occurring at a silver electrode, now getting thinner.

It is obvious that (9.1) and (9.2) as well as (9.3) and (9.4) represent identical chemical reactions, only evolving in different directions. This is an interesting observation: lowering an electrode potential causes a reduction current, e.g., according to (9.1) and increasing the potential of this electrode causes the flow of the current to reverse its direction, resulting in a oxidation current, according to (9.3). Obviously, there must exist an electrode potential for which the (net) current is zero! Now what is this potential and how is it related to the chemical reaction of (9.1) and (9.3) and the concentrations of the species involved? This relation between the concentration of chemical species, participating in a redox reaction (= a reaction involving the reduction or oxidation of species) is found by Nernst[1].

Let us first generalize (9.1) into:

$$ox + ne \rightleftarrows red \tag{9.5}$$

where $n$ is the number of electrons involved in the redox reaction. In the case where there is neither oxidation nor reduction current, (9.5) represents a true equilibrium as indicated by the double arrow. Let us assume that this equilibrium occurs at electrode potential $E$ (always with respect to the potential of the reference electrode, being 0 Volt per definition for a NHE). If we would force the electrode to a potential $E'$, we would have to use an amount of work to oxidize (if $E' > E$) $n$ mol of red into ox that is equal to

$$n(E' - E)F \quad [\text{VC/mol}] = [\text{J/mol}] \tag{9.6}$$

in which $F$ equals Faraday's constant[2], expressing the number of Coulombs[3] that is present in one mole of electrons: $1.6022 \cdot 10^{-19} \times 6.0220 \cdot 10^{23} = 96\,484$ C/mol. The ratio of the concentrations of red and ox ions in that case is derived from Boltzmann[4] statistics, stating that

$$e^{-\alpha/RT} \tag{9.7}$$

equals the fraction of species (in moles) having an extra energy of at least $\alpha$. $RT$ [J/mol] is the product of the gas constant $R$ (= 8.314 J/(mol·K) and temperature $T$. In our case, the species obtain this extra energy by the increased electrode potential. Now, the new ratio in red and ox concentrations can be expressed by substituting the amount of extra work given by (9.6) into (9.7), defining this ratio:

$$\left(\frac{[ox]}{[red]}\right)_{new} = e^{-n(E'-E)F/RT} \tag{9.8}$$

where [species] represents the concentration of the species in mol/dm$^3$. It will be no surprise, suggesting that this identity was already valid for the original equilibrium

---

[1] Walter Nernst (1864–1941), German physical chemist, winner of the Nobel prize in Chemistry in 1920.
[2] Michael Faraday (1791–1867), British physicist.
[3] Charles Augustin Coulomb (1736–1806), French physicist and mathematician.
[4] Ludwig Boltzmann (1844–1906), Austrian theoretical physicist.

situation of (9.5) with the assumed electrode potential $E$. In that case with non-existent $E'$ (=0), (9.8) becomes

$$\left(\frac{[ox]}{[red]}\right)_{original} = e^{nEF/RT} \tag{9.9}$$

or

$$E = \frac{RT}{nF} \ln\left(\frac{[ox]}{[red]}\right) \tag{9.10}$$

dropping the subscript "original". This assumed concentration-dependent electrode potential $E$ has to be added to the standard potential of an electrode, $E^0$, being the experimentally determined electrode potential under standard conditions for a specific red/ox couple (unity concentrations of 1 mol/dm$^3$, at 101.3 kPa and $T = 298$ K). The values of $E^0$ are different for each red/ox couple and can be found in tables.

In conclusion, we derive the Nernst Law for any electrochemical reaction of the form $ox + n_e \rightleftarrows red$ being

$$E_{Nernst} = E^0 + \frac{RT}{nF} \ln\left(\frac{[ox]}{[red]}\right)$$

or

$$E_{Nernst} = E^0 + \frac{2.303 \cdot RT}{nF} \log\left(\frac{[ox]}{[red]}\right) \tag{9.11}$$

By definition, a red or ox species, present in solid form or in gaseous form at atmospheric pressure gets unity concentration.

This important equation (9.11) is the basis of potentiometry, i.e., measuring the electrode potential with respect to a reference electrode as a function of the concentration of some chemical species in the electrolyte at (almost) zero current.

### Example 9.1 From electrode potential to concentration

A silver electrode is immersed in an aqueous solution containing $AgNO_3$. $AgNO_3$ dissolves easily in water. The potential of that electrode is 0.70 V, measured with respect to a Normal Hydrogen Reference Electrode at room temperature. What is the concentration of $Ag^+$, present in the solution?

The standard electrode potential of the redox reaction $Ag^+ + e \leftrightarrow Ag$ is $E^0 = 0.80$ V (as tabled in many textbooks, e.g., in [1]). Now substituting the relevant parameters in equation (9.11) yields:

$$0.70 = 0.80 + \frac{2.303 \cdot 8.314 \cdot 298}{96\,484} \cdot \log\frac{[Ag^+]}{[1]},$$

**Table 9.1** *Overview of electrode types for potentiometry [2]*

| Electrode type | Example |
|---|---|
| Redox | Pt electrode in $Fe^{3+}/Fe^{2+}$-containing solution |
| First kind | Pb electrode in $Pb^{2+}$-containing solution |
| Second kind | Ag/AgCl electrode in $Cl^-$-containing solution |
| Membrane or ISE (Ion Selective Electrode) | Electrode in which a potential difference occurs across a membrane (on top of the electrode); pH glass electrode, $NO_3^-$ electrode |

or $[Ag^+] = 2.0 \cdot 10^{-2}$ mol/dm$^3$ or 20 mM, being 20 milliMolar, meaning $2.0 \cdot 10^{-2}$ mol/dm$^3$.

Many types of electrodes exist, each requiring a detailed description, not given in this introductory chapter. All potentiometric electrodes, however, can be grouped according to their kind, as summarized in Table 9.1.

Basically, the concentration of a certain ion can be determined by measuring the electrode potential with respect to a reference electrode and using Nernst's law, but only when one specific redox couple is present in the solution, like the examples mentioned in Table 9.1 for redox electrodes or electrodes of the first or second kind. When more redox couples are present, some mixed potential results, which does not represent the concentration of one specific ion. In all these cases extra membranes have to be engineered and added to the electrode in order to obtain specific information of the concentration of just one ionic species. This results in the last class of electrodes, mentioned in Table 9.1: the ion selective electrodes (ISE).

Some examples:

$$Fe^{3+} + e \rightarrow Fe^{2+}, \quad E_{Nernst} = 0.77 + \frac{RT}{F} \ln\left(\frac{[Fe^{3+}]}{[Fe^{2+}]}\right), \text{ at Pt electrode, redox type;}$$

$$Pb^{2+} + 2e \rightarrow Pb, \quad E_{Nernst} = -0.13 + \frac{RT}{2F} \ln([Pb^{2+}]), \text{ at Pb electrode, first kind;}$$

$$Cl_2 + 2e \rightarrow 2Cl^-, \quad E_{Nernst} = 1.36 + \frac{RT}{2F} \ln\left(\frac{[Cl_2]}{[Cl^-]^2}\right)$$

$$= 1.36 - \frac{RT}{F} \ln([Cl^-]), \text{ at Pt electrode, redox type.}$$

Basic instrumentation for the measurement of the electrode potential with respect to the reference electrode only consists of a good voltmeter with a very high input impedance ($\gg 1$ M$\Omega$), or an impedance transformer (a "follower") with a high input impedance ($\gg 1$ M$\Omega$) and a simple voltmeter (see section 7.1). The main concern for the electronic part of the sensor system is to avoid noise, caused by the high source impedance of the electrochemical cell.

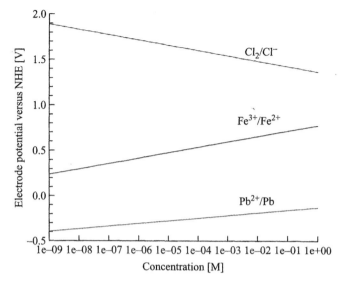

**Figure 9.4** *The electrode potentials of several redox couples as a function of the concentration of the varied species, $Cl^-$, $Fe^{3+}$ and $Pb^{2+}$ at room temperature, illustrating Nernst law, equation (9.11)*

When measuring properly, the three given examples yield the following electrode potentials when the $Fe^{3+}$, the $Pb^{2+}$ and the $Cl^-$ concentration, respectively, is varied (see Figure 9.4). Remember that species present in solid form or in gaseous form at atmospheric pressure get unity concentration in the Nernst equation, which is here the case for the Pb and the $Cl_2$ concentration. Also, it is assumed that the $Fe^{3+}$-concentration is varied during constant $Fe^{2+}$-concentration in Figure 9.4.

## 9.3. Amperometry

Amperometry[5] is a technique of measuring the concentration of an electro-active species via the electrical current that occurs due to the electrochemical reaction of that species at an electrode. Consider a possible redox reaction, e.g.,

$$Ag^+ + e \rightleftarrows Ag \qquad (9.12)$$

at a silver electrode in an electrolyte, containing $Ag^+$-ions.

When we now apply a potential step to the Ag-electrode in negative direction (lowering the electrode potential with respect to some reference electrode), it is already explained in Section 9.2 that a reduction current will occur (as shown in

---

[5] After André Marie Ampère (1775–1836), French physicist and mathematician.

Figure 9.3(b)) resulting in a reduction reaction of the equilibrium reaction (9.12):

$$Ag^+ + e \rightarrow Ag \qquad (9.13)$$

What is happening is that silver ions from the electrolyte are now reduced to silver atoms and are deposited at the silver electrode, growing thicker. Due to this process, there are fewer $Ag^+$-ions near the electrode surface than in the bulk of the solution, causing a gradient in $Ag^+$-concentration; from the bulk (higher concentration) to the electrode (lower concentration). Such a gradient causes mass transport by diffusion: movement of a species under the influence of a concentration gradient.

As a result of this mass transport, the $Ag^+$-ions reach the electrode surface at a certain rate (= number of ions per second) and are subsequently reduced to Ag atoms. When we now express the number of ions in moles and normalize this rate per unit area, then the so-called flux of $Ag^+$-ions, $J_{Ag^+}$, in mol/(m$^2$s) is obtained. Clearly, from (9.13), this flux of silver ions at the electrode surface is related to the reducing current, $i_{red}$, through the electrode. This current can be easily expressed in the flux of $Ag^+$-ions, $J_{Ag+}$, by multiplying this flux with the actual electrode area $A$ and Faraday's constant $F$:

$$i_{red} = -J_{Ag^+} A \cdot F \quad [C/s] = [A] \qquad (9.14)$$

All this is schematically shown in Figure 9.5 in a simplified representation.

Now let us summarize and generalize. The following general reduction reaction occurs at an electrode with a suitable potential:

$$ox + ne \rightarrow red \qquad (9.15)$$

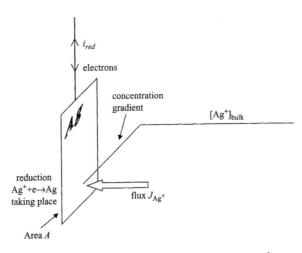

**Figure 9.5** *Impression of the flux caused by the concentration gradient resulting from the reduction of $Ag^+$-ions at the electrode*

The depletion of ox at the electrode surface causes a concentration gradient of ox, leading to mass transport by diffusion. This results in a flux of ox, $J_{ox}$ that is related to the reduction current, $i_{red}$, through the electrode with area $A$ according to

$$i_{red} = -nFAJ_{ox} \quad \text{or} \quad J_{ox} = -\frac{i_{red}}{nFA} \tag{9.16}$$

Equation (9.16) is called *Faraday's law* stating that the amount of species involved in the redox reaction, transported by the flux in the electrolyte, is proportional to the amount of charge, transported through the electrode (=electrical current).

### Example 9.2 From current to flux

An electric reducing current of 10 μA passes a silver electrode, immersed in an aqueous solution containing plenty of $AgNO_3$ (easily dissolving in water). The electrode area in contact with the solution is 1 mm². What is the flux of $Ag^+$ ions, $J_{Ag^+}$, that reaches the electrode and how many silver ions are reduced to silver per second?

Using (9.16) yields: $J_{Ag^+} = (10 \cdot 10^{-6})/(1 \cdot 96484 \cdot 1 \cdot 10^{-6}) = 1.04 \cdot 10^{-4}$ mol/m²s. In 1 s on a 1 mm² electrode, $J_{Ag^+} \cdot 1 \cdot A \cdot N_A = 1.04 \cdot 10^{-4} \cdot 1 \cdot 1 \cdot 10^{-6} \cdot 6.022 \cdot 10^{23} = 6.24 \cdot 10^{13}$ $Ag^+$ ions are being reduced.

Now how can we relate the measured $i_{red}$ to the original concentration of ox in (9.15)? In first approximation, this can be derived and expressed quite easily: $i_{red}$ can be measured and is via (9.16) related to the flux $J_{ox}$. The driving "force" for $J_{ox}$ is the concentration gradient of ox near the electrode, so we only have to find an expression for this gradient. For that, Figure 9.6 is used, being a more strict visualization of Figure 9.5.

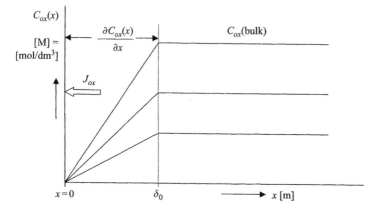

**Figure 9.6** *Illustration of the flux $J_{ox}$ caused by the gradient $\partial C_{ox}/\partial x$, present in the diffusion layer with thickness $\delta_0$*

The flux $J_{ox}$ is linearly related to the concentration gradient via a proportionality constant, $D_{ox}$:

$$J_{ox} = -D_{ox}\frac{\partial C_{ox}(x)}{\partial x} \quad [\text{mol/m}^2\text{s}] \qquad (9.17)$$

The minus sign is obvious from Figure 9.6. $D_{ox}$ is the diffusion coefficient, the ease with which the ox particles travel through the electrolyte, and is expressed in m$^2$/s. Combining (9.16) and (9.17) yields an expression relating the measured current $i_{red}$ to the concentration gradient:

$$i_{red} = -nJ_{ox}AF = nFAD_{ox}\frac{\partial C_{ox}(x)}{\partial x} \quad [\text{A}] \qquad (9.18)$$

Equation (9.18) is generally valid and is called Fick's first law of diffusion[6].

Now, we make a few assumptions in order to find a simple analytical expression for the gradient.

1. The slope of the gradient, $\partial C_{ox}/\partial x$, is linear, as drawn in Figure 9.6.
2. The thickness of the layer in which diffusion takes place is fixed, $\delta_0$ in Figure 9.6.
3. The potential of the electrode is such that $C_{ox}(x=0) = 0$, i.e., the concentration of ox at the electrode surface drops to zero.

In that case, (9.18) becomes

$$i_{red,limiting} = nFAD_{ox}\frac{(C_{ox}(bulk) - C_{ox}(x=0))}{\delta_0} = \frac{nFAD_{ox}}{\delta_0}C_{ox}(bulk) \qquad (9.19)$$

This is an important equation. It expresses the so-called limiting current for a given species at a certain concentration. It is called "limiting," because a larger current for the specific reduction reaction of (9.15) is not possible at a fixed bulk concentration $C_{ox}$(bulk): $J_{ox}$ is at its maximum value. Note from (9.19), that apart from the variable to be measured, $C_{ox}$(bulk), the diffusion coefficient $D_{ox}$ also determines the limiting current. This value can be found in appropriate tables, or the curve $i_{red} = f(C_{ox}(\text{bulk}))$ can be experimentally determined with known concentrations of $C_{ox}$, i.e., the sensor can be calibrated. Obviously, a curve as in Figure 9.7 will then be the result.

A successful example of an amperometric sensor that is based on the relatively simple equation (9.19), is the oxygen sensor ($O_2$-sensor) as introduced by Clark[7]. An electrochemical cell, consisting of a Pt disk working electrode and a Ag ring counter electrode in an inner electrolyte solution, as shown in Figure 9.8(a) and (b), is closed by a thin $O_2$-permeable membrane.

---

[6] Adolf Eugene Fick (1831–1879), German physiologist, mathematician and physicist.
[7] Leland C. Clark, medical biologist, developed the heart-lung machine and invented the Clark Oxygen Electrode.

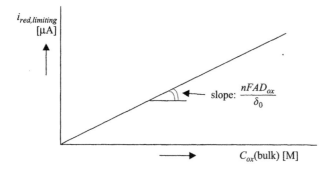

**Figure 9.7** *Illustration of the relation between the limiting current and the bulk concentration*

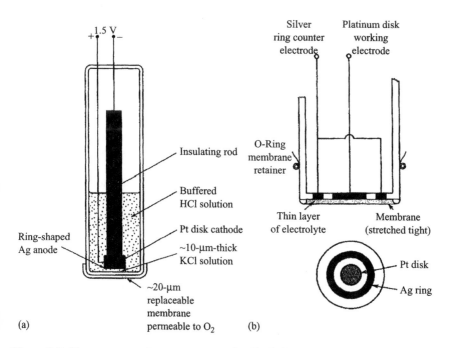

**Figure 9.8** *The amperometric oxygen sensor: the Clark electrode: (a) cross-sectional view of the total cell; (b) a detailed view of the top, showing the membrane and the working and counter electrode*

The *Pt* working electrode is kept at a potential of −0.7 V versus the Ag/AgCl electrode functioning as working as well as reference electrode. Oxygen can arrive at this working electrode when it travels through the membrane (slowly, due to the low diffusion coefficient of $O_2$ in the membrane) and the thin layer of electrolyte (fast, due to the high diffusion coefficient of $O_2$ in water). At the electrode, the

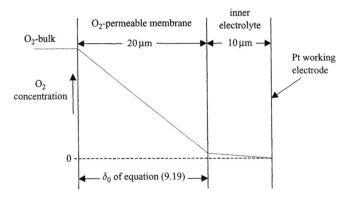

**Figure 9.9** *Illustration of the $O_2$-concentration gradient in a Clark electrode, mainly present in the membrane and determined by its thickness $\delta_0$*

oxygen will immediately and completely be reduced at $-0.7$ V according to

$$O_2 + 4e + 4H^+ \rightarrow 2H_2O$$

Due to the difference in diffusion coefficient in the membrane and the electrolyte, and the fact that the electrolyte film is even thinner than the membrane, it is reasonable to assume that the $O_2$-concentration gradient (determining the flux and the reduction current, according to (9.19)) is only present in the membrane and is virtually absent in the thin electrolyte film, as schematically shown in Figure 9.9.

This makes the Clark electrode successful: the carefully engineered membrane fixes the thickness $\delta_0$ in which the $O_2$-gradient is present. Figure 9.10(a) shows the importance of choosing the correct potential of the Pt working electrode. When the potential is too negative, other interfering redox reactions start to occur, resulting in an increased current. When the potential is not negative enough, not all $O_2$ is completely reduced, causing a partial collapse of the current. The Clark electrode must operate, therefore, at a potential in the plateau region of the current. Only then the measured current truly represents the limiting current according to (9.19). By now, it should be clear that this limiting current increases for higher $O_2$-concentrations, as shown in Figure 9.10(b).

Equation (9.19) is not generally valid because of the assumptions we had to make in order to derive (9.19) from (9.18). The weakest assumption is number 2, that of the assumed constant diffusion layer thickness $\delta_0$. In reality this layer in which the concentration gradient is present becomes thicker in time, making $\delta$ a function of time: $\delta(t)$. What happens is schematically shown in Figure 9.11.

Clearly, the slope of the concentration profile

$$\frac{\partial C_{ox}(x)}{\partial x} = \frac{(C_{ox}(bulk) - C_{ox}(x=0))}{\delta(t)}, \quad \text{with } C_{ox}(x=0) = 0 \quad (9.20)$$

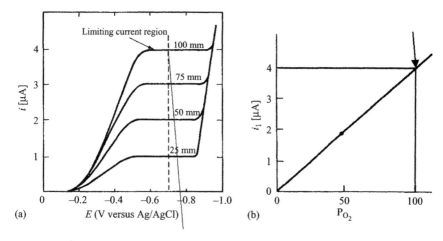

**Figure 9.10** *(a) Current versus electrode potential curves for 4 different $O_2$-concentrations, clearly showing the limiting current region at the appropriate electrode potential. "mm" stands for "mmHg" and 1 mmHg = 133.3 Pa; (b) the limiting current (plateau value) plotted as a function of the partial $O_2$-pressure in %, $P_{O_2}$*

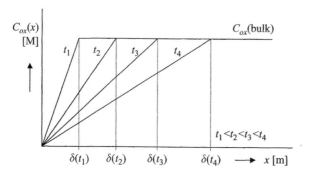

**Figure 9.11** *A more realistic impression of the processes occurring at a current-carrying electrode: the concentration profile develops in time; the diffusion film thickness increases in time*

decreases for increasing $t$, because $\delta(t)$ increases. Consequently, according to (9.18) and (9.19) the flux and the measured current decreases with time, as shown in Figure 9.12.

Unfortunately, there is no longer one constant limiting current for each bulk concentration anymore. Assumption number 1, stating that $\partial C_{ox}/\partial x$ is thought to be linear is also not true in reality, but the error thus introduced is only a constant factor of $2/\sqrt{\pi}$ (ca. 13%). The real, curved concentration profile is shown in Figure 9.13 together with the assumed linear profile.

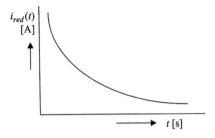

**Figure 9.12** *The increasing diffusion film thickness causes a decrease in the slope of the concentration gradient (Figure 9.11), resulting in a decreased flux and electrode current as illustrated here*

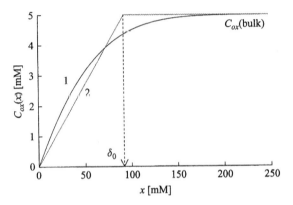

**Figure 9.13** *Illustration of a minor correction: discontinuous gradients in one medium do not occur in reality. Curve 2 is nice for modelling, curve 1 is closer to reality*

In order to get rid of the errors, introduced by the assumptions, we need a more rigorous description. For that, some extra mathematics is required, starting with the equation that describes the conservation of matter: the continuity equation:

$$\frac{\partial C_{ox}(x,t)}{\partial t} = -\frac{\partial J_{ox}(x,t)}{\partial x} \qquad (9.21)$$

This equation represents the change in concentration per second being equal to the change in flux per meter; [mol/(s·m³)] and can easily be illustrated with Figure 9.14. The change in concentration at $x$ (in compartment $\Delta x$) can only be caused by the difference in flux into and flux out of this compartment:

$$\frac{\partial C_{ox}(x,t)}{\partial t} = \frac{J_{ox}(x,t) - J_{ox}(x + \Delta x, t)}{\Delta x}$$

Letting $\Delta x$ approach zero results in (9.21). Note that $\partial J_{ox}(x,t)/\partial x$ has dimensions of mol/(m³s), being a change in concentration per unit time, as required. Combining

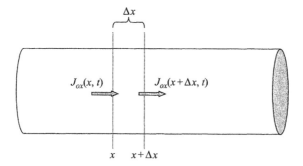

**Figure 9.14** *Illustration used for the determination of the continuity equation (9.21), leading to Fick's second law of diffusion, equation (9.22)*

(9.17) and (9.21) yields Fick's second law of diffusion:

$$\frac{\partial C_{ox}(x,t)}{\partial t} = D_{ox}\left(\frac{\partial^2 C_{ox}(x,t)}{\partial x^2}\right) \tag{9.22}$$

If we now presume a planar electrode (e.g., a platinum disk) and an unstirred solution [4], we can calculate the diffusion-limited current, $i_{red}$, using equation (9.22) and the following boundary conditions:

$$C_{ox}(x, t=0) = C_{ox}(bulk) \tag{9.23}$$

$$\lim_{x \to \infty} C_{ox}(x,t) = C_{ox}(bulk) \tag{9.24}$$

$$C_{ox}(x=0, t) = 0 \quad \text{for } t > 0 \tag{9.25}$$

The initial condition (9.23) merely expresses the homogeneity of the solution before the experiment starts at $t = 0$, and the semi-infinite condition (9.24) is an assertion that regions sufficiently distant from the electrode are unperturbed by the experiment. The third condition (9.25), expresses the surface condition after the (negative) potential step, resulting in sudden, complete depletion of ox-particles at the electrode surface.

The solution of (9.22), using conditions (9.23)–(9.25) can be obtained after Laplace transformation, resulting in a mathematical expression $C_{ox}(x,t)$. This expression (not given here) can be substituted in (9.23), producing the current-time response:

$$i(t) = nFAC_{ox}\sqrt{\frac{D_{ox}}{\pi t}} \tag{9.26}$$

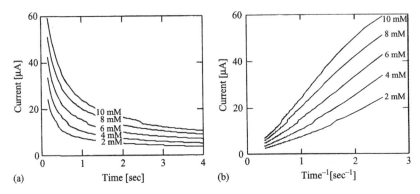

**Figure 9.15** *Chronoamperometric $H_2O_2$ measurement results: (a) the current relaxation curves at a Pt electrode as a function of the $H_2O_2$ concentration; (b) the same data but plotted on a non-linear time axis*

which is known as the Cottrell equation[8]. As with every equation, having its own name, this is an important one. The meaning and practical use of the Cottrell equation (9.26) for amperometric concentration determination will be illustrated with the following example [3]. In Figure 9.15(a) several current versus time responses are shown for different concentrations of hydrogen peroxide, $H_2O_2$. The reduction reaction involved is

$$H_2O_2 + 2H^+ + 2e \rightarrow 2H_2O \quad (9.27)$$

This reaction, and thus the reduction current as shown in Figure 9.15(a) occur at a Pt working electrode of 0.5 cm$^2$ when stepping from 0 to $-0.5$ V (versus a so-called Ag/AgCl reference electrode).

To show that we are really dealing with the phenomena as predicted by Cottrell, we plot the curves of Figure 9.15(a) again, but now against $1/\sqrt{t}$, because (9.26) then predicts straight lines:

$$\frac{\partial i(t)}{\partial (1/\sqrt{t})} = nFA\sqrt{\frac{D_{ox}}{\pi}} C_{ox} \quad [As^{1/2}] \quad (9.28)$$

with different slopes for each different $H_2O_2$ concentration. Figure 9.15(b) confirms this prediction. The slopes of these curves are plotted in Figure 9.16, now as a function of the applied $H_2O_2$ concentration, giving the calibration curve for this sensor.

---

[8] Frederick Gardner Cottrell (1877–1948), American analytical chemist; the equation was published in 1902.

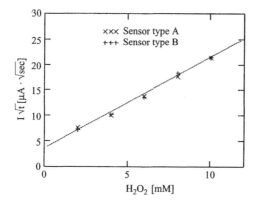

**Figure 9.16** *Measured calibration curve for chronoamperometric responses for two different Pt electrodes*

From the Cottrell equation, the diffusion coefficient can be determined as a verification:

$$D_{H_2O_2} = \pi \left(\frac{slope}{nFA}\right)^2 = 1.62 \cdot 10^{-9} \text{ m}^2/\text{s}$$

which is equal to the theoretical value of $1.6 \cdot 10^{-9}$ m$^2$/s.

In conclusion: this type of amperometric sensing appears to be a reliable method for the detection of the hydrogen peroxide concentration. This approach can also be applied for other redox couples, present in the solution, as long as they are alone. When, however, other redox-active species are also present, concentration information about one ion specifically can only be obtained by adding a selective membrane to the electrode.

## 9.4. Electrolyte conductivity

### 9.4.1. *Faradaic and non-faradaic processes*

Two types of processes occur at electrodes [1]. One kind comprises those just discussed, in which charges (e.g., electrons) are transferred across the metal-solution interface. This electron transfer causes oxidation or reduction to occur. Since these reactions are governed by Faraday's law (i.e., the number of species involved in the redox reaction, transported by the flux in the electrolyte, is proportional to the amount of charge, transported through the electrode (=electrical current); they are called faradaic processes. Electrodes at which faradaic processes occur are sometimes called charge transfer electrodes. Under some conditions a given electrode-solution interface will show a range of potentials where no charge

**Table 9.2** *Modes of ionic mass transport with their driving forces*

| Modes | Driving force |
|---|---|
| Diffusion | Difference in concentration, concentration gradient |
| Migration | Difference in electric potential, potential gradient, electric field |
| Convection | Difference in density, stirring |

transfer reactions occur because such reactions are thermodynamically or kinetically unfavourable. These processes are called non-faradaic processes. Although charge does not cross the interface under these conditions, external currents can flow (at least transiently) when the potential changes. We discuss now the case of a system where only non-faradaic processes occur: electrolyte conductivity sensing.

*Electrolyte conductivity* (EC) *sensing* nowadays is a well-established and much practised technique of measuring. After an introduction of the basic concept of EC sensing, the relevant equations to describe conductivity are presented. An additional topic of this section is to treat several options for increasing the reliability of EC sensing.

Electrolyte conductivity (EC) is an expression for the mobility and concentration of ions in an aqueous solution as a result of an electric field [4]. For the determination of EC, a voltage difference must be present between two conducting electrodes, placed in the solution. This difference in potential results in an electric field between the electrodes causing the mobile ions to migrate. The resulting ionic mass transport manifests itself at the conducting electrodes as an electronic current, known or measurable.

Ionic mass transport can take place by three different means, each governed by a different driving force, as indicated in Table 9.2. It will be clear that the only mode of mass transport involved in EC is migration.

### 9.4.2. Theoretical background of EC sensing

**Relevant equations and definitions [4]**
The underlying ionic transport mode of EC is migration. The driving force of migration is a potential gradient, $dV/dx$, or electric field $E$:

$$E = \frac{dV}{dx} \quad [\text{V/m}] \tag{9.29}$$

This electric field imposes an electric force, $F_e$, on every charge particle, i.e., every ion:

$$F_e = |z_i| q E \quad [\text{N}] \tag{9.30}$$

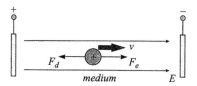

**Figure 9.17** *The force balance experienced by a charged particle under the influence of an applied electric field, E, in a viscous medium, resulting in a constant travelling speed, v*

with $|z_i|$ the valence of ion $i$, and $q$ the electronic charge. The ion will accelerate until the frictional drag (due to the viscosity of the medium) exactly counter balances the electric force. Then, the ion travels at a constant velocity, $v$. At this velocity, the ion experiences a drag force, $F_d$, equal to

$$F_d = 6\pi \eta r v \quad [\text{N}] \tag{9.31}$$

with $\eta$ the viscosity of the medium and $r$ the ionic radius. This situation is depicted in Figure 9.17.

At constant velocity, $F_e = F_d$, and from (9.30) and (9.31) the mobility of the ion $i$, $\mu_i$, is defined:

$$\mu_i = \frac{v}{E} = \frac{|z_i|q}{6\pi \eta r} \quad [\text{m/s per V/m} \equiv \text{m}^2/(\text{Vs})] \tag{9.32}$$

The resulting ionic mass transport flux, $J_i$, is

$$J_i = -\mu_i C_i \frac{dV}{dx} \quad [\text{mol}/(\text{m}^2\text{s})] \tag{9.33}$$

with $C_i$ the concentration of ion $i$ [mol/m$^3$]. The resulting electronic current density, due to all $n$ ions that are present in the solution is

$$J = F \frac{dV}{dx} \sum_{i=1}^{n} |z_i| \mu_i C_i \quad [\text{A/m}^2] \tag{9.34}$$

When the electric field $E = dV/dx$ is linear between two electrode plates with electrode area $A$ and distance $\ell$, as depicted in Figure 9.18, the electronic current can be obtained:

$$i = JA = FA \frac{\Delta V}{\ell} \sum_{i=1}^{n} |z_i| \mu_i C_i \quad [\text{A}] \tag{9.35}$$

with $\Delta V$ the potential difference that is present between the electrodes.

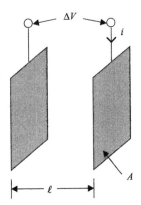

**Figure 9.18** *Illustration of two plan-parallel electrodes with electrode area A and distance $\ell$. The cell-constant for this conductivity sensor is $k = \ell/A$ $[m^{-1}]$*

### Example 9.3 Ion velocity due to migration

What is the maximum velocity of a silver ion in an aqueous solution migrating between two plan-parallel electrodes placed $\ell = 1$ mm apart with 0.1 $V_{RMS}$ over these electrodes?

The mobility of $Ag^+$-ions is tabled in text books: $\mu_{Ag^+} = 6.43 \cdot 10^{-8}$ $m^2/sV$ at room temperature. $V_{electrode} = 0.1$ $V_{RMS}$ means $\Delta V_{electrode} = 0.141$ $V_{top}$.

Using equations (9.29) and (9.32) the maximum velocity $v = \mu_{Ag^+} \cdot E = \mu_{Ag^+} \cdot (\Delta V_{top}/\ell) = 6.43 \cdot 10^{-8} \cdot (0.141/1 \cdot 10^{-3}) = 9.1 \cdot 10^{-6}$ m/s $= 9.1$ μm/s.

Now, the conductance $G$ can be expressed as

$$G = \frac{1}{R} = \frac{i}{\Delta V} = \frac{FA}{\ell} \sum_{i=1}^{n} |z_i| \mu_i C_i \quad [\Omega^{-1}] = [\text{S}]\text{iemens} \qquad (9.36)$$

where $R$ is the resistance. The conductance $G$ is a variable that can be measured, and depends on the type and dimensions of the electrodes. To be independent on these system parameters, the conductivity $\kappa$ can be defined as

$$\kappa = \frac{1}{\rho} = Gk \quad [\Omega^{-1} m^{-1}] \qquad (9.37)$$

with $\rho$ $[\Omega m]$ the resistivity, and $k$ $[m^{-1}]$ the system parameter, called the cell-constant. In the case of two plan-parallel electrodes shown in Figure 9.18 with area $A$ and distance $\ell$, $k = \ell/A$ as can be derived from (9.36) and (9.37). So, from these equations, the expression for the conductivity, $\kappa$, is

$$\kappa = F \sum_{i=1}^{n} |z_i| \mu_i C_i \quad [\Omega^{-1} m^{-1}] \qquad (9.38)$$

For solutions of simple, pure electrolytes (i.e., one positive and one negative ionic species, like KCl or CaCl$_2$) the conductivity can be normalized with respect to its equivalent concentration of positive (or negative) charges, $C_{eq}$, from which the equivalent conductivity $\Lambda_{eq}$ can be obtained:

$$\Lambda_{eq} = \frac{\kappa}{C_{eq}} \quad [\Omega^{-1}\text{m}^2/\text{mol of charge}] \tag{9.39}$$

with $C_{eq} = |C_i|z_i$. From (9.38) and (9.39), $\Lambda_{eq}$ can be expressed as

$$\Lambda_{eq} = F(\mu_+ + \mu_-) \tag{9.40}$$

with $\mu_+$, $\mu_-$ the cationic and anionic mobility, respectively. This equation leads to the definition of equivalent ionic conductivity, $\lambda_{i,eq}$:

$$\lambda_{i,eq} = F\mu_i \quad [\Omega^{-1}\text{m}^2/\text{mol of charge}] \tag{9.41}$$

Note, that following the same line of reasoning a molar conductivity, $\Lambda_m$, and a molar ionic conductivity, $\lambda_{i,m}$, can be defined:

$$\Lambda_m = \frac{\kappa}{C} \quad [\Omega^{-1}\text{m}^2/\text{mol of species}] \tag{9.42}$$

with $C$ [mol/m$^3$] the concentration of the added species and

$$\lambda_{i,m} = F\mu_i z_i \quad [\Omega^{-1}\text{m}^2/\text{mol of species}] \tag{9.43}$$

**Sensing of EC**
For a proper understanding of practical EC determination, it is necessary to introduce the electrode-solution interface before continuing.

In Figure 9.19 a complete, two-electrode EC sensor is schematically depicted. As explained, a voltage applied between the two electrodes would result in an electric current due to the migrating ions as a result of the electric field that is caused. It will be explained that EC can not be successfully determined by a DC voltage over the electrodes.

*Scenario 1*
The applied DC voltage is so high that redox reactions at the electrodes will occur and current flows via $R_{diffusion}$ through the faradaic path of Figure 9.19. Possible reactions in a solution of NaCl in water are, e.g.:

at the anode, +, $\quad 2\text{Cl}^- \rightarrow \text{Cl}_2 + 2e$

at the cathode, −, $\quad 2\text{H}_2\text{O} + 2e \rightarrow 2\text{OH}^- + \text{H}_2$

In this case, however, the concentrations of mobile ions contributing to the EC (Cl$^-$ and OH$^-$, in this example), changes at the proximity of the electrode, giving

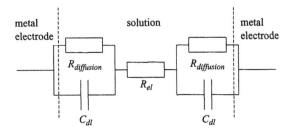

**Figure 9.19** *Schematic model of a two-electrode EC sensor. $R_{el}$ represents the measured electrolyte resistance, $R_{diffusion}$ and $C_{dl}$ the (simplified) faradaic pathway electrode resistance and the non-faradaic pathway electrode double layer capacitance, respectively*

rise to mass transport by diffusion. Thus, a proper EC determination in which only transport by migration is allowed, is no longer possible.

*Scenario 2*
The applied DC voltage is low enough to avoid any redox reactions. In this case, however, the charge on the conducting electrode that is present due to the applied voltage will almost immediately be counteracted by an equal amount of ionic charge, of opposite sign, in the solution at the direct proximity of the electrode. Of course, it is these two charges that form the well-known double layer capacitance, also present in Figure 9.19. But now, the potential applied over the electrode, with respect to the solution, will be present over $C_{dl}$ only: no longer over the solution itself, thus inhibiting any migration due to the absence of the driving force; the electric field.

From this discussion, it may be clear that EC can only be determined properly by applying an AC voltage over the electrode with such an amplitude that redox processes are avoided, and with such a frequency that the impedance of $C_{dl}$ is of the order of, or much smaller than, the $R_{el}$ to be determined, to allow a practical and precise EC determination.

### 9.4.3. Extra reliability

Reliability of EC sensing means the unambiguous determination of $R_{el}$ of Figure 9.20.

Theoretically, this can be accomplished by carefully taking into account the electrode-solution interface impedance, schematically depicted by $R_{diffusion}$ and $C_{dl}$ of Figure 9.20. This impedance, however, is by no means a constant and ideal element, but varies with the concentration of the electrolyte, its composition, and also with the electrical potential and frequency over the electrode-solution interface. Therefore, a practically more feasible approach is to design the EC sensor in such a way that the electrode-solution interface impedance influences the total

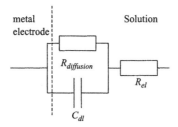

**Figure 9.20** *Simplified representation of the electrode-solution interface impedance, including the electrolyte resistance, $R_{el}$*

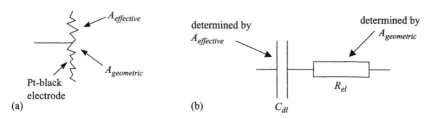

**Figure 9.21** *(a) A Pt-black electrode can be characterized by its geometric electrode area, $A_{geometric}$ and the real area, including the surface area of all micro-structures, $A_{effective}$; (b) increasing the roughness of the electrode surface decreases the impedance of the double layer capacitance, leaving the electrolyte resistance unaffected, resulting in a more precise determination of $R_{el}$ at a given measurement frequency*

impedance of the sensor (as shown in Figure 9.19) as minimally as possible. Two options are elaborated in the next sub-sections.

### Use of Pt-black electrodes

By a proper (electro)chemical treatment, the otherwise shiny surface of a Pt electrode can become black due to the thus obtained cauliflower-like structure. The resulting rough surface can be characterized by its original geometric electrode surface area, $A_{geometric}$ and by its effective surface area, $A_{effective}$, as schematically depicted in Figure 9.21.

Remember, that at proper operation, $R_{diffusion}$ of Figure 9.20 is absent, because only migrational effects via $C_{dl}$ are involved in EC determination. Now, the effect of the Pt black electrode can be easily explained. The electrolyte resistance, $R_{el}$ of Figure 9.21(b), is determined by the electrode area $A_{geometric}$ and is not affected by the rough surface and remains constant at constant electrolyte concentration. The double layer capacitance, $C_{dl}$, however, is determined by the rough electrode surface and thus by $A_{effective}$. As the impedance of $C_{dl}$ ($= 1/j\omega C_{dl} \div 1/A_{effective}$) dramatically decreases with a likewise dramatically increasing $A_{effective}$ with respect to $A_{geometric}$, due to the Pt-black formation, the impedance of $C_{dl}$ can virtually be neglected with respect to $R_{el}$ at a suitable measuring frequency. Thus, the goal

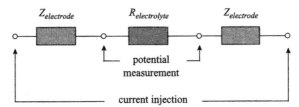

**Figure 9.22** *A typical electrical representation of a 4-points EC probe, showing the separate current injection and potential sensing electrodes and, more importantly, the position of the unfavourable electrode impedance, $Z_{electrode}$, with respect to the potential sensing electrodes*

as stated at the start of this section, an unambiguous determination of $R_{el}$, by decreasing the effect of the interface impedance, is obtained.

However, by fouling and/or degeneration of the initial rough Pt surface, the ratio $A_{effective}/A_{geometric}$ gradually decreases, thereby lowering the favourable effect of the treatment. Moreover, the Pt black electrode is mechanically rather weak, making (mechanical) cleaning not very possible. Therefore, periodic regeneration of the Pt black electrode is necessary in order to keep its favourable behaviour.

**The conventional 4-points EC measurement**
A well-known alternative to the previously described approach is the 4-points measuring method, as schematically depicted in Figure 9.23 (see Section 7.1.4).

By separating the current injection electrodes from the voltage measurement electrodes, the voltage drop over the electrode impedance, $Z_{electrode}$ in Figure 9.22, is no longer present in the measured potential. Moreover, as there are no longer strict demands on the current electrodes, they can be designed very small.

Also this method, however, has some disadvantages. Firstly, the design of the probe is more complicated due to the increased number of electrodes and their interrelation. Secondly, the design and implementation of the electronics is also a bit more complicated. Thirdly, the signal-to-noise ratio of the whole sensor plus electronics is a critical and difficult to tackle parameter, due to the high impedance of the potential sensing electrodes.

**Example 9.4 EC sensing with an interdigitated electrode pair**
One of the common manifestations of an EC sensor is the so-called interdigitated conductivity electrode [4]. Especially in small (chemical micro-)systems, the planar structure is an advantage. Moreover, this device can be fabricated in a clean room with usual photolithographic techniques. A typical lay out of such a device is depicted in Figure 9.23(a).

The overall dimensions are $1 \times 1$ mm$^2$. In reality, the device consists of 9 fingers in total. In order to minimize the interfering effect of the double layer impedance, the operating

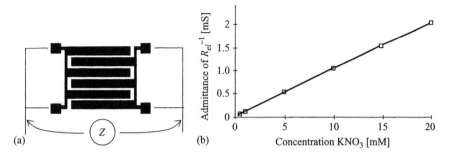

**Figure 9.23** *(a) Impression of an interdigitated electrolyte conductivity electrode; (b) admittance versus potassium-nitrate concentration, measured at 200 kHz. The range from 0.5 to 20 mM $KNO_3$ is equivalent to 0.072 to 2.9 mS/cm*

frequency had to be chosen at 200 kHz. The result of a set of measurements in different potassium nitrate ($KNO_3$) concentrations is shown in Figure 9.23(b). Here, the reciprocal value of $R_{el}$ is plotted as a function of the concentration, because then a linear dependence is expected (and obtained). The calculation of the cell constant of such an interdigitated structure is a mathematically complicated task, not repeated here [5]. The calculated cell constant of this device is 1.28 cm$^{-1}$ and the experimentally determined value is 1.24 cm$^{-1}$.

## 9.5. Further reading

Allen J. Bard, Larry R. Faulkner, *Electrochemical Methods: Fundamentals and Applications*, New York [etc.]: Wiley, 2nd ed. (2001); ISBN: 0-471-04372-9 hbk. This excellent book is not a standard textbook concerning electrochemical sensors as such, but a very fundamental and at the same time highly readable book about all possible electrochemical aspects of electrode–solution interfaces and much more.

## 9.6. Exercises

1. What is the dimension of $RT/F$? What is the quantity of $RT/F$ at room temperature ($T = 298\,\text{K}$)?

2. Why is measuring the electrode potential using a voltmeter with relatively low input impedance not wise in potentiometry?

3. What is the potential difference between a silver electrode and an aluminium electrode, immersed in an aqueous solution at room temperature containing $50 \cdot 10^{-3}$ mol/dm$^3$ $AgNO_3$ and $20 \cdot 10^{-3}$ mol/dm$^3$ $Al(NO_3)_3$?

4. What is the flux of electrons at a distance of 1 meter from an electron emitter, spraying electrons homogeneously in all directions? A current of 1 µA is applied to the emitter.

5. An $O_2$-sensitive Clark electrode is used in an aqueous solution in which 5 millimol/dm$^3$ of $O_2$ is dissolved. The measured limiting current is 4 µA. The thickness of the $O_2$-permeable membrane is 20 µm and the surface area of the Pt-electrode is 1 mm$^2$. Calculate the diffusion coefficient $D_{O_2}$ of oxygen in the membrane.

6. Referring to Example 9.2, what is the electrochemical deposition rate of Ag on the electrode (expressed in thickness increase per time)? Additional data: 1 mol Ag weighs 0.108 kg and 1 m$^3$ Ag weighs $7.9 \cdot 10^3$ kg.

7. Conductivity is temperature-dependent. What parameter causes this temperature dependence and will the conductivity increase or decrease when raising the temperature?

8. Express the capacitance $C$ and the resistance $R$ of the configuration of Figure 9.18, using its cell-constant, in the case of a block of material placed between the electrodes with relative permittivity $\varepsilon_r$ and resistivity $\rho$. In addition, derive the expression for the $RC$-product. What is the dimension of $RC$?

9. Check the experimentally determined curve shown in Figure 9.23(b) for [KNO$_3$] = 10 mM, knowing that $\mu_{K^+} = 7.62 \cdot 10^{-8}$ m$^2$/sV, $\mu_{NO_3^-} = 7.40 \cdot 10^{-8}$ m$^2$/sV at $T = 298$ K and the cell-constant is $k = 1.24$ cm$^{-1}$.

# Chapter 10

# Imaging Instruments

The purpose of imaging instruments is to measure aspects of the 2- or 3-dimensional geometrical structure of physical objects. Imaging instruments use radiation as the carrier of information. For example, optical imaging (e.g. photography) uses visible light.

An imaging instrument typically exists of a radiator (illuminator) and an image formation device (camera). Radiant energy is emitted into space. Eventually, it reaches the surfaces of objects where it interacts with the material. In the case of reflection, each surface patch acts as a new radiator that emits energy. Other examples of light-material interactions are absorption and transmission. The image formation device intercepts part of the energy and distributes it over a plane. The resulting spatial distribution of this radiant energy is what we call the *image* of the object.

However, the image, as defined above, is still in the physical domain of radiant energy, e.g. the optical domain. In order to process the image we may want to convert it to another domain, for instance, the electrical domain (e.g. video) or the digital domain. Especially, a digital representation of the image is of interest because such a representation paves the way to digital image processing and computer vision techniques.

The diversity of applications of imaging systems is wide. The following list gives an impression, but is not complete:

- Navigation and traffic control: object detection and tracking.
- Safety and surveillance systems: identification, detection and tracking of persons.
- Robotics and automation: measurement of position, displacement, orientation and rotation of recognized objects.
- Industrial applications: quality control of products.
- Medical imaging systems for diagnosis and surgery.

The chapter starts with a discussion on general principles of imaging instruments (Section 10.1). Visible light (the optical domain) is not the only physical modality that can be used to form an image. Section 10.1.1 gives an overview of other modalities. It appears that the wavelength of the radiation is a parameter that controls many of the properties of the radiation including:

- Resolving power and diffraction (Section 10.1.2)
- Spectral radiation (Section 10.1.3)
- Quantum effects (Section 10.1.4)
- Light-material interactions (Section 10.1.5).

Section 10.1.6 discusses some general principles of image formation.

Section 10.2 focuses on optical imaging techniques using visible light as the physical modality, including a short review of multi-spectral imaging, e.g. colour. Section 10.3 discusses some advanced image instruments used for various applications. Section 10.4 concludes the chapter with a short introduction to digital image processing and computer vision techniques.

Details of the topics introduced in this chapter can be found in [1]–[5].

## 10.1. Imaging principles

### 10.1.1. Physical modalities

Visible light is an example of EM (electromagnetic) radiation consisting of alternating electric and magnetic fields. Taken together these fields form an EM wave. An important parameter of an EM wave is its frequency $f$. In free space, an EM wave propagates at the speed of light, $c = 2.9979 \cdot 10^8$ ms$^{-1}$. The *wavelength* $\lambda$ is inversely related to the frequency $\lambda = c/f$. This is another parameter that is used to classify the EM wave.

Visible light occupies a small fraction of the full spectrum of wavelengths. This spectrum is enormous: from femtometers ($10^{-15}$ m) to meters (Figure 10.1). Consequently, besides optical imaging (using visible light as the physical modality), many other possibilities for image formation exist such as:

- Gamma rays: SPET (single photon emission tomography)
- X rays: X-ray imaging and CT (computed tomography)
- Infrared radiation: thermal imaging
- Microwaves: radar and SAR imaging (synthetic aperture radar).

Besides EM radiation, two other types of radiation exist: *acoustic radiation* and *particle radiation*. Acoustic waves need a medium which can be locally deformed causing local variations of pressures that propagate through the medium. The speed of propagation depends on the density of the medium and its stiffness (see Section 8.10). Since the speed does not depend on the frequency, the wavelength is again inversely proportional to the frequency.

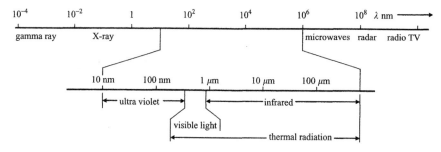

**Figure 10.1** *The electromagnetic spectrum*

*Acoustic radiation* is applied in ultrasound imaging systems (e.g. medical imaging), and underwater SONAR[1] systems (vessel detection and tracking). In water at 20° C, the speed of sound is about 1500 ms$^{-1}$. A frequency of 2 MHz (commonly used in medical imaging) corresponds to a wavelength of about 0.75 mm. At a frequency of 10 kHz (used in conventional SONARs) the wavelength is about 15 cm.

*Particle radiation* consists of a flow of particles, e.g. electrons (beta radiation) or positrons. Due to the dual nature of waves and particles (Section 10.1.4), particle radiation can also be associated with waves. The frequency of a particle with energy $E$ and momentum $p$ is given by the Bohr's frequency condition $hf = E$ in which $h = 6.6262 \cdot 10^{-34}$ Js is Planck's constant. The wavelength follows from "de Broglie wave number relation" $p\lambda = h$. An electron microscope using electrons with energies of 30 keV operates at wavelengths of about $7 \cdot 10^{-3}$ nm.

### 10.1.2. Resolving power

The wavelength of the physical modality of an imaging system is of particular interest. The resolving power (i.e. the ability of the imaging instrument to discern the details of an object) is limited to dimensions that are of the order of the wavelength of the radiation. For instance, an ultrasound imaging system operating at 2 MHz cannot have a resolving power above roughly 1 mm.

An important phenomenon related to the wavelength of radiation is *diffraction*. A distant point source, imaged by a perfect optical lens (Figure 10.2), appears as a bright spot surrounded by side lobes. If the lens is rotationally symmetrical with diameter $D$, the shape of an image of the distant point is:

$$E_d(x, y) = C \left( \frac{2J_1(\pi r D/f\lambda)}{\pi r D/f\lambda} \right)^2 \quad \text{with } r = \sqrt{x^2 + y^2} \quad (10.1)$$

---

[1] SONAR: SOund Navigation And Ranging.

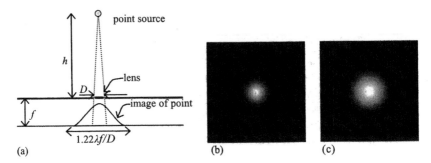

**Figure 10.2** *The diffraction limited image of point source using a theoretical perfect lens: (a) geometrical set-up (h ≫ λ); (b) image of the point source. The central part is called the Airy disk; (c) logarithmic view of (b) to enhance the rings that surround the bright central region*

$J_1(\cdot)$ is the Bessel function of the first kind and the first order. Here, $f$ is the focal length (not the frequency). Figure 10.2 shows the so-called *point spread function* of the perfect lens.

Diffraction is explained by Huygens' principle which states that each point of a wave front can be regarded as the origin of an elementary point source of a spherical wave. This, applied to the aperture of the lens, creates a system of inline point sources of waves. Each point source has its own distance to a point in the image plane. The different time delays of each source cause an interference pattern.

The resolving power of a lens is defined as the angular distance between two remote points which can just be discerned. For ideal lenses, i.e. diffraction limited lenses, this angular distance follows from (10.1):

$$\varphi = \frac{1.22\lambda}{D} \tag{10.2}$$

The resolving power of practical lenses is usually much worse than indicated by (10.2). But anyhow decreasing the wavelength favours the resolving power of an imaging system.

Diffraction occurs with each wave-related modality. Ultrasound imaging systems suffers from the same type of blurring due to the directivity of its bundle (Section 8.10). Radar systems behave similarly. The resolving power of electron microscopes can be very high because the wavelength is of the order of 10 pm.

### 10.1.3. Spectral EM radiation

Radiant energy is composed of various components, each with its own wavelength (Figure 10.3). Therefore, if needed, the radiant energy, flux, or any radiometric quantity can be decomposed into components per unit wavelength. The distribution of the radiant energy over the wavelength is called the *spectral radiant energy*

$U'(\lambda)$. The total radiant is the sum of the contributions of each wavelength:

$$U = \int_0^\infty U'(\lambda)d\lambda \tag{10.3}$$

An example is the thermal emission of a black body. Due to the thermal molecular motion a body spontaneously emits EM radiation. The spectral radiant emittance of a black body is given by Planck's law:

$$E'_s(\lambda) = \frac{2\pi c^2 h}{\lambda^5 (\exp(hc/\lambda kT) - 1)} \text{Wm}^{-2} \text{ m}^{-1} \tag{10.4}$$

$T$ is the temperature. $k$ is Boltzmann's constant $k = 1.380\,622 \times 10^{-23} \text{JK}^{-1}$. A black body is a Lambertian radiator. Real objects emit less than a black body does.

Figure 10.3 shows the spectral irradiance from the sun both outside and inside the atmosphere (at sea level). Outside the atmosphere, the spectral irradiance matches that of a black body radiator at a temperature of 5900 K. At sea level, part of the energy is absorbed due to chemical components in the atmosphere.

Continuous planar waves with a fixed relation between the phases of its components (so-called *coherent* radiation) do not often occur. Natural sources emit trains

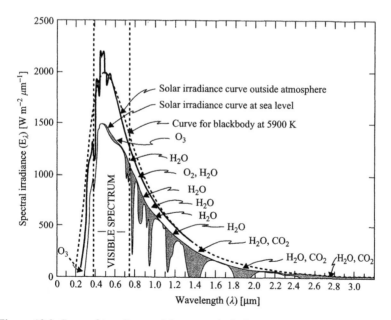

**Figure 10.3** *Spectral irradiance of the sun and of a black body radiator at 5900 K. [Source: RCA Electro-Optics Handbook, Technical Series EOH-11, Lancaster 1974]*

of short-duration waves. These trains hold random relations between their phase differences (*incoherent* radiation).

Also, the orientations of the electric and magnetic fields of the trains are often random. However, *polarization filters* can be used to confine the orientations to one linear plane. The radiation is then said to be *linearly polarized*.

### 10.1.4. The discrete nature of radiation

The dual nature of waves and particles, mentioned above, also means that waves can be associated with a continuous flow of particles. According to Einstein's relation the smallest unit of energy of a wave is $E = hf$. Therefore, radiant energy is quantized to multiples of $hf = hc/\lambda$. For visible light such a quantum is called a *photon*.

**Example 10.1 Number of particles of light emitted by a lamp**
We reconsider Example 2.3. A lamp has a radiant flux of $P = 5$ W. A table top is located below the lamp at a height of 1 m. The irradiance at the table top just below the lamp is $E_d = 0.4$ W/m² (Example 2.3). Suppose that the wavelength of the radiation is $\lambda = 600$ nm $= 6 \times 10^{-7}$m. The number of photons that the lamp emits per second is $P\lambda/hc \approx 1.5 \times 10^{19}$. Let us consider a small surface patch of the table top just below the lamp. The area of a surface patch is, say, $A = 1$ mm². The number of photons that hits that surface patch is $AE_d\lambda/hc \approx 1.2 \times 10^{12}$ per second.

In Example 10.1 we assumed that the radiation can be described by a single wavelength. But, as mentioned above, radiation often has a spectrum of wavelengths. The radiant flux $P'(\lambda)d\lambda$ within a narrow wavelength band $d\lambda$ can be converted to a flow of quanta: $n(\lambda) = \lambda P(\lambda)d\lambda/hc$. The total flow of quanta is found as the integral $n = \int n(\lambda)d\lambda$.

If a surface is exposed to a radiant flux $P$ (with wavelength $\lambda$) during a finite period of time $T$, the total number of received quanta is $N = TP\lambda/hc$. This quantity, however, is only a mean value. The real number of received quanta is a random variable with an expectation of $N$. The random variable has a Poisson distribution (Section 3.2.1). The corresponding standard deviation is $\sqrt{N}$. Therefore, the signal-to-noise ratio (in dB) is

$$SNR = 10^{10}\log N \qquad (10.5)$$

**Example 10.2 Signal-to-noise ratio of a photodiode**
Suppose that the irradiance at the table top in Example 10.1 is measured by means of a lens and photodiode (Figure 10.4). The irradiance of the table top below the lamp is about $E_d = 0.4$ W/m².

The distance from the lens to the table top, measured along the direction at which the lens points, is $H = 2$ m. The diameter of the lens is $D = 12$ mm. The lens is mounted such

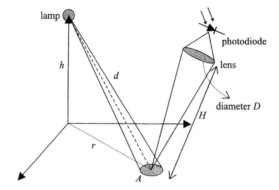

**Figure 10.4** *Irradiance of a photodiode*

that the table top is in focus with the photosensitive area of the diode. We assume that the photosensitive area corresponds to an area $A = 0.25$ mm² of the table top. The table top itself can be regarded as a Lambertian reflector. Furthermore, it is given that the table top reflects 10% of its incident irradiance. The purpose is to calculate the signal-to-noise ratio due to quantum limitations of the irradiance of the sensor.

The surface is a Lambertian reflector. Therefore, the radiance $L$ is given by equation (2.12): $L = E_s/\pi = 0.1 E_d/\pi = 0.04/\pi$ W/m² sr. The intensity of the surface in the direction of the lens is $I = AL$. The solid angle, seen by the surface and formed by the lens is $\Omega = \pi D^2/(4H^2)$. Then, if the transparency of the lens is 100%, the radiant flux incident on the photodiode is $P_{diode} = I\Omega = AL\pi D^2/(4H^2) = 9 \times 10^{-14}$ W. If $\lambda = 600$ nm, the number of photons (photon flow) is about $n = 2.7 \times 10^5$ per second.

In digital photography, for example, the integration time (shutter time) could be $T = 10$ ms. The mean number of photons is $N = Tn = 2700$. The signal-to-noise ratio becomes $\sqrt{2700} \approx 52$, equivalent to 34 dB. In practice, this could be much lower because of transmission losses in the lens, and because the DQE (detective quantum efficiency, i.e. the fraction of photons that are actually converted into free electrons) is less than 100%. Dark current noise and thermal noise may also affect the SNR.

The conclusion is that the discrete nature of radiation amounts to the noise in the image. This phenomenon is observed especially in low power situations, high bandwidths (small integration time), or short wavelengths.

## 10.1.5. Light-material interactions

When a wave traverses a medium, two types of interaction between the wave and the material can happen:

- Volume related interaction (absorption and scattering)
- Surface related interaction (reflection and refraction).

The latter happens when changes of material properties occur over a distance which is much smaller than the wavelength.

**Absorption**

The property of interest here is the attenuation coefficient $\mu(x, y, z)$. When a ray propagates along a (straight) line $\ell$ through space, its intensity is partly absorbed, and the corresponding loss of energy is converted to heat. Measured over an infinitesimal distance $ds$ along $\ell$ the absorption is proportional to its intensity, and proportional to $ds$. Thus, if $s$ is the running arc length of the ray, then $I(s + ds) - I(s) = -\mu(\ell(s))I(s)ds$. This leads to the following integral:

$$I(s) = I(0) \exp\left(\int_{t=0}^{s} \mu(\ell(t))dt\right) \qquad (10.6)$$

$\mu(\ell(s))$ is the attenuation coefficient at a position on the line $\ell$ at a distance $s$ from its origin.

An example of where absorption is important is X-ray imaging. In daily life, we observe the effects of absorption during a hazy day.

**Scattering**

Scattering is a phenomenon where the individual photons undergo a change of direction due to collisions with other particles. The energies of the photons may change as well, and thus the wavelengths may change. Scattering is a complex process. The effect on the incident radiation is that it loses energy. This can be described analogously to (10.6). However, unlike absorption, the scattered energy is not lost, but still remains active as radiant energy. One of the effects that might occur, for instance in X-ray imaging, is a veil over the image. Such a veil complicates the detection of objects.

**Reflection and refraction**

When a ray hits a discontinuity of the media, part of the radiant energy is transmitted, part of it is absorbed, and part is reflected. The transmission occurs according to Snells' law (Figure 10.5(a)):

$$\frac{\sin \varphi_i}{\sin \varphi_r} = \frac{n_r}{n_i} \qquad (10.7)$$

$n_i$ and $n_r$ are indices of refraction. Refraction is the basis of operation of optical lenses.

There are two physical principles to be applied to reflection: specular reflection and diffuse reflection. These principles are depicted in Figure 10.5. If a surface is smooth (on a scale larger than the wavelength), the reflection is specular, i.e. mirror-like $\varphi_o = \varphi_i$. The reflectivity (the fraction of incident energy that is reflected) depends on $\varphi_i$, $\varphi_r$ and the polarization of the incident wave. If the surface is rough, but with facets that are smooth (Figure 10.5(b)), then the energy is reflected "grossly" in the

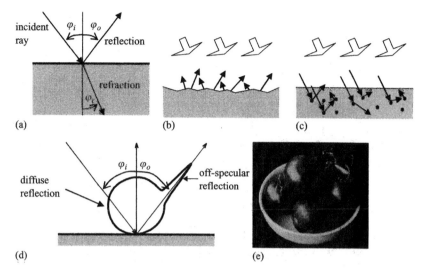

**Figure 10.5** *Reflection and refraction: (a) specular reflection and refraction on a mirror-like surface; (b) off-specular reflection at a rough surface modelled as a facetted mirror-like surface; (c) diffuse reflection due to scattering; (d) the reflectance map of a combined diffuse and off-specular reflecting surface; (e) the tomatoes and the apple in this image show both diffuse and off-specular reflection*

direction $\varphi_o$. Due to the facets the reflected energy is distributed around $\varphi_o$. This is called off-specular reflection.

Diffuse reflection occurs when transmitted energy is scattered beneath the surface due to the interaction with particles (Figure 10.5(c)). Part of the energy is scattered back. The orientation of scattering is arbitrarily, but in a direction perpendicular to the surface the path back to the surface is shorter. Therefore, the attenuation in that direction is smallest, and the reflection is strongest. In fact, this reflection is proportional to $\cos \varphi$. Thus, the reflection is Lambertian (diffuse).

Figure 10.5(d) shows how the intensity of a reflected wave depends on $\varphi$. The figure shows both diffuse reflection and off-specular reflection. The off-specular reflection is seen in the graph as a peak in a direction opposite to the incident angle of the illumination. Figure 10.5(e) is an example of an image of such objects. The off-specular reflections cause glossy spots (so-called *highlights*). Their positions are directly related to the location of the illuminator in the scene. Specular and off-specular reflection is often independent of the wavelength. Therefore, the glossy spots take the colour of the illuminator. In contrast, diffuse reflection gives rise to images which are less dependent on the locations of the illuminators, and which reflect the colour of the object.

Another difference between the two types of reflections is that (off-)specular reflection maintains the polarization of a waveform. In contrast, diffusely reflected

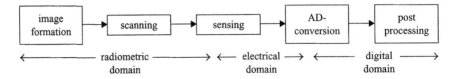

**Figure 10.6** *Functional structure of an image device*

waveforms are not polarized, regardless of whether the incident radiation is polarized or not.

### 10.1.6. General layout of an imaging device

An imaging device consists of at least two functional parts (Figure 10.6): *image formation* and *image sensing*. The image formation is a physical process that ends up with the irradiance at a surface (called the *image plane*). The purpose of image sensing is to transform the image from the domain of the physical modality (e.g. visible light) to the electrical domain, i.e. a voltage or a current.

An optional third functional part is *scanning*. The 2D image plane is examined in a systematic pattern so as to transform the spatial distribution of the radiant energy on the image plane into a time dependent signal. Typical patterns are raster-like (TV), polar-like (RADAR), sector-like (ultrasound imaging) and random access (CMOS camera).

The last two parts in Figure 10.6 are also optional. Some imaging devices produce digital output. For that, AD-conversion is needed. A typical example is digital video. Furthermore, advanced imaging devices need further processing to obtain the desired result.

Perhaps, the simplest device is a photocell that detects the presence or absence of light. Here, the image formation is trivial. A lamp, sometimes in addition to a lens, suffices. In this case we have a point measurement. Therefore, no scanning is needed, and a simple photo resistor can do the sensing.

At the other end of the spectrum, we have complex, active imaging devices such as RADAR systems and flying spot scanners to obtain range images (distance maps), or CT scanners to obtain 3D images of the interior of the human body.

### 10.2. Optical imaging

Optical imaging, as an image based measurement system, is perhaps the most widely used imaging technique. Illuminators, lenses (for image formation), and cameras (scanning and sensing) are relatively cheap components. In addition, the potentials of optical imaging are large, and many fields of applications are covered.

## 10.2.1. *Illumination techniques*

The quality of an image based measurement system often depends on the choice of illumination. For opaque objects, reflection is the main principle of light-material interaction (Section 10.1.5). Usually, the radiant energy reflected from objects has a diffuse and an off-specular component. Which type of reflection is beneficial depends on the application. If we want to measure geometric parameters (shape, size, position, orientation), then diffuse reflection is preferred since in that case off-specular reflection is often a disturbing factor. However, scratch detection on mirror-like surfaces (e.g. steel plate) profits mostly from specular reflection.

There are various methods used to separate the diffuse reflection from the specular one. One method is based on the choice of illumination. If the illumination is diffuse (such as during daylight on a cloudy day), then the diffuse reflection component will prevail. Such a situation can be created in a measurement set-up by the arrangement of a number of illuminated diffuser screens around the object (Figure 10.7(a)).

Specular reflection can be obtained by using a collimator that projects a bundle of light coming from one direction (Figure 10.7(b)). If the direction of the specular reflected ray is not in the line of sight of a camera, then no specular light will be

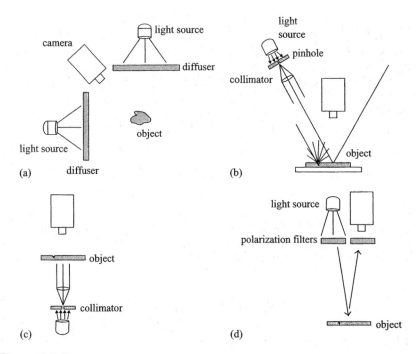

**Figure 10.7** *Illumination techniques: (a) diffuse illumination using diffuser; (b) dark field illumination using collimated (bundled) light; (c) background illumination using either a collimator (shown) or a diffuser; (d) diffuse illumination using two polarization filters*

captured by the camera. Thus, if the surface produces much specular reflection, and relatively little diffuse reflection, then the image of that surface will be dark. However, surface patches with much diffuse reflection (e.g. scratches on a glossy plate) will produce bright spots in the image. This technique is called *dark field illumination*.

Another technique is *background illumination* (Figure 10.7(c)). Here, the objects are illuminated from the rear. The image contains the silhouette of the object. This is useful, for instance, for object recognition based on the shape as defined by the silhouette.

The polarization of light is another technique used to produce only diffuse reflection components (Figure 10.7(d)). The incident light is polarized using a polarization filter. The reflected light consists of a diffuse component and an off-specular component. The polarization is maintained in the off-specular component, but not in the diffuse component. Therefore, we can remove the off-specular component by means of a second polarization filter. The orientation of this filter should be perpendicular to that of the first filter. The diffuse component will pass the second filter although half of its energy will be lost due to the polarization.

### 10.2.2. Perspective projection

Image formation in the optical domain takes place by means of a lens. The process can be modelled using a *perspective projection model* (also called *pinhole camera model*) (see Figure 10.8). The 2D image plane with coordinates $(x, y)$ coincides with the $XOY$-plane of a 3D world coordinate system $(X, Y, Z)$. The optical axis coincides with the $Z$-axis, and the lens is located at a distance $f$ from the image plane. The lens itself is modelled by a pinhole; that is, a non-transparent plane with a tiny hole at its origin.

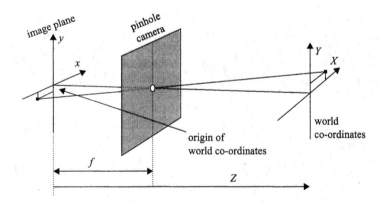

**Figure 10.8** *Perspective projection as a model for the image formation with a lens*

Suppose that there is a surface patch of an object in the world coordinate system at a position $(X, Y, Z)$. The surface patch emits light in many directions, and also in the direction of the pinhole. The ray in this direction shines through the pinhole and hits the image plane at a position $(x, y)$ (in image coordinates):

$$x = \frac{-f}{Z - f} X \qquad y = \frac{-f}{Z - f} Y \qquad (10.8)$$

This relation is called the *perspective projection*. It is a geometric relation that associates each 3D point $(X, Y, Z)$ in world coordinates with a 2D point $(x, y)$ in image coordinates. The relation is fundamental for the measurement of position, orientation and shape of objects, and also for stereometry.

The irradiance of the image at $(x, y)$ depends on many factors, e.g.:

- The radiant intensity of the surface patch at $(X, Y, Z)$ in the direction of the lens.
- The diaphragm of the lens.
- The orientation of the optical axis with respect to the position of the surface patch.
- Possible losses due to absorption.

The image of the ray will always be blurred due to the inevitable diffraction (Section 10.1.2). Other phenomena that lead to blurring are an imperfect lens, motion blurring, atmospheric blurring, a defocused lens, and so on.

### 10.2.3. Optical scanning and sensing devices

The result of optical image formation is radiant flux that is spatially distributed over the image plane. The task of scanning and sensing is to access each position on the image plane, and to convert the local irradiance to a voltage. If optical point sensors (Section 7.5) are used, such as photoconductors or photodiodes, the scanning must be done entirely mechanically. CCD and CMOS cameras consist of a mosaic of photosensitive cells. Such a cell is called a *pixel* (picture element). The scanning of pixels is accomplished electronically.

**The CCD**

A CCD (*charge coupled device*) is an integrated circuit for the acquisition of images. It consists roughly of three parts (Figure 10.9):

- An array (mosaic) of charge-storage elements.
- A mechanism for the transport of electrical charge.
- An output circuitry.

The charge-storage elements are arranged either linearly (the line-scan CCD), or according to an orthogonal grid (the area CCD). First, we discuss the line-scan CCD.

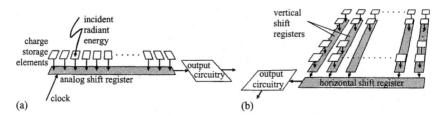

**Figure 10.9** *Charge coupled devices: (a) line-scan CCD; (b) area CCD*

Each charge-storage element (pixel) has a surface layer that is photosensitive. At a given moment, under control of a clock signal, an electronic shutter can be opened. From that moment on, the photons that are incident on the surface charge the element. Each detected photon emits a free electron. With that, the accumulated charge in the element is proportional to the radiant flux at the surface integrated over time.

After the period of integration, the electronic shutter closes. From that moment on, the transport mechanism comes into action. The charges in the elements are locked parallel into an analogue shift register. Subsequently, under control of another clock signal, the charges in the register are shifted out serially. The charge that appears at the output of the shift register is pushed into the output circuitry. A charge-voltage converter finally produces an output voltage that is proportional to the charge.

The ultimate result is that with consecutive clock pulses of the shift register the output voltage, as a time dependent signal, is proportional to the spatial distribution of the radiant energy at the linear array.

The area CCD is provided with a more advanced transport mechanism. The elements are connected to the parallel input of a parallel bank of vertical, analogue shift registers. The serial outputs of the vertical shift registers are connected to the parallel input of a horizontal shift register. Reading the device takes place by means of a vertical sync pulse that shifts the charges of the vertical registers one position. The charges at the output of these registers are pushed into the horizontal shift register. Next, a fast sequence of horizontal sync pulses shifts the charges of the horizontal register into the output circuitry. This process (one vertical sync pulse followed by a sequence of fast horizontal sync pulses) repeats until the last horizontal row has been processed.

The voltage output of an area CCD encodes the irradiance of each pixel by scanning each row from left to right. The rows themselves are scanned either interlaced, that is, first the odd rows and then the even rows (conforming to television standards), or progressively, i.e. row by row. The (conventional) video CCD cameras produce an analogue signal that complies with a television standard. Digital CCD cameras are equipped with an AD converter which encodes the pixel values into a digital video standard (Example 10.3).

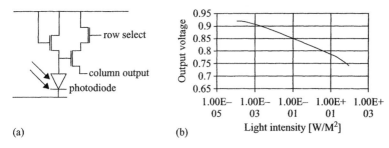

**Figure 10.10** *CMOS Imager: (a) CMOS pixel architecture; (b) logarithmic response*

The line-scan CCD finds its application in optical scanners. With a mechanical scan (in a direction orthogonal to the array of elements) photographs, prints and drawings can be digitally scanned and copied to files in digital formats. The area CCD is more suitable for the acquisition of images. They are applied in video cameras and digital "still picture" cameras.

**The CMOS camera**
The CMOS camera consists of a 2D array of photodiodes whose outputs can be addressed by means of MOSFETs. Figure 10.10(a) shows the architecture of a single pixel. If the "row select" has been activated, the voltage of the photodiode appears at the column output. This enables the addressing of each pixel in random order.

The voltage of a photo diode holds a logarithmic relation with its leakage current. The leakage current itself depends linearly on the incident radiant flux. Figure 10.10(b) shows the logarithmic response of such a device. The most important source of noise is the dark current of the device (i.e. the leakage current that flows in the absence of light). This current is additional to the *signal current* that flows due to the radiation. Due to the quantization of current to multiples of electron per second, the total current obeys the Poisson statistics (Chapter 3). Therefore, the standard deviation of a current $i$ is $\sigma_i = (2eiB)^{1/2}$ where $e$ is the charge of an electron, and $B$ is the bandwidth of the system. Another source of noise is due to the discrete nature of light (10.1.4). Unlike the CCD there is no integration time. However, we may consider the reciprocal of the bandwidth $1/B$ as a measure of integration time.

Compared with the CCD, the CMOS has a large dynamic range. Another possible advantage is that the pixels in a CMOS camera are randomly accessible.

### 10.2.4. *Multi-spectral imaging and colour vision*

Many objects have their own typical colour. Examples are found in the food industry where products like green peas, French beans, tomatoes, and so on are bound to their colours. In other applications, the colour of the objects is related to a physical state of the object. The obvious example is the temperature of a body (Section 10.1.3). Therefore, the assessment of these colours may provide much information about the objects.

The wavelength domain adds another dimension to the concept of an image. The spectral irradiance at the image plane now becomes a function, not only of space, but also of wavelength: $E'_d(x, y, \lambda)$. The additional information stems from the fact that the irradiance at a given position $(x, y)$ in the image plane results from a reflected ray somewhere on a surface patch of the imaged object. The reflected ray itself results from two factors: the illumination of the patch, and the reflectivity. Hence, the irradiance at the image plane can be written as the product of two factors:

$$E'_d(x, y, \lambda) = f_{illum}(x, y, \lambda) f_{obj}(x, y, \lambda) \tag{10.9}$$

$f_{illum}(\cdot)$ represents the irradiance of the surface patch. $f_{obj}(\cdot)$ represents the reflectivity of the surface patch in the direction of the lens. As such, $f_{obj}(\cdot)$ depends on the geometry of both the object and the camera, and on the optical properties of the object.

The introduction of spectral images raises the question of how to sample in the wavelength domain. Usually, the spatial domain needs dense sampling in order not to lose the resolving power. Image sizes varies from, say, $100 \times 100$ pixels up to $3000 \times 2000$ or more. In order not to push up the memory requirements too far it is important to keep the wavelength sampling as sparse as possible.

The actual number of samples needed depends on the smoothness of $E'_d(x, y, \lambda)$ with respect to $\lambda$. The radiation emitted by some chemical substances consists of a number of narrow spectral lines at various wavelengths. The representation of these spectra needs dense sampling. Fortunately, for most natural materials the spectral reflectivity is rather smooth within the visible range. Often, about three samples suffice provided that the spectral sensitivity of each sample is selected appropriately.

Taken together, the relative sensitivities should cover the whole visible range in order not to lose spectral information. Figure 10.11(a) shows the relative sensitivities for a particular CCD. The resulting channels are shown in Figure 10.11(b), (c) and (d).

Suppose that the image is sampled with $N$ spectral bands. The sampled image at a position $(x, y)$ is denoted by $g_n(x, y)$ where $n = 1, \ldots, N$ is the index to the $n$-th spectral band. For each band there is a relative spectral sensitivity $h_n(\lambda)$. The pixel values are found to be:

$$g_n(x, y) = c \int h_n(\lambda) E'_d(x, y, \lambda) d\lambda \tag{10.10}$$

$c$ is a constant that depends on the area of the pixel, and the shutter time.

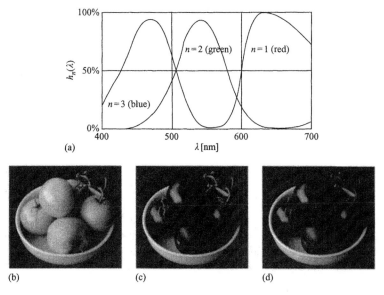

**Figure 10.11** *Multi-spectral imaging: (a) relative spectral sensitivities of the SONY ICX084AK CCD; (b) red channel; (c) green channel; (d) blue channel*

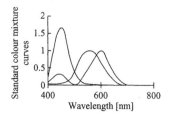

**Figure 10.12** *Standard colour mixture curves of the CIE*
*[Source: CIE publication Tech. 15, Paris 1971]*

## Colour vision

Colour is a psycho-physiological concept. The perception of colour is a complex human activity that takes place both in the retina of the eye and in the brain. Despite the various attempts that have been undertaken to model this activity, these models are still the subject of dispute. The simplest model relies on the assumption that the retina of the eye consists of a mosaic of three different types of receptors that are sensitive to three different parts of the electromagnetic spectrum. Although these parts overlap considerably they are associated with red, green and blue, respectively.

The three degrees of freedom of this model have led to the generally accepted convention that colour can be represented numerically by three quantities called the *tristimulus values*. For that purpose, in 1931, the "Commission Internationale de l'Éclairage" (CIE) has proposed to use a set of three *standard colour mixture curves*, $\bar{x}(\lambda)$, $\bar{y}(\lambda)$ and $\bar{z}(\lambda)$ shown in Figure 10.12. The tristimulus value of a radiator with spectral radiant flux $P'(\lambda)$ is[2]:

$$X = \int_0^\infty \bar{x}(\lambda)P'(\lambda)d\lambda \quad Y = \int_0^\infty \bar{y}(\lambda)P'(\lambda)d\lambda \quad Z = \int_0^\infty \bar{z}(\lambda)P'(\lambda)d\lambda \quad (10.11)$$

In fact, $Y$ is proportional to the luminous flux (Section 2.3.5) of the radiator.

The two *chroma coordinates* $(x, y)$ are defined in terms of the tristimulus values as ratios:[2]

$$\begin{aligned} x &= X/(X+Y+Z) \\ y &= Y/(X+Y+Z) \end{aligned} \quad (10.12)$$

The corresponding colour representation is indicated with CIE $xyY$-space. Thus, this space contains one luminous coordinate, and two chroma coordinates. The $xyY$-space (or equivalently the $XYZ$-space) is the world wide basic standard for colours.

The three degrees of freedom in the human perception of colour imply that colour display devices (monitors, projectors, printers) need only three primary colours. A weighted mixture of these primaries can match almost any arbitrary colour. Consequently, all colour acquisition devices (cameras, scanners) need three spectral sensitivities. Usually, these sensitivities are associated with red, green and blue (Figure 10.11). The so-called *RGB* tristimulus values are obtained by normalizing each output channel such that the resulting three values have a range of [0,1].

Usually, the sensitivities of a device do not match the standard colour mixture curves. A *RGB* colour representation is device-dependent. For a faithful reproduction of a wide range of colours, the primary colours of the display device must be adapted to the *RGB* representation of the acquisition device. It needs calibration.

### Digital colour representation
The enormous consumer market for video and photography has lead to a large range of affordable consumer products. Multi-spectral imaging devices for scientific and industrial applications often share the same technology of these consumer products in order to come to economic solutions. Consequently, *RGB* representations of a pixel are often used in image based measurement systems. However, the same

---

[2] The world coordinate system in a geometric domain and the tristimulus values in the colour domain are both conventionally denoted by $(X, Y, Z)$. Unawareness of this might lead to confusion. However, the usage is too well established to introduce a different notation. The context should make clear which domain is meant.

consumer market has also lead to all kinds of image data types aiming at compression so as to reduce the memory requirements or the bandwidth. For measurement applications compression is undesirable.

### Example 10.3 Sony DFW-V500 digital colour camera

The Sony DFW-V500 is a digital camera that adopts the IEEE 1394-1995 standard. The camera uses spectral sensitivities as shown in Figure 10.11(a). The specifications mention that the camera is able to present 30 frames (images) per second in VGA (640 × 480 pixels) mode and in Yuv (4:2:2) format.

"Yuv" means that $RGB$ is converted to digital $Yuv$ according to the following equation:

$$\begin{bmatrix} Y \\ u \\ v \end{bmatrix} = \text{int}\left( \begin{bmatrix} 16 \\ 128 \\ 128 \end{bmatrix} + \begin{bmatrix} 65.481 & 128.553 & 24.966 \\ -37.797 & -74.203 & 112 \\ 112 & -93.786 & -18.2142 \end{bmatrix} \begin{bmatrix} R^{0.45} \\ G^{0.45} \\ B^{0.45} \end{bmatrix} \right)$$

(10.13)

The first step is the so-called *gamma correction*, i.e. $R^{0.45}$, $G^{0.45}$ and $B^{0.45}$. This is done to match the logarithmic nature of the human eye. The second step is the linear transform of the gamma corrected $RGB$-space. The first component $Y$ carries information about the luminance, i.e. the "black and white" image. The two other components, $u$ and $v$, carry the chromatic information. In essence, they specify $B - Y$, and $R - Y$. The last step is a linear mapping of each component, and a consecutive quantization to 8 bits. The $Y$ component gets a maximal excursion of 219 with an offset of $+16$. With that, the range of $Y$ is [16, 219]. $Y = 16$ represents black. The range of $u$ and $v$ components is mapped to $[-112, +112]$ with a binary offset code (so that $+128$ corresponds to zero). The reason for not using the full range of a byte is that this creates a footroom and headroom that can be used for transmission of additional information.

"(4:2:2)" means that a bandwidth compression using subsampling has been applied. For every two pixels on a row, the two $Y$ bytes are transmitted, only one $u$ byte and one $v$ byte are sent, thus implementing a compression ratio of 1:1.5.

Figure 10.13 demonstrates the $Yuv$ representation in (4:2:2) format. The orginal $RGB$ images are shown in Figure 10.11. Small degradations are seen near the vertical edges in the image.

## 10.3. Advanced imaging devices

Optical imaging is cheap, and it provides a rich source of information about the geometry and radiometry of objects. Nevertheless, there are many other types of imaging devices each with its own specific application field, and with its own features. The following list gives an impression.

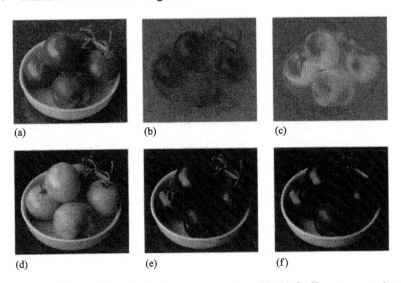

**Figure 10.13** *Image degradation due to compression: (a)–(c) the Yuv representation in 4:2:2 format. The original RGB is shown in Figure 10.11; (d)–(f) the reconstructed RGB representation*

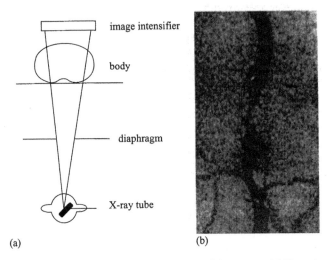

**Figure 10.14** *X-ray image formation for medical diagnosis: (a) X-ray imager; (b) coronary angiography for measurement of the constriction in blood vessel*

### X-ray imaging

Here, the image formation is based on the attenuation coefficient $\mu(X, Y, Z)$ of material for X-rays. When an X-ray passes through a body (Figure 10.14) along a line $\ell$, it is absorbed according to equation (10.6). $\ell$ is the line that originates from the X-ray source, and ends up on the image plane at the image position $(x, y)$.

$I_0$ is the radiant intensity of the source in the direction of $(x, y)$. $E_d(x, y)$ is the irradiance at $(x, y)$. In fact, the X-ray image is a shadow of the internal organs of the body.

The purpose of the image intensifier in Figure 10.14 is to convert the X-ray image to visible light, and to intensify the irradiance. X-ray imaging systems find applications in medical diagnostic systems and in material science. The advantage of the technique is that it is non-invasive and inexpensive with respect to CT and NMR.

**Computer tomography**
A disadvantage of X-ray imaging is that the shadows of different organs may pile up. That makes it difficult to interpret the image. Computer tomography (CT) is an extension of X-ray imaging that overcomes this problem. The goal is to determine the attenuation coefficient $\mu(X, Y, Z)$ for each 3D position. Initially, the $Z$-position is kept constant, $Z = Z_0$, so that the target is to obtain a *slice* $\mu(X, Y, Z_0)$. If needed, the procedure can be repeated for other values of $Z$ in order to obtain a full 3D reconstruction of the interior of the body.

Figure 10.15 shows the set-up. An array of detectors is lined up along an arc (say 512 detectors over 45°). The X-ray source and the detectors are placed in a plane perpendicular to the $Z$-axis. Each detector receives a radiant flux according to

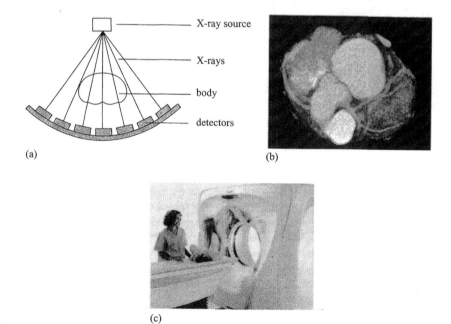

**Figure 10.15** *CT scanner (courtesy of Philips Medical Systems): (a) geometrical arrangement; (b) coronary; (c) Philips MX8000 IDT CT scanner*

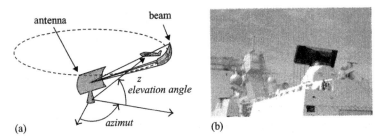

**Figure 10.16** *Radar: (a) polar scan pattern of a radar; (b) polar scanning radar system [courtesy of Thales Nederland]*

(10.6). This gives 512 equations to recover $\mu(X, Y, Z_0)$. In general, this is not enough data to make the reconstruction. However, the combination of X-tube and detectors can be rotated a little around the body, giving rise to another set of 512 equations. This process continues until enough data have been acquired to make an accurate reconstruction of $\mu(X, Y, Z_0)$. There are various alternatives for the method of reconstruction (e.g. back projection).

**Range imaging**
A range image is a depth map of the surroundings of the imaging device. Suppose that $x, y$ are coordinates in the image plane, and $z(x, y)$ is the range image, then $z(x, y)$ has the meaning of "distance" from the image plane to the nearest object seen in a direction that is determined by $x, y$.

For instance the scanning method in a *radar system* (Figure 10.16(a)) is polar-like. The variables $x$ and $y$ are associated with the azimuth angle and the elevation angle. $z$ is the distance from an object to the origin of the camera's coordinate system. Some radar systems realise the scanning mechanically by rotating the antenna. Random access scanning can be realised by phased-array radars. The distance $z$ can be measured by a "time-of-flight" measurement of an electromagnetic burst that is sent by the antenna, reflected by the object, and received back by the antenna.

Another range imaging device is the *flying spot scanner* (Figure 10.17). Here, a spot illuminator (laser) scans the 3D spaces by means of a rotating mirror system. Scanning is carried out with two independent rotations over two different axes, i.e. a polar scan. The light beam intersects the surface of an object at a certain 3D position. This is seen in the image of a pinhole camera as a projected spot on the image plane. By back projecting this spot along a ray through the pinhole and calculating where this ray intersects the light beam of the scanner the 3D position of the object can be reconstructed.

The pattern of light is not necessarily restricted to that of a spot. If a (planar) stripe of light is projected into space, it is seen on the surface of the object as a curve in 3D space, which can also be reconstructed from an observation in the image

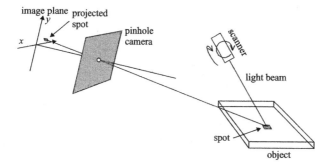

**Figure 10.17** *Flying spot scanner*

plane. With that, the mechanical scan can be restricted to 1D, i.e. a rotation over a single axis.

One can even project a 2D light pattern, so-called structured light, on the surface of the object, thus avoiding mechanical scanning. But in that case, the 3D reconstruction is ambiguous. To solve that ambiguity, it is necessary to project multiple light patterns.

Another method to get range data is stereo vision which uses two cameras (Figure 10.18). Suppose that in the first image a point feature has been localized at some position in that image. This point feature must come from some surface patch of the object in the scene. That patch must be located somewhere on a ray (straight line) defined by the point feature in the image and the pinhole position of the first camera. (Figure 10.18(a)). The surface patch also causes a point feature in the second image. Suppose that we can find two point features, one in the first image, and one in the second image, that stem both from the same surface patch in the scene (finding these two points is the so-called correspondence problem). Then the 3D position of the surface patch can be reconstructed by retracing the two rays from the two corresponding point features back through the two pinholes. The interception point of the two rays must be the position of the surface patch.

Figure 10.18(d) is a range image obtained from a left image (Figure 10.18(b)) and a right image, Figure 10.18(c). The reconstruction is only successful if sufficient points can be found in one image that can be associated with corresponding points in the other image. Such is the case if the local contrast in the image meets some constraints. For instance, corner-like point features can easily be associated. At locations without contrast, the correspondence problem cannot be solved. Stereo vision fails at these locations.

**Imaging with other physical modalities**
Up to now, the discussion of imaging devices was restricted to that with electromagnetic waves as the physical modality. Such a restriction is not necessary. Other modalities include the magnetic domain and the acoustic domain.

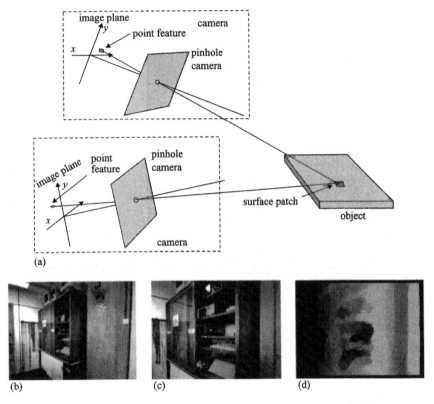

**Figure 10.18** *Stereo vision: (a) geometry of a stereo vision system; (b) left image; (c) right image; (d) range image*

An example of the former is NMR (*nuclear magnetic resonance*) imaging. Here, the energy of atomic particles (e.g. protons) is temporarily raised to a higher state by means of a radio frequency magnetic pulse. The process in which the particles fall back to their lower energy state induces transient pulses with a specific resonance frequency (called the Larmor frequency). These pulses can be measured using external coils. The process has a time constant that is specific for the type of material. The spatial variability can be resolved by placing the object in a permanent, space dependent magnetic field. Such a field influences the Larmor frequency. With that, a one-to-one correspondence can be set-up between the space coordinates and the Larmor frequency. Figure 10.19 is an example of an NMR image.

Another physical modality is the acoustic domain. An acoustic wave is cast into the object of interest. At the boundary between neighbouring tissues the transition of acoustic impedance induces a reflection of the acoustic wave. Therefore, a time-of-flight measurement reveals the internal structure of the object (Section 8.10). Figure 10.20 provides an example.

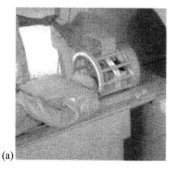 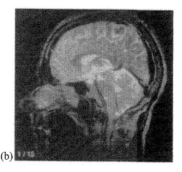

**Figure 10.19** *NMR imaging [courtesy of Philips Medical Systems Best]: (a) Philips Gyroscan ACS-NT magnetic resonance imaging system; (b) NMR Image of the sagittal brain*

**Figure 10.20** *Fetal medicine ultrasound imaging*

## 10.4. Digital image processing and computer vision

The image of an object can capture the required information about the object. However, usually, the measurement process does not end with image acquisition. The images need to be processed in order to obtain the measurement result. Image processing must give us the required description of the object. Often, the description is in terms of parameters of the object: size, position, orientation, shape, colour, etc. In other applications, the object needs to be classified, or even identified.

In some applications the processing occurs manually. A typical example is medical diagnostics. Despite many efforts of the "computer vision" community in the last two decades to automate the interpretation of medical images, it is still the expertise of the radiologists and other medical specialists that is decisive. In these cases, image processing is merely a tool that is used in a semi-automated, interactive system.

Nevertheless, many other applications exist where the automated processing of images is successful, sometimes even better than human visual inspection. Such an automated process is often referred to as "computer vision".

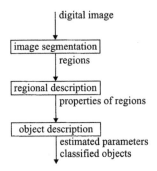

**Figure 10.21** *Functional structure of a computer vision system*

Computer vision often consists of three steps (Figure 10.21):

- *Image segmentation*
  The goal is to separate the image plane into non-overlapping, meaningful regions (or segments). The connotation of "meaningful" is that the regions should bear a one-to-one correspondence with the object's parts that are visible in the scene.
- *Regional description*
  The regions obtained from image segmentation have all kind of properties such as shape, colour, position and orientation. These properties can be estimated from the image data. Also, the mutual relations between regions (larger-than, above, etc.) can be determined.
- *Object description*
  The final step is to cast the regional description into a final object description. Examples of tasks included here are event detection, object recognition, statistical analysis, etc.

The concepts mentioned above will be illustrated by means of an example.

**Example 10.4 Recognition of electronic components**
The measurement problem is as follows. Electronic components consisting of resistors, capacitors and transistors are placed on a conveyor belt in order to separate them. The vision task is twofold. First, we have to detect and to localise each component. Second, each detected component should be assigned a class label "resistor", "capacitor", or "transistor". For that purpose, a CCD camera system is available (Figure 10.22). An example of an image is shown in Figure 10.23(a).

In this kind of application, it is often assumed that the height of the objects is small relative to the distance to the camera. In addition, it is assumed that the image plane and the object plane are parallel. In that case, the relationship between a point on the surface of the object and the corresponding point in the image plane is given by equation (10.8). $Z$ and $f$ are two system parameters that must be obtained by a proper calibration procedure.

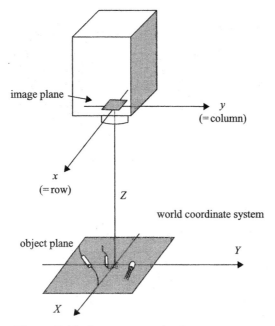

**Figure 10.22** *Camera system for object recognition*

If the image and object planes are not parallel equation (10.8) is still valid, but now Z is not constant anymore. We can still use the images, but the calibration should account for perspective projection. Even if nonlinear lens distortion takes place, i.e. small deviations from the pinhole model, the image can be used provided that calibration based on a suitable nonlinear model takes place.

**Image segmentation**
The objects are seen as regions in the image plane. The goal of image segmentation is to reveal these regions. Often, image segmentation is one of the most difficult and crucial subtasks in a computer vision system. However, if the illumination of the object is selected carefully (Section 10.2.1), a simple threshold operation can separate the objects from the background. If the grey level of an image pixel is denoted by $f(x, y)$, then the threshold operation is:

$$g(x, y) = \begin{cases} 1 & \text{if } f(x, y) > threshold \\ 0 & \text{elsewhere} \end{cases} \quad (10.14)$$

The next step is to find the individual objects in the image. Sometimes, we can assume that there is no contact between objects. A white island in the image corresponds to exactly one object. Such an island is called a particle or *connected component*. The operation that identifies each particle and assigns a unique label to it is called *component labelling*.

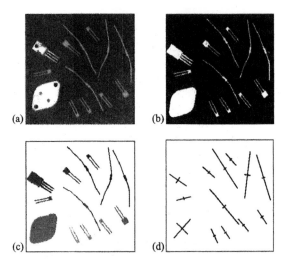

**Figure 10.23** *Image processing: (a) original; (b) segmented image; (c) individual particles; (d) regional description: position, orientation, principal components*

**Example 10.5  Recognition of electronic components (continued)**
Figure 10.23(b) shows the result of thresholding applied to the image in Figure 10.23(a). The objects are represented by the set of white pixels. Figure 10.23(c) shows the result of component labelling.

**Regional description**
The purpose of regional description is to extract features from the regions found (or particles) that provide much information about these particles. Examples are:

*Position*
The position of a particle in the image plane can be defined as the centre of gravity:

$$\bar{x} = \frac{1}{N} \sum_{i=1}^{N} x_i \qquad \bar{y} = \frac{1}{N} \sum_{i=1}^{N} y_i \qquad (10.15)$$

where $\{(x_i, y_i) | i = 1, \ldots, N\}$ are the $x, y$ coordinates of the $i$-th pixel of the set of $N$ pixels that makes up a particular particle.

*Size measures*
The size of a particle can be defined in various ways. Obvious choices are the area and the perimeter. The area can be defined as the number of pixels that make up the particle. A definition of perimeter is the number of boundary pixels of the particle.

*Orientation and shape parameters*
Shape parameters are geometric parameters that are independent of the position and the orientation of the particle. Shape parameters often used are the principal

components of the particle. Such a parameter can be derived as follows. If we would like to have a measure of the extension of the particle along the $x$-axis, then we could define the so-called second order moment:

$$\sigma_x^2 = \frac{1}{N} \sum_{i=1}^{N} (x_i - \bar{x})^2 \tag{10.16}$$

In fact, this definition is quite analogous to the one of the variance of a random variable (Chapter 3). The second order moment $\sigma_x^2$ is a measure of the dispersion of the shape measured over the $x$-axis.

Unfortunately, $\sigma_x^2$ is certainly not independent of the orientation of the particle. If we rotate the particle, then the set $x_i, y_i$ rotates accordingly, and $\sigma_x^2$ will change. In fact, $\sigma_x^2$ will depend on the rotation angle $\varphi$. Hence, we should write $\sigma_x^2(\varphi)$.

The first principal value $\sigma_1^2$ is defined as the maximum of $\sigma_x^2(\varphi)$ if we vary $\varphi$ over an angle of $180°$. If we denote the angle for which that maximum is obtained by $\varphi_1$, then

$$\sigma_1^2 = \max_{0 < \varphi < 180°} \sigma_x^2(\varphi) = \sigma_x^2(\varphi_1) \tag{10.17}$$

$\varphi_1$ is called the first principal direction. The direction for which $\sigma_x^2(\varphi)$ reaches its minimum is called the second principal direction $\varphi_2$. This angle follows directly from $\varphi_1$ as it can be proven that $\varphi_2 = \varphi_1 + 90°$. The second principal value is found as $\sigma_2^2 = \sigma_x^2(\varphi_2)$. Hence, $\sigma_2^2$ measures the extension of the particle in a direction orthogonal to the first principal direction.

**Example 10.6 Recognition of electronic components (continued)**
All these concepts are depicted graphically in Figure 10.23(d). For each particle in Figure 10.23(c), a cross has been constructed with its position at the centre of gravity of the particle, with its orientation according to the first principal direction, and with the lengths of the segments according to the square roots of the two principal values.

**Object recognition**
The last step is often the association of particles with the types of objects. (Their positions and orientations are already determined in the preceding step.) In simple problems, the objects can easily be classified by using some features derived from the particles, e.g. the area, the perimeter, the principal components, the colour, etc. These features might form a cluster in the feature space. Each cluster corresponds exactly to one type of object. Recognition boils down to determining which cluster corresponds well to the observed features of an object. This process is called *pattern classification*.

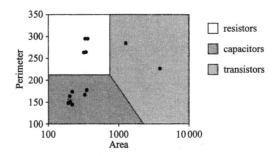

**Figure 10.24** *Pattern classification*

**Example 10.7 Recognition of electronic components (continued)**
Figure 10.24 shows the feature space (area versus perimeter). Each object from Figure 10.23 is seen here as a point with the corresponding area and perimeter. It appears that the objects fall into three well-separated clusters in the feature space. From this we conclude that it is easily possible to divide the space into three compartments, and assume that future objects will also fall in the same compartments.

## 10.5. Further reading

[1]  G. Saxby, *The Science of Imaging – An introduction*, IOP Publ., Bristol, 2002.
[2]  G.W. Stimson, *Introduction to Airborne Radar*, Second Edition, Scitech Publishing, Mendham, NJ, 1998.
[3]  D.A. Forsyth, J. Ponce, *Computer Vision – A modern approach*, Prentice-Hall, Upper Saddle River, NJ, 2003.
[4]  L.G. Shapiro, G.C. Stockman, *Computer Vision*, Prentice-Hall, Upper Saddle River, NJ, 2001.
[5]  R. Klette, K. Schlüns, A. Koschan, *Computer Vision – Three-dimensional data from images*, Springer-Verlag, Singapore, 1998.
[6]  R. Hartley, A. Zisserman, *Multiple View Geometry in Computer Vision*, Cambridge University Press, Cambridge, 2000.

The topic of imaging is wide. Each modality (optical, X-ray, radar, acoustic, etc.) has an overwhelming amount of literature. Within the various modalities many fields of specialization exist, for instance image formation, image acquisition, and so on. The literature for each of these fields is too numerous to mention. However, an overview of the various principles of image formation, acquisition and processing is given in [1]. An introduction to image-based measurement systems is provided in [2], but the discussion in this book is limited to the use of optical images. The topics of X-ray imaging and ultrasound medical imaging are dealt with in [3]. Radar is covered by [4]. Sonar technology is discussed in [5]. This book focuses on underwater acoustic systems in nautical applications.

Saxby's book gives a quick overview of various imaging-formation techniques. It is mentioned here because it is a first introduction for the non-specialist. The book by Stimson, about radar technology is also suitable for the non-specialist. An introductory text for 3D computer vision can be found in the book by Klette et al. The subject is treated at a more skilled level in the book of Hartley and Zissermann. Modern approaches to computer vision and digital image processing techniques can be found in the books by Forsyth and Ponce, and Shapiro and Stockman.

## 10.6. Exercises

1. A CCD camera is used to measure the length of a line segment. The line segment is located in a plane that is perpendicular to the optical axis of the camera. The lens is described as a pinhole model with a focal distance of 6 mm. The distance from the pinhole to the line segment is 490 mm. The pitch (distance between pixels of the CCD) is 0.006 mm. Digital image processing is used to locate the two end points of the line. The measured pixel coordinates of these points are (100, 130) and (130, 170), respectively. Calculate the length of the line segment.

2. The pinhole model is often used to describe the image formation using a lens. Show that with a small modification the pinhole model is also suitable to describe X-ray imaging.

3. Which of the following statements are not correct?
a. The advantage of a CMOS camera with respect to a CCD camera is that the former is random addressable, and the latter is not.
b. The main principle of a RADAR system to measure range data is based on Doppler frequency shifts.
c. The principle of CT medical imaging is based on differences between X-ray attenuation coefficients of biological tissues.
d. The principle of ultrasound imaging is based on differences between acoustic impedances of the tissues.

4. Which imaging system is most appropriate to expose welding processes?
a. A CCD camera
b. An X-ray imaging system
c. A flying spot camera
d. A CMOS camera

# Chapter 11

# Design of Measurement Systems

A measurement *system* may be considered as a set of interrelated components, of which the input is a physical process and the output consists of data reflecting specific properties of this process. Often, the mental view of a measurement *instrument* is that of a box with one or more input terminals, some analogue and digital signal processing elements and a display to show the measurement results. Thus, a measurement instrument can be considered as the physical realisation of a system performing the measurement process. However, such a realisation does not necessarily have the appearance of a single box. It may consist of a set of components distributed over a certain area, possibly extending to a space larger than the earth.

During the nineties, the growth of computer technology had a great impact on advances in measurement instruments. The traditional mechanical instruments were replaced by all-electronic instruments; signal processing was included and the development in telecommunication and communication protocols allowed easy interconnection of measurement instruments worldwide. Consequently, the terminology has been shifted from "instruments" to "systems", the latter including the whole range from embedded (micro) sensors to Internet based data acquisition. Further expansion in computer science has resulted in the substitution of many instrument functions traditionally implemented in hardware by software (virtual instrumentation, Section 11.2.2).

Obviously, the design of a measurement system is not a trivial task. Even for the measurement of a single parameter, a variety of measurement principles is available, and for each principle many sensor types, measurement configurations and different ways of signal processing and data handling are available. This leads to an almost infinite number of possible measurement systems for each measurand. In this chapter, we will review some essential steps in the design and decision process.

## 11.1. General aspects of system design

### 11.1.1. *Design levels*

The design of a measurement system (actually of any technical system) is a very complex mental activity. In many books on this topic, the design process is decomposed into a number of basic steps, with feedback loops and iteration sequences. Most of these studies distinguish at least three major levels:

- Task definition
- Concept generation
- Evaluation.

Obviously, any design should start with a description of the requirements the system should finally be able to meet. Important specifications of a measurement system concern:

- Information handling performance (range, accuracy, bandwidth, sensitivity, noise level etc.);
- Technical performance (dimensions, power consumption etc.);
- Environmental conditions (temperature, pressure, shock resistance, radiation hardness etc.);
- Economically related aspects (development time, expected lifetime, reliability, costs etc.).

Some of these items may be specified with concrete figures, and have hard limitations. Others are less restrictive ("it would be nice if..."). Therefore, the designer must consider proper weighting factors attributed to each of the requirements set by the prospective user.

The level of concept generation is the most creative part of the design process. It is recommended in this stage to generate as many scenarios and concepts as possible, and group them in some order of confidence (which might be a rather subjective one). All of these concepts are subjected to an initial evaluation, to sort out the most promising ones (but never throwing away the remaining ones: they could be of use in a later development stage). Now the selected scenarios are evaluated thoroughly and in much more detail. The outcome of this process is compared with the results of the task definition phase, from which follows a decision based on the smallest difference between the requirements and the predicted system performance.

Unfortunately, the procedure as described above is highly simplified. Major practical problems arise from uncertainty due to insufficiently known system behaviour and sparse knowledge of weighing factors characterizing the requirements. A possibly ill-defined market further complicates the decision-making: there are always clients popping up with special wishes, for instance concerning environmental conditions, measurement speed or instrument size.

Consequently, a final decision cannot be taken after a single course through the design sequence described here. Jumping back to either the conceptual stage or even the task definition stage might be necessary. Usually, a final decision is made only after a number of iterations.

Despite all the uncertain factors, formalization of the design process does make sense because it helps the designer to follow a well structured path, leaving room for creativity and forcing detailed analysis, two essential elements for a successful design.

In complex systems, the design process can be further subdivided into different hierarchical levels. Examples of such levels are:

- System architecture and external functionality (what it does)
- Internal functional structure (how it does)
- Realization (with what it does).

For each of these levels the design cycle as described before may apply. The architectural level involves the aspects of the system that are of direct concern of the user, for instance the choice between a fully centralized system or a system with distributed intelligence (see Section 1.3). The external functionality specifies the input-output relations of the system. The internal functional structure deals with the methods that are used to accomplish the external functionality. It involves internal subtasks, functions, and processes (the type of a filter, its bandwidth, its order etc., the type of an ADC and its number of bits, the type of signal transmission, etc.). Finally, the realization can be set: what is the most appropriate sensor type (piezoelectric or electrostatic in case of an ultrasound system; a laser diode or an LED in case of an optical system); what should be the length or the strength of a particular mechanical part, what kind of material should be applied?

Even during the last stage of material or component selection, jumping back to a higher level of the design process can become necessary when entering into a dead end, due to for instance unexpectedly high costs or lack of availability of particular components.

The final appearance of a measurement system depends largely on the application for which it is intended. Roughly, we can distinguish three major application areas:

1. Research and development
2. Industrial production
3. Consumer products.

This division is oversimplified but it clearly illustrates the differences in design approach. Measurement systems in the first group of applications are, generally, highly specific, (partly) custom made and expensive. The user interface is often self-made and complicated; the same applies to the data processing part. User and designer are often the same person or belong to the same team; anyhow, there is a short "distance" between designer and user.

Measurement systems for consumer products are often embedded in that product. The measuring device is almost invisible; there is hardly any interface to the user (in particular as part of an embedded control system). It should be extremely cheap (for mass production), and the life time is usually short (adapted to that of the product). Since these systems are produced in large quantities production cost is a major item during the design. There is a relatively long distance between user and manufacturer.

Examples of measurement devices for the high volume market are the acceleration sensor for airbags, thermometers in a freezer or washing machine, sensing devices in toys and other "smart" products (for instance an autonomous vacuum cleaner), optical and ultrasonic sensors in intruder alarms, and many more. The design time of such systems is also extremely short, because it is linked to the product lifetime and a demanding market for innovative products. For this category of applications, a fully automated design process could be of high value.

The category of measurement systems for industrial production lies somewhere between these two extreme application fields: research (from a single piece) and consumer product (millions). They are characterized by a high reliability (a failure means loss of production) and a rather simple user interface. Usually there is a close contact between user and manufacturer, especially in the design phase of the system. Military and medical applications have their own position in this spectrum, which is also reflected in the design process.

### 11.1.2. *Design automation*

Considering the design process as a concatenation of sub-designs, each consisting of a series of subsequent observations and evaluations, with decision points and feedback loops, one might be tempted to think about a fully automated design machine. Inputs are the requirements and constraints with associated weighing factors, as set by the client or user. The core of the machine is a huge database containing all possible measurement principles; strategies; all commercially available sensors with their specifications; all available electronic components (amplifiers, filters, ADC's and DAC's). The output is a list of candidate measurement systems ranked according to the degree of matching to the set of requirements, as obtained by a simple search through the whole object space.

To a certain extent, this would be a feasible concept. However, this is not as trivial as it seems to be. For example, consider the measurement of speed. It is not enough to state that the measurand is just speed. The object of which the speed has to be measured should also be specified. It makes quite a difference when it concerns a car on a road, or a flying bird, or an electron in a particle accelerator. If the final aim is to monitor the speed of cars on a highway in the Netherlands, then it is essential to know that, according to the Dutch law, the accuracy of the equipment should be better than 3% (where "accuracy" has to be specified as well). Therefore, not only the measurands, but also the objects and their environment have to be specified.

Design of Measurement Systems 313

**Table 11.1** *Different sensor types with their ISA specification standard*

| Sensor-type | ISA-standard |
|---|---|
| Piezo-electrical acceleration sensor | S37.2 |
| Strain-gauge as linear acceleration sensor | S37.5 |
| Piezo-electrical pressure sensor | S37.10 |
| Strain-gauge as pressure sensor | S37.3 |
| Potentiometric pressure-sensor | S37.6 |
| Potentiometric displacement sensor | S37.12 |
| Strain gauge as force sensor | S37.8 |

The expert system should be able to correctly and unambiguously interpret these specifications.

The success of a fully or partly automated design machine also depends on the availability of knowledge about system parts and their specifications. Again, we consider just an example: a transducer, being an important part of a measurement system. The Instrument Society of America (ISA) has defined specification schemes for a number of transducer types: the SP37 standard [1]. Table 11.1 shows some transducer types covered by this ISA standard. A specification scheme contains a number of required and supplemental fields, comprising particular properties of the transducer as well as procedures how to determine these. Adaptation of transducer suppliers to such a scheme facilitates the comparison of transducer performance and hence the selection process.

## Example 11.1 Specification of a potentiometric displacement transducer
The way to specify a potentiometric displacement transducer is published in a leaflet of 21 letter-format pages [2]. Items are:

- Drawing symbol (Figure 11.1)
- Specification characteristics
- Individual acceptance tests and calibrations
- Qualification test procedures.

*Drawing symbol*
"The drawing symbol for a potentiometric displacement transducer is a square with an added equilateral triangle, the base of which is the left side of the square.
Subscripts: A = Angular, L = Linear.
The letter 'D' in the triangle designates 'displacement' and the subscripts denote the second modifier, as symbolized by '$D_A$'.

A variable resistance of length X symbolizes the potentiometer. The lines from it and to it and to ground represent the electric leads or terminations."

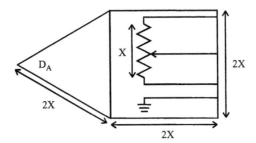

**Figure 11.1** *Drawing symbol of the potentiometric displacement transducer [2]*

*Performance characteristics*
"Range: Expressed as '_____ to _____ radians/millimetres (degrees/inch)'.[1]
Linearity: Expressed as '_____ % ± _____ % (VR)' (VR = Voltage Ratio)."

*Calibration and test procedures*
"Results obtained during the calibration and testing shall be recorded on a data sheet in Section 7, Figure 7, of this standard. Calibration and testing shall be performed as defined in ISA-S37.1 unless otherwise specified.
Insulation resistance: Verify the Dielectric Withstand Voltage using sinusoidal ac voltage test with all transduction element terminals (or leads) paralleled and tested to case and ground pin."

*Qualification test procedures*
"The tests and procedures of Section 5, Calibration and Test procedures, shall be run to establish reference performance during increasing (and decreasing) steps of 50, 60, 80, 100, 80, 60, 50, 40, 20, 0, 20, 40, 50% VR minimum."

Despite the exploding growth of the Internet, online information about transducer specifications is still far from complete, and turns out to be quickly obsolete. For the time being, an innovative design requires the assistance of sensor experts; the way towards fully automated designing is still a very long one.

## 11.2. Computer based measurement systems

### 11.2.1. *The role of the PC in measurement*

The classical form of a stand-alone measurement instrument is a case containing signal processing hardware, a user interface and a number of terminals to connect the instrument to the measurement object. When the instrument is intended for non-electrical measurands, a separate sensor head with connecting cable to the instrument itself facilitates access to the measurement object. The user interface

---

[1] To our regret, we observe that even the ISA seems unable to conform unambiguously to SI-units. The consequences of misunderstandings regarding units can be quite serious.

may be a simple numerical display and some control buttons for on-off switching and range setting.

More sophisticated, versatile instruments have a menu driven control panel allowing the user to configure the instrument according to specific measurement functions. The number of options remains restricted to those implemented by the manufacturer. The flexibility of such instruments is mainly due to the embedded software, performing dedicated processing of the signals obtained.

As an example, consider the measurement of the resistance between two terminals of some electrical system. Depending on the application, we can think of a simple handheld multimeter set in resistance measurement mode. The measured value appears on an LCD, with an uncertainty as specified by the manufacturer. On the other side of the spectrum stands a computer controlled impedance analyzer, offering four-point measurements (Section 7.1.4), the presentation of a complete frequency characteristic (in various modes, for instance modulus and phase; real and imaginary parts; polar plot) over a wide frequency range (from almost zero up to several GHz) and many more features. However, even such an instrument is restricted to impedance measurements only.

When simultaneously measuring a multitude of different parameters (for instance temperature, atmospheric pressure, humidity, wind speed and wind direction in a weather station), a complete instrument for each of these parameters could be an option. Obviously, to minimize costs, materials and power consumption, a different system architecture is preferred, in which particular functions are shared, for instance according to the structure of Figure 1.6. In this and similar cases the instrument could consist of a data acquisition card, containing programmable preamplifiers, a multiplexer and an AD converter. The card is inserted in a PC that contains the associated software for the card and possibly other packages for proper data processing and presentation. The market offers a multitude of such data acquisition cards, which vary with respect to the number of input channels, resolution and sampling rate.

Once having introduced the PC as part of a measuring instrument, we now consider some additional features of this approach. The PC substantially increases the flexibility of the measurement system. Huge amounts of data can be stored on disk for complex, offline data processing or just for archiving purposes. With appropriate software statistical analysis on measurement data can be performed, trends in a particular quantity analyzed and the correlation between two or more quantities investigated. Graphical representation facilitates quick evaluation of single experiments and the visualization of complex processes on a single display, eliminating the need for reading individual displays for each measurand. Figure 11.2 shows the presentation of a sequence of measured temperatures and simultaneously the moving average over 10 consecutive measurements.

The functionality of the PC can further be extended with a function generator (for specific test signals), a filter (in the digital domain), a controller (using the output

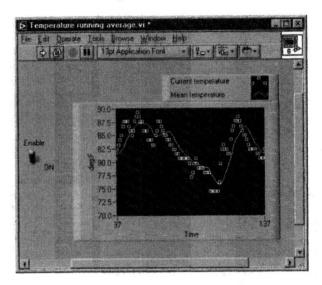

**Figure 11.2.** *Example of a virtual instrument for measuring the temperature and the mean of the last 10 measurements (in LabVIEW[1])*

channels of the acquisition card) and much else. With these extensions, the PC has taken over several instruments in a traditional measurement (and control) system. Moreover, since the PC is a freely programmable device, measurement structures and procedures can be set up and controlled in an arbitrary way.

A further step in the development of increasing flexibility is the application of object oriented programming to the design of a measurement system. Since many functions are already implemented as software in the PC, configuring a measurement system using graphical symbols (icons) is a small step to what is called *virtual instrumentation*. This topic will be discussed in a separate section.

### 11.2.2. Virtual instruments

Classical computer programming is based on a declarative approach, in which commands, according to some language, are written into a *program file*. Former programmable measurement instruments used this type of programming.

This programming technique may be called command-driven. However, a measurement system works data driven: the data is or should be processed immediately when produced. Several vendors have defined "visual languages", for example, the Simulink part of MATLAB, HP-Vee and LabVIEW. Icons representing specified actions to be performed are placed on a "worksheet". These icons have also input and output terminals. Connecting graphically the output terminal of one icon to the

---

[1] LabVIEW is a product of National Instruments.

Design of Measurement Systems   317

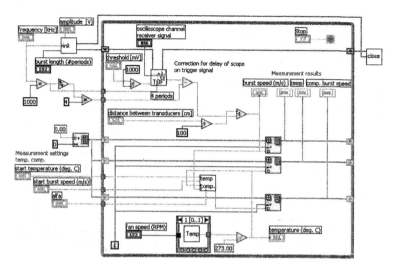

**Figure 11.3** *Wiring diagram of a virtual measurement system (LabVIEW) for the determination of the time-of-flight in an acoustic flow measurement system*

input terminal of another defines a data-stream. In this way, a complete network of instruments and their connections can be simulated. Real or simulated data are connected to the inputs of the icons as if they were real instruments. After starting the execution of the program and a logic check of the network, the icons perform corresponding actions on the input data and the resulting output data appears at the output terminal of the icon.

Figure 11.3 shows, just as an illustration, the wiring diagram of a system that determines the time-of-flight in an acoustic flow measurement system. The big grey box represents a loop, indicating that the measurement sequence is repeated several times. The icons on the left side are user controlled input parameters. Outputs are burst speed, temperature and again the burst speed but compensated for temperature effects (see Section 8.7.4). This system is used by students to investigate the effect of all kinds of parameters on the performance of a flow measurement system using acoustic time-of-flight [3].

Although the program is data driven, the PC-architecture is still command driven. This means that all input terminals of the icons have to be polled continuously. Moreover, a PC contains a processor that executes all tasks consecutively: no two actions can be executed at the same time. This limits the possibilities of carrying out real time tasks. Dedicated PC-boards with their own processing hardware considerably alleviate this problem.

In the next section, we present, merely as an illustration, examples of three different design stages: a selection process, an uncertainty evaluation and a LabVIEW application.

## 11.3. Design examples

### 11.3.1. *Selection of a level sensor*

Design methods have evolved over time, from purely intuitive (as in art) to formal (managerial). The process of sensor selection is somewhere in between: it is an act of engineering where advanced tools for simulating system behaviour support the design. The basic attitude is (still) the expertise, contained in the minds of people and acquired through long experience.

**Task definition**
Sensor selection means meeting requirements. Unfortunately, these requirements are often not known precisely or in detail, in particular when the designer and the user have different professions. The first task of the designer is to get as much information as possible about the intended application of the measurement instrument, all possible conditions of operation, the environmental factors, and the specifications with respect to quality, physical dimensions and costs.

The list of demands should be exhaustive. Even when not all items are relevant, they must be indicated as such. This will leave more room to the designer, and minimizes the risk to be forced starting all over again. The list should be made in such a way as to enable unambiguous comparison with the final specifications of the designed instrument. Once the designer has a complete idea about the future use of the instrument, the phase of the conceptual design can start.

**Selecting the measurement principle**
Once the tasks have properly defined, the measurement principle has to be considered. This step in the design process is illustrated by the example of a measurement for just a single, static quantity: the amount of fluid in a container. The first question to be answered is in what units the amount should be expressed: volume or mass. This may affect the final selection result. Figure 11.4 shows five measurement principles.

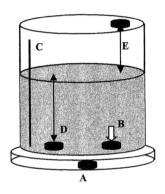

**Figure 11.4** *Various possibilities to measure the amount of fluid*

A. the tank is placed on a balance, to measure its total weight;
B. a pressure gauge on the bottom of the tank;
C. a gauging-ruler from top to bottom with electronic read out;
D. a level detector on the bottom, measuring the column height;
E. a level detector from the top of the tank, measuring the height of the empty part.

Obviously, many more principles can be found to measure a quantity that is related to the amount of fluid in the tank.

In the conceptual phase of the design as many principles as possible should be considered, even unconventional ones. Based on the list of demands it should be possible to find a proper candidate principle from this list, or at least to delete some of the principles, on an argued base. For instance, when the fluid should remain in the tank during the measurement, principles based on volume flow or mass flow are excluded. If the tank contains a chemically aggressive fluid, a non-contact measurement principle is preferred, placing principles B, C and D lower on the list. A level gauge (C) may exhibit hysteresis because liquid drops may remain attached to the electrodes in the empty part of the tank. Further, method A can possibly be eliminated in this stage because of high costs. The conceptual design ends up with a set of principles with pros and cons, ranked according to the prospects of success.

**Selecting the sensing method**
After having specified a list of candidate principles, the next step is to find a suitable sensing method for each of them. In the example of Figure 11.4 we will further investigate principle E, a level detector placed at the top of the tank. Again, a list of the various possible sensor methods is made. This may look as follows:

(a) a float, connected to an electronic read-out system;
(b) optical time-of-flight measurement;
(c) optical range measurement;
(d) electromagnetic distance measurement (radar);
(e) acoustic time-of-flight measurement; etc.

As in the conceptual phase, these methods are evaluated using the list of demands, so not only the characteristics of the sensing method but also the properties of the measurement object (liquid level) and the environment should be taken into account. Method (a) has the advantage of simplicity and robustness: the float is connected through a lever to some angular encoder. To extend the range, the lever can be replaced by a wire (cable actuated position sensors are available on the market in a wide variety of types, ranges and application areas). The acoustic ToF method is also a good candidate because of being contact-free, a property that might be useful for very hot or aggressive liquids. The same holds for optical and electromagnetic methods. When comparing acoustic and optical ToF methods, the specified accuracy is an important criterion. Optical ToF require high-speed electronics in particular when measuring small distances; acoustic ToF methods behave better in this respect, but temperature changes may reduce the measurement accuracy.

Obviously, at this stage it is also important to consider methods to reduce such environmental factors (Section 3.6). Anyhow, this stage ends up with a list of candidate sensing methods with merits and demerits with respect to the requirements.

**Decision**
The final step is the selection of the components that make up the sensing system. Here it has to be decided between a commercially available system and the development of a dedicated system. The major criteria are costs and time: both are often underestimated when own development is considered.

In this phase of the selection process, sensor specifications become crucial. Sensor providers publish specifications in data sheets or on the Internet. However, the accessibility of such data is still poor, making this phase of the selection process critical and time consuming, particularly for non-specialists in the sensor field.

Computer aided sensor selection programs are under development and partly realized, but up to now their use is limited. A prerequisite for a general, successful tool facilitating sensor selection is the continuous availability of sensor data on the Internet, in a much more standard format, and regularly updated (see also Section 11.1.2).

Evidently, the example of the level sensor is highly simplified. Usually, the selection process is not that straightforward. Since the sensor is often just one element in the design of a complex technical system, strong and frequent interaction with other design disciplines is necessary.

### 11.3.2. Evaluation of the uncertainty of a rain gauge

An essential step in the design process is the analysis and evaluation of the selected system concepts. For measurement instruments, many software tools are available now to assist the designer during this stage. However, a prerequisite is a proper model of the system to be evaluated.

In this example we consider a rain gauge, consisting of a measuring beaker (upper area $A = 500 \text{ mm}^2$), placed on an electronic scale (actually a pressure sensor). The output of the scale is connected to a read-out unit that interfaces the sensor output to a numerical display (Figure 11.5). The weight of the collected drops is measured every 15 minutes. Our goal is to evaluate the total uncertainty of this design, given a particular type of pressure unit. The following processes contribute (most) to the measurement uncertainty:

- the quantized character of rain (drops)
- the specifications of the pressure sensor (Table 11.2)
- the quantization error of the ADC in the read-out unit (or the numerical display).

We will evaluate the individual uncertainty due to each of these errors, express them in units of mm, and combine them according to the rules given in Chapter 3.

**Figure 11.5** *A simple rain meter*

**Table 11.2** *Selected specifications of the pressure sensor with read-out unit*

| | Symbol | Min | Typ. | Max | Unit |
|---|---|---|---|---|---|
| Pressure range | $P$ | 0 | – | 10 | kPa |
| Full scale span | $V_{FS}$ | 24 | 25 | 26 | mV |
| Offset and temp. effect | $V_{off}$ | −0.1 | – | 0.1 | mV |

**Rain drops**

Rain is characterized by a Poisson distribution (Section 3.2). We investigate the uncertainty due to this quantization process. First we estimate the average volume of a single drop (which could be done experimentally). Suppose the volume is $V = 100\,\text{mm}^3$. Suppose after 15 minutes the water level in the beaker is $x = 20\,\text{mm}$. This corresponds to $N = 100$ drops. Now we can derive the uncertainty due to the quantization process of the drops (Section 3.2.2):

$$\lambda = 100$$
$$\sigma_n = \sqrt{\lambda} = 10$$
$$\sigma_e = 2\,\text{mm}$$

This is an approximate result. Actually, the uncertainty due to the quantization could be a little more because of the variations in the volumes of different drops. However, we will ignore this effect.

**ADC: round off error (quantization error)**

Both the display and the internal ADC give rise to quantization errors, with a uniform distribution. Suppose the resolution of the display matches that of the ADC, and the least significant digit on the display is 1 mm. The resulting uncertainty follows from:

$$\Delta x = 1\,\text{mm}$$
$$\sigma_e = \frac{\Delta x}{\sqrt{12}} = \frac{1}{\sqrt{12}}\,\text{mm} \approx 0.3\,\text{mm}$$

## Offset and drift of the electronics

Table 11.2 gives only an indication of the temperature effect on the offset; apparently the offset can be adjusted by the user when starting the measurements. Since we do not know the temperature, we assume the offset to be uniformly distributed between the minimum and the maximum values. The resulting uncertainty (in mV) is:

$$\Delta V_T = 0.2\,\text{mV}$$

$$\sigma_T = \frac{\Delta V_T}{\sqrt{12}} = \frac{0.2}{\sqrt{12}} \approx 0.06\,\text{mV}$$

## Combining uncertainties

Using the model of Figure 11.6 we combine these three errors to find the total uncertainty. First we express all uncertainties in mm at the output of the system, and next we add these contributions according to the rules outlined in Section 3.4.3 (assuming the system is linear). This procedure is illustrated with the error budget in Figure 11.7. The total uncertainty (expressed in the standard deviation) is 3.1 mm. Note that the quantization error in the reading is not significant in this example.

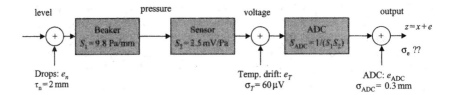

Figure 11.6 *Error model of the rain meter*

| Error source | Uncertainty | Sensitivity | Contribution to variance | |
|---|---|---|---|---|
| Quantum effect (drops) | $\sigma_{e_n} = 2\,mm$ | $S_n = 1$ | $(2 \cdot 1)^2 = 4\,mm^2$ | |
| Temperature offset | $\sigma_{e_T} = 60\,\mu V$ | $S_T = 0.04\,mm/\mu V$ | $(60 \cdot 0.04)^2 = 5.76\,mm^2$ | |
| Round-off error | $\sigma_{e_{ADC}} = 0.3\,mm$ | $S_{ADC} = 1$ | $(0.3 \cdot 1)^2 = 0.09\,mm^2$ | |
| Combined uncertainty | | | $9.85\,mm^2$ | $(9.85)^{0.5} = 3.1\,mm$ |

Figure 11.7 *Error budget of the rain meter*

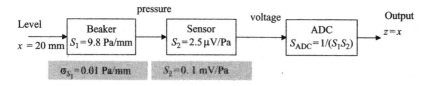

**Figure 11.8** *Error evaluation of the pressure sensor*

## Gain error of the read-out unit

The last step in this evaluation is the estimation of the error due to the pressure sensor. We consider two possible causes of errors:

- the sensitivity of the pressure sensor (specified by the manufacturer)
- the uncertainty in the gravitation.

The procedure is illustrated with the error model in Figure 11.8.

Again, we express these uncertainties in terms of the standard deviation at the output. Since the system is non-linear (the sensitivity error is a multiplicative error), we apply the combination rules as given in Section 3.4.3 for non-linear relationships:

$$\sigma_e^2 = \left(\frac{S_2 x}{S}\right)^2 \sigma_{S_1}^2 + \left(\frac{S_1 x}{S}\right)^2 \sigma_{S_2}^2$$

$$= \left(\frac{x}{S}\right)^2 (S_2^2 \sigma_{S_1}^2 + S_1^2 \sigma_{S_2}^2)$$

and applied to the rain meter:

$$\sigma_e \approx \frac{x}{24}\sqrt{0.0025^2 + 0.98^2} \approx 0.04 x$$

Note that the value of this uncertainty type depends on the measured value itself.

## 11.3.3. Design using virtual instruments

This last example illustrates the use of virtual instruments in the design of a rather complicated measurement system: the measurement of the state-of-charge of a battery. The system uses LabVIEW[2] as basis for the virtual instrument. It describes a real-time evaluation system for an algorithm, which calculates the state-of-charge (SoC) in percent, as well as the standby time left and the talk time left for a portable application. With the system we evaluate the accuracy of the algorithm. A more detailed presentation of this algorithm can be found in [4].

---

[2] LabVIEW is a product of National Instruments.

## Definitions

In order to efficiently discuss this section, some of the common terms used in the battery SoC industry have to be defined. Further, we give some battery-related definitions concerning the SoC application. For more information the reader is referred to [4], from which most of the material presented here is taken.

| | |
|---|---|
| *Ampere-Hour (Ah)* | *A measurement of electric charge computed as the integral product of current (in amperes) and time (in hours).* |
| *Cell* | *The basic electrochemical unit used to generate electrical energy from stored chemical energy or to store electrical energy in the form of chemical energy. A cell consists of two electrodes in a container filled with an electrolyte.* |
| *Battery* | *Two of more cells connected in an appropriate series/parallel arrangement to obtain the required operating voltage and capacity for a certain load. The term battery is also frequently used for single cells.* |
| *EMF* | *The electromotive force of the battery.* |
| *C-rate* | *A charge or discharge current equal in amperes to the rated capacity in Ah. Multiples larger or smaller than the C-rate are used to express larger or smaller currents. For example, the C-rate is 1100 mA in the case of an 1100 mAh battery, whereas the C/2 and 2C rates are 550 mA and 2.2 A respectively. In the real world, a cell does not maintain the same rated capacity at all C rates.* |
| *Cycle Life* | *The number of cycles that a cell or battery can be charged and discharged under specific conditions, before the available capacity in Ah fails to meet specific performance criteria. This will usually be 80% of the rated capacity.* |
| *Cut-off voltage* | *The lowest operating voltage of a cell at which the cell is considered depleted. Also often referred to as end-of-discharge voltage or final voltage.* |
| *Self-discharge* | *The recoverable loss of useful capacity of a cell on storage due to internal chemical action. This is usually expressed in a percentage of the rated capacity lost per month at a certain temperature, because self-discharge rates of batteries are strongly temperature-dependent.* |
| *State-of-Charge (SoC)* | *The charge (in Ah) that is present inside the battery.* |
| *Remaining capacity* | *The charge (in Ah) that is available to the user under the valid discharge conditions.* |
| *Remaining time of use (or Time Left)* | *The estimated time that the battery can supply charge to a portable device under the valid discharge conditions before it will stop functioning.* |

Design of Measurement Systems    325

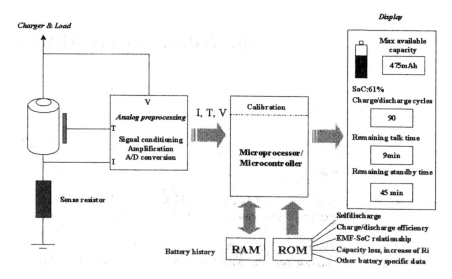

**Figure 11.9** *General functional architecture of a possible SoC indication system*

## System structure

The SoC measurement method and the computational model based on the corrected SoC definition must be simple, convenient, practical and reliable. Figure 11.9 shows an example of a possible SoC indication system.

The battery may include a plurality of battery cells connected in series, each of the battery cells having at least two terminals. The circuit board may include an analogue-to-digital converter (ADC) for converting a voltage ($V$) drop between at least two connection pins to a digital signal and also for converting the measured analogue values of the voltage and temperature ($T$). A microprocessor/microcontroller (in which the SoC algorithm is stored) determines the SoC of the battery system based on these digitized signals. Two types of memory are needed. Basic battery data, such as the amount of self-discharge as a function of $T$ and the discharging efficiency as a function of $I$ and $T$, is stored into and read from the ROM. In the ROM memory we can also store the EMF-SoC relationship or other battery specific data. The RAM is used to store the history of use, such as the number of charge/discharge cycles, which can be used to update the maximum battery capacity. Each part (software algorithm or hardware device) of this system will influence the final accuracy in the SoC indication (inaccuracy in $V$, $T$ and $I$ measurements will give an inaccuracy in the final SoC).

The evaluation of the algorithm requires a real battery and a real current/voltage source/meter. The mobile phone itself is simulated, including the display and the control switches. The user is able to control the source by virtual control buttons on the virtual user interface (UI) control panel (Figure 11.10).

**Figure 11.10** *The user interface (UI)*

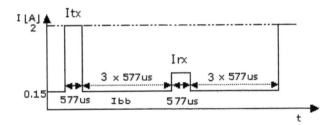

**Figure 11.11** *The GSM current pulse*

The source/meter is a Keithley model 2420, able to simulate the mobile phone current consumption with currents from $\pm 100$ pA to $\pm 3$ A, and voltage measurements from $\pm 1$ $\mu$V to $\pm 60$ V. The estimated values presented to the user are SoC (expressed in %) and remaining time of use $t_{rem}$ (in hour, minutes and seconds), under all valid charge and discharge conditions of the battery.

During talk mode the discharge current simulates the shape of a real GSM current (Figure 11.11). $I_{tx}$ represents the transmitter current (depending on the power or the distance between the mobile phone and the base station), $I_{rx}$ is the receiver current, and $I_{bb}$ the base band current. $I_{tx}$ can be changed by changing the value of $P_{out}$, as shown in Figure 11.10.

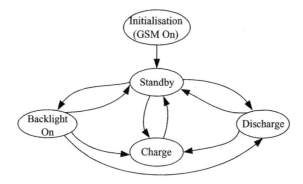

**Figure 11.12** *State transition of a mobile phone*

## States of the system
The implemented algorithm operates in five different states:

- Initialization state
- Standby state
- Charge state
- Discharge state
- Backlight on state.

The standby state of the simulated mobile phone corresponds to the equilibrium state of the algorithm. In other words, on one hand the user is always able to switch into a certain state via the UI control panel and on the other hand the algorithm determines the transitions according to parameter values. The state transitions of the mobile phone (which can be switched by the user) can be described by a flow diagram as shown in Figure 11.12.

## The initialization state
Every time at the switching on of the system, it starts with the initialization state. It reads the voltage and displays (by finding the value with the EMF curve) automatically the relative SoC. It initializes also the source/meter. From this state the system switches automatically into the standby state, except if we choose via the UI the system to charge or discharge.

## The standby state
In this state, the source/meter is drawing a small current and the SoC is calculated using the EMF-curve. By reading the voltage we find immediately the SoC. From this state the system is able to switch into charge-, discharge-, backlight on mode.

## The charge state
In this state, the source/meter is in either CC (constant current) mode or in CV (constant voltage), and the SoC is calculated by the coulomb counting method (current measurements and integration). At the end of charging, the system switches

automatically to standby state, but the user is always able to order this transition before charging ends.

### The discharge state
In this state the source/meter pulls power out of the battery. The power consumption is similar to the power drawn by a real mobile phone (see the GSM pulse from Figure 11.11). After discharge, and without user intervention, the system returns to standby state.

### The backlight on state
In this state the source/meter is pulling a little more current than in the standby state. The SoC is still calculated by the EMF curves, because the current is still under a certain limit. The system is staying in this state for about 5 seconds (arbitrary chosen value) before it comes back into the standby state. During these 5 seconds, all other transitions remain possible.

### Connections for the LabVIEW setup
Figure 11.13 shows the setup diagram, which describes the connections between the battery, the computer (via the LabVIEW card) and the source/meter. A GPIB card is used to connect the measurement systems to the computer. In our case we use the GPIB card for connecting the source/meter to the computer.

The SoC algorithm is programmed in LabVIEW so the evaluation environment is a computer with LabVIEW running on it. Via the UI we can decide the state in which the algorithm should go. The source/meter, which simulates the mobile phone, is controlled completely under LabVIEW.

Figure 11.14 shows the various external connections. A safety box is inserted to ensure the battery is never operated in an unsafe region. The battery manufacturer

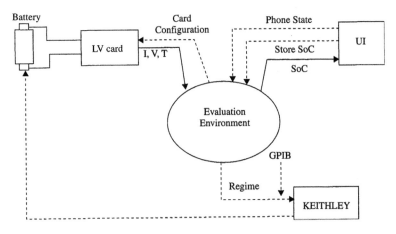

**Figure 11.13** *The LabVIEW Setup*

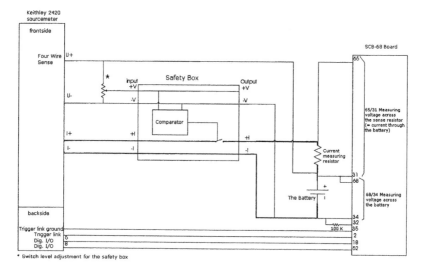

**Figure 11.14** *LabVIEW setup connections*

determines the region within which it is considered safe to use a battery. Outside the safe region, destructive processes may start to take place.

The SCB-68 board in Figure 11.14 is a shielded board with 68 screw terminals for easy connection to National Instruments 68-pin products. The resistor of $100\,\text{k}\Omega$ is used to inhibit voltage peaks.

The battery current is measured by measuring the voltage drop across a current measuring resistor. For calculating the value of this resistor we consider that the maximum current (2 A in our case) times the resistance should be the full scale of the ADC. In our case the range of the ADC is 0–10 V so $R = \frac{10}{2} = 5\,\Omega$ (see also exercise 1 of Chapter 7). This resistor must have a power rating of at least $20\,\text{W}\,(2^2 \cdot 5)$ to prevent destruction during discharge with high power.

This (partly) virtual system enables accurate prediction of the performance over a wide range of conditions, before the system itself has been realised. All measurements are executed automatically. Since many measurements are required for a full evaluation of the performance, it is worthwhile to spend some time in the preparation of such a versatile test set-up.

## 11.4. Further reading

B.S. Blanchard, W.J. Fabrycky, *Systems Engineering and Analysis*, Prentice Hall, 1990; ISBN 0-13-880758-2.
A comprehensive book on the design of technical systems. It is organized in six parts: an introduction, with background and definitions; system design processes; tools for system

analysis; design for operational feasibility; system engineering management, and a final part on applications. Each chapter ends with questions and problems. One of the appendices shows a list of design criteria.

L. Finkelstein, K.T.V. Grattan, *Concise Encyclopedia of Measurement and Instrumentation*, Pergamon Press, 1994; ISBN 0-08-036212-5.
As a contrast to the book by Blanchard, this book presents in pages 68–72 a very brief but nevertheless clear overview of design principles for instrumentation systems, providing quickly insight in the major philosophy on this topic.

# Appendices

**A. Scales**

*A.1. Moh's scale of hardness*

| Hardness | Mineral |
|---|---|
| 1 | Talc |
| 2 | Gypsum |
| 3 | Calcite |
| 4 | Fluorite |
| 5 | Apatite |
| 6 | Orthoclase |
| 7 | Quartz |
| 8 | Topaz |
| 9 | Corundum |
| 10 | Diamond |

## A.2. Beaufort scale for wind force

| Force | Speed (km/h) | Name | Condition (sea) | Condition (land) |
|---|---|---|---|---|
| 0 | <2 | Calm | Sea like a mirror | Smoke rises vertically |
| 1 | 1–5 | Light air | Ripples only | Smoke drifts and leaves rustle |
| 2 | 6–11 | Light breeze | Small wavelets (0.2 m), crests have a glassy appearance | Wind felt on face |
| 3 | 12–19 | Gentle breeze | Large wavelets (0.6 m), crests begin to break | Flags extended, leaves move |
| 4 | 20–29 | Moderate breeze | Small waves (1 m), some whitecaps | Dust and small branches move |
| 5 | 30–39 | Fresh breeze | Moderate waves (1.8 m), many whitecaps | Small trees begin to sway |
| 6 | 40–50 | Strong breeze | Large waves (3 m), probably some spray | Large branches move, wires whistle, umbrellas are difficult to control |
| 7 | 51–61 | Near gale | Mounting sea (4 m) with foam blown in streaks downwind | Whole trees in motion, inconvenience in walking |
| 8 | 62–74 | Gale | Moderately high waves (5.5 m), crests break into spindrift | Difficult to walk against wind. Twigs and small branches blown off trees |
| 9 | 75–87 | Strong gale | High waves (7 m), dense foam, visibility affected | Minor structural damage may occur (shingles blown off roofs) |
| 10 | 88–102 | Storm | Very high waves (9 m), heavy sea roll, visibility impaired. Surface generally white | Trees uprooted, structural damage likely |
| 11 | 103–118 | Violent storm | Exceptionally high waves (11 m), visibility poor | Widespread damage to structures |
| 12 | >118 | Hurricane | 14 m waves, air filled with foam and spray, visibility bad | Severe structural damage to buildings, wide spread devastation |

## A.3. Richter and Mercalli scales for seismic activity

The Richter scale measures the energy of an earthquake by determining the size of the greatest vibrations recorded on a seismogram.

| Richter | Energy (kg TNT) | Description |
|---|---|---|
| 0 | 0.6 | – |
| 1.0 | 20 | – |
| 2.0 | 600 | Smallest quake people can normally feel |
| 3.0 | $2 \cdot 10^4$ | Most people near epicenter feel the quake |
| 4.0 | $6 \cdot 10^5$ | A small fission atomic bomb |
| 5.0 | $2 \cdot 10^7$ | A standard fission bomb |
| 6.0 | $6 \cdot 10^8$ | A hydrogen bomb |
| 7.0 | $2 \cdot 10^{10}$ | Major earthquake (about 14 every year) |
| 8.0 | $6 \cdot 10^{11}$ | Largest known: 8.9 in Japan and in Chile/Ecuador |
| 9.0 | $2 \cdot 10^{13}$ | Roughly the world's energy usage in a year |

The Mercali scale measures an earthquake according to the observable results or effects the damage caused or the sensations described by people.

| Magnitude | Observable results and effects |
|---|---|
| I | Most people do not notice, animals may be uneasy, can be detected by a seismograph |
| II | Hanging objects sway back and forth |
| III | Many people feel the movement, parked cars may rock |
| IV | Doors, windows, and shelves may rattle, people indoors can feel movement |
| V | Light furniture moves, pictures fall off walls, objects fall from shelves |
| VI | Nearly everyone feels movement, light furniture falls over, windows may crack |
| VII | Some people fall over, walls may crack |
| VIII | Heavy furniture falls over, some walls crumble |
| IX | Many people panic, some buildings collapse, dams crack |
| X | Railroad lines are bent, most buildings are damaged, roads crack |
| XI | Bridges collapse, buried pipes break, most buildings collapse |
| XII | All manmade structures are destroyed |

## B. Units and quantities

### B.1. Derived units with special names and symbols

| Derived quantity | Name | Symbol | Expression in terms of other SI units | Expression in terms of SI base units |
|---|---|---|---|---|
| Plane angle | Radian | rad | – | $m \cdot m^{-1} = 1$ |
| Solid angle | Steradian | sr | – | $m^2 \cdot m^{-2} = 1$ |
| Frequency | Hertz | Hz | – | $s^{-1}$ |
| Force | Newton | N | – | $m \cdot kg \cdot s^{-2}$ |
| Pressure, stress | Pascal | Pa | $N/m^2$ | $m^{-1} \cdot kg \cdot s^{-2}$ |
| Energy, work | Joule | J | $N \cdot m$ | $m^2 \cdot kg \cdot s^{-2}$ |
| Power, radiant flux | Watt | W | J/s | $m^2 \cdot kg \cdot s^{-3}$ |
| Electric charge | Coulomb | C | – | $s \cdot A$ |
| Electric potential difference | Volt | V | W/A | $m^2 \cdot kg \cdot s^{-3} \cdot A^{-1}$ |
| Capacitance | Farad | F | C/V | $m^{-2} \cdot kg^{-1} \cdot s^4 \cdot A^2$ |
| Electric resistance | Ohm | $\Omega$ | V/A | $m^2 \cdot kg \cdot s^{-3} \cdot A^{-2}$ |
| Electric conductance | Siemens | S | A/V | $m^{-2} \cdot kg^{-1} \cdot s^3 \cdot A^2$ |
| Magnetic flux | Weber | Wb | $V \cdot s$ | $m^2 \cdot kg \cdot s^{-2} \cdot A^{-1}$ |
| Magnetic flux density | Tesla | T | $Wb/m^2$ | $kg \cdot s^{-2} \cdot A^{-1}$ |
| Inductance | Henry | H | Wb/A | $m^2 \cdot kg \cdot s^{-2} \cdot A^{-2}$ |
| Celsius temperature | Degree celsius | °C | – | K |
| Luminous flux | Lumen | lm | $cd \cdot sr$ | $m^2 \cdot m^{-2} \cdot cd = cd$ |
| Illuminance | Lux | lx | $lm/m^2$ | $m^2 \cdot m^{-4} \cdot cd = m^{-2} \cdot cd$ |
| Activity (of a radionuclide) | Becquerel | Bq | – | $s^{-1}$ |
| Absorbed dose | Gray | Gy | J/kg | $m^2 \cdot s^{-2}$ |
| Dose equivalent | Sievert | Sv | J/kg | $m^2 \cdot s^{-2}$ |
| Catalytic activity | Katal | kat | – | $s^{-1} \cdot mol$ |

## B.2. Decimal system of prefixes

| Factor | Symbol | Prefix | Factor | Symbol | Prefix |
|---|---|---|---|---|---|
| $10^{24}$ | Y | yotta | $10^{-1}$ | d | deci |
| $10^{21}$ | Z | zetta | $10^{-2}$ | c | centi |
| $10^{18}$ | E | exa | $10^{-3}$ | m | milli |
| $10^{15}$ | P | peta | $10^{-6}$ | $\mu$ | micro |
| $10^{12}$ | T | tera | $10^{-9}$ | n | nano |
| $10^{9}$ | G | giga | $10^{-12}$ | p | pico |
| $10^{6}$ | M | mega | $10^{-15}$ | f | femto |
| $10^{3}$ | k | kilo | $10^{-18}$ | a | atto |
| $10^{2}$ | h | hecto | $10^{-21}$ | z | zepto |
| $10^{1}$ | da | deka | $10^{-24}$ | y | yocto |

## B.3. Prefixes for binary multiples

| Factor | Name | Symbol | Origin | Derivation |
|---|---|---|---|---|
| $2^{10}$ | kibi | Ki | kilobinary: $(2^{10})^1$ | kilo: $(10^3)^1$ |
| $2^{20}$ | mebi | Mi | megabinary: $(2^{10})^2$ | mega: $(10^3)^2$ |
| $2^{30}$ | gibi | Gi | gigabinary: $(2^{10})^3$ | giga: $(10^3)^3$ |
| $2^{40}$ | tebi | Ti | terabinary: $(2^{10})^4$ | tera: $(10^3)^4$ |
| $2^{50}$ | pebi | Pi | petabinary: $(2^{10})^5$ | peta: $(10^3)^5$ |
| $2^{60}$ | exbi | Ei | exabinary: $(2^{10})^6$ | exa: $(10^3)^6$ |

## C. Expectations, variances and covariance of two random variables

### C.1. The expectation of a single random variable

Suppose $\underline{a}$ is a random variable with probability density $f_{\underline{a}}(a)$. Let $g(\underline{a})$ be a function of $\underline{a}$. The expectation of $g(\underline{a})$ is defined as:

$$E\{g(\underline{a})\} \stackrel{def}{=} \int_{a=-\infty}^{\infty} g(a) f_{\underline{a}}(a) da \qquad (C.1)$$

Examples are:

The mean: $g(\underline{a}) = \underline{a}$  $\quad E\{\underline{a}\} = \mu_a = \int_{a=-\infty}^{\infty} a f_{\underline{a}}(a) da$

The mean square: $g(\underline{a}) = \underline{a}^2 \quad \mathrm{E}\{\underline{a}^2\} = \int_{a=-\infty}^{\infty} a^2 f_{\underline{a}}(a) da$

The variance: $g(\underline{a}) = (\underline{a} - \mu_a)^2 \quad \mathrm{Var}\{\underline{a}\} = \int_{a=-\infty}^{\infty} (a - \mu_a)^2 f_{\underline{a}}(a) da$

A special case occurs when a new variable $\underline{b}$ is formed as follows:

$$\underline{b} = \alpha \underline{a} + \beta \quad \Rightarrow \quad \begin{array}{l} \mu_b = \alpha \mu_a + \beta \\ \mathrm{Var}\{\underline{b}\} = \alpha^2 \mathrm{Var}\{\underline{a}\} \end{array} \tag{C.2}$$

A corollary of equation (C.2) is that the standard deviation of $\underline{b}$ is found as $|\alpha|\sigma_a$.

## C.2. Joint and marginal density of two random variables

Suppose $\underline{a}$ and $\underline{b}$ are two random variables with probability density $f_{\underline{a}}(a)$ and $f_{\underline{b}}(b)$. The probability that $\underline{a} \leq a$ and simultaneously $\underline{b} \leq b$ is called the *joint distribution function*: $F_{\underline{a},\underline{b}}(a,b) \stackrel{def}{=} \Pr(\underline{a} \leq a \text{ and } \underline{b} \leq b)$. The *marginal distribution functions* $F_{\underline{a}}(a)$ and $F_{\underline{b}}(b)$ can be derived from $F_{\underline{a},\underline{b}}(a,b)$ as follows:

$$\begin{array}{l} F_{\underline{a}}(a) = \Pr(\underline{a} \leq a) = \Pr(\underline{a} \leq a \text{ and } \underline{b} \leq \infty) = F_{\underline{a},\underline{b}}(a, \infty) \\ F_{\underline{b}}(b) = \Pr(\underline{b} \leq b) = \Pr(\underline{a} \leq \infty \text{ and } \underline{b} \leq b) = F_{\underline{a},\underline{b}}(\infty, b) \end{array} \tag{C.3}$$

The *joint* probability *density function* is defined:

$$f_{\underline{a},\underline{b}}(a,b) \stackrel{def}{=} \frac{\partial^2 f_{\underline{a},\underline{b}}(a,b)}{\partial a \partial b} \tag{C.4}$$

The *marginal probability density function* can be derived from the joint probability density:

$$\begin{array}{l} f_{\underline{a}}(a) = \int_{b=-\infty}^{\infty} f_{\underline{a},\underline{b}}(a,b) db \\ f_{\underline{b}}(b) = \int_{a=-\infty}^{\infty} f_{\underline{a},\underline{b}}(a,b) da \end{array} \tag{C.5}$$

In the special case where $f_{\underline{a},\underline{b}}(a,b) = f_{\underline{a}}(a) f_{\underline{b}}(b)$, the random variables $\underline{a}$ and $\underline{b}$ are called *independent*.

## C.3. The expectation of a function of two random variables

Suppose $g(\underline{a}, \underline{b})$ is a real function of the two random variables $\underline{a}$ and $\underline{b}$. The expectation of $g(\underline{a}, \underline{b})$ is defined as follows:

$$E\{g(\underline{a}, \underline{b})\} \stackrel{def}{=} \int_{a=-\infty}^{\infty} \int_{b=-\infty}^{\infty} g(a,b) f_{\underline{a},\underline{b}}(a,b) db\, da \tag{C.6}$$

Specifically, let $g(\underline{a}, \underline{b})$ be the sum of two functions of $\underline{a}$ and $\underline{b}$, i.e. $q(\underline{a})$ and $r(\underline{b})$, respectively: $g(\underline{a}, \underline{b}) = q(\underline{a}) + r(\underline{b})$. Then:

$$\begin{aligned} E\{q(\underline{a}) + r(\underline{b})\} &= \int_{a=-\infty}^{\infty} \int_{b=-\infty}^{\infty} (q(a) + r(b))\, f_{\underline{a},\underline{b}}(a,b) db\, da \\ &= \int_{a=-\infty}^{\infty} \int_{b=-\infty}^{\infty} q(a) f_{\underline{a},\underline{b}}(a,b) db\, da \\ &\quad + \int_{a=-\infty}^{\infty} \int_{b=-\infty}^{\infty} r(b) f_{\underline{a},\underline{b}}(a,b) db\, da \\ &= \int_{a=-\infty}^{\infty} q(a) f_{\underline{a}}(a) da + \int_{b=-\infty}^{\infty} r(b) f_{\underline{b}}(b) db \\ &= E\{q(\underline{a})\} + E\{r(\underline{b})\} \end{aligned} \tag{C.7}$$

Thus, the addition and the expectation operator are distributive.

A corollary of (C.7) is the following expression for the variance:

$$\begin{aligned} \mathrm{Var}\{\underline{a}\} &= \int_{a=-\infty}^{\infty} (a - \mu_a)^2 f_{\underline{a}}(a) da \\ &= \int_{a=-\infty}^{\infty} (a^2 - 2a\mu_a + \mu_a^2) f_{\underline{a}}(a) da \\ &= \int_{a=-\infty}^{\infty} a^2 f_{\underline{a}}(a) da - \int_{a=-\infty}^{\infty} 2a\mu_a f_{\underline{a}}(a) da + \int_{a=-\infty}^{\infty} \mu_a^2 f_{\underline{a}}(a) da \\ &= E\{\underline{a}^2\} - \mu_a^2 \end{aligned} \tag{C.8}$$

Thus, the variance can be expressed in the mean square and the expectation of $\underline{a}$.

## C.4. The expectation and variance of the sum of two random variables

We form the weighted sum of two random variables as follows:

$$\underline{c} = \alpha \underline{a} + \beta \underline{b} \tag{C.9}$$

A corollary of equations (C.2) and (C.7) is the following expression for the expectation of $\underline{c}$:

$$E\{\underline{c}\} = \alpha E\{\underline{a}\} + \beta E\{\underline{b}\} = \alpha \mu_a + \beta \mu_b \qquad (C.10)$$

The calculation of the mean square of $\underline{c}$ is more involved:

$$\begin{aligned} E\{\underline{c}^2\} &= \int_{a=-\infty}^{\infty} \int_{b=-\infty}^{\infty} (\alpha a + \beta b)^2 f_{\underline{a},\underline{b}}(a,b) db\, da \\ &= \int_{a=-\infty}^{\infty} \int_{b=-\infty}^{\infty} (\alpha^2 a^2 + \beta^2 b^2 + 2\alpha\beta ab) f_{\underline{a},\underline{b}}(a,b) db\, da \\ &= \int_{a=-\infty}^{\infty} \alpha^2 a f_{\underline{a}}(a) da + \int_{b=-\infty}^{\infty} \beta^2 b^2 f_{\underline{b}}(b) db \\ &\quad + \int_{a=-\infty}^{\infty} \int_{b=-\infty}^{\infty} 2\alpha\beta ab f_{\underline{a},\underline{b}}(a,b) db\, da \\ &= \alpha^2 E\{\underline{a}^2\} + \beta^2 E\{\underline{b}^2\} + 2\alpha\beta E\{\underline{a}\,\underline{b}\} \end{aligned} \qquad (C.11)$$

The term $E\{\underline{a}\,\underline{b}\}$ is the so-called second order cross moment.

For the variance of $\underline{c}$ we find a similar result:

$$\text{Var}\{\underline{c}^2\} = \int_{a=-\infty}^{\infty} \int_{b=-\infty}^{\infty} (\alpha a + \beta b - \alpha\mu_a - \beta\mu_b)^2 f_{\underline{a},\underline{b}}(a,b) db\, da \qquad (C.12)$$

$$= \int_{a=-\infty}^{\infty} \int_{b=-\infty}^{\infty} (\alpha(a-\mu_a) + \beta(b-\mu_b))^2 f_{\underline{a},\underline{b}}(a,b) db\, da \qquad (C.13)$$

$$= \alpha^2 \text{Var}\{\underline{a}^2\} + \beta^2 \text{Var}\{\underline{b}^2\} + 2\alpha\beta E\{(\underline{a}-\mu_a)(\underline{b}-\mu_b)\} \qquad (C.14)$$

The term $E\{(\underline{a}-\mu_a)(\underline{b}-\mu_b)\}$ is called the *covariance* between $\underline{a}$ and $\underline{b}$.

# References

**Chapter 1**

[1] *Encyclopaedia Britannica*, 2000.

[2] L. Finkelstein, K.T.V. Grattan, *The Concise Encyclopedia of Measurement and Instrumentation*, Pergamon Press (1994), ISBN 0-08-036212-5, p. 201.

[3] DIN1319 (1995).

[4] Chr. Freyherr von Wolff, *Auszüge aus den Anfangsgründen aller Mathematischen Wissenschaften zu bequemerem Gebrauche der Anfänger (1772)*; citation from [5].

[5] D. Hofmann, *The Role of Measurement for Innovation and Society*, in: K.Kariya, L.Finkelstein (eds.), *Measurement Science, A discussion*, Ohmsha, Japan (2000), ISBN 4-274-90398-2, pp. 65–73.

[6] R. Kurzweil, *Das Zeitalter der künstlichen Intelligenz*, Carl Hanser Verlag (1993); citation from [5].

[7] R.E. Zupko, *Revolution in Measurement: Western European Weights and Measures Since the Age of Science*, The American Philosophical Society, Philadelphia, USA, (1990).

[8] R.E. Young, T.J. Glover, *Measure for Measure*, Sequoia Publ., Littleton, Colorado, USA (2000), ISBN 1-889796-00-X.

[9] P.K. Stein, *The Role of the Individual in Measurement Society*, in: K.Kariya, L.Finkelstein (eds.) *Measurement Science, A discussion*, Ohmsha, Japan (2000), ISBN 4-274-90398-2, pp. 103–112.

**Chapter 2**

[1] *Physics World*, November 1999, p. 22.

[2] T.J. Quinn, *The Kilogram: The Present State of our Knowledge*, IEEE Trans. on Instrumentation and Measurement, 40, 2 (April 1991), pp. 81–85.

[3] A.J. Daub, *Meten met Maten: Vademecum van veertig eeuwen (Measuring by Measures, Handbook of 40 Centuries)*, De Walburg Pers, Zutphen, 1979.

[4] B.N. Taylor, *The Possible Role of the Fundamental Constants in replacing the Kilogram*, IEEE Trans. on Instrumentation and Measurement, 40, 2 (April 1991), pp. 86–91.

[5] B.N. Taylor, P.J. Mohr, *The Role of Fundamental Constants in the International System of Units (SI): Present and Future*, IEEE Trans. on Instrumentation and Measurement, 50, 2 (April 2001), pp. 563–567.

[6] S.I. Park, H.K. Hong, *Development of 10-V Josephson Series Arrays*, IEEE Trans. on Instrumentation and Measurement, 52, 2 (April 2003), pp. 512–515.

[7] B.W. Petley, *The Fundamental Physical Constants and the Frontier of Measurement*, Bristol [etc.]: Hilger (1985), ISBN: 0-85274-427-7.

[8] A.J. Blundell, *Bond Graphs for Modelling Engineering Systems*, Ellis Horwood Publishers, Chichester, United Kingdom, 1982.

[9] S.Middelhoek, S.A. Audet, *Silicon Sensors*, Academic Press, 1989, ISBN 0-12-495051-5.

[10] J. Kohlmann, R. Behr, T. Funck, *Josephson Voltage Standards*, Meas. Sci. Technol. 14 (2003), pp. 1216–1228.

[11] B. Jeckelmann, B. Jeanneret, *The Quantum Hall Effect as an Electrical Resistance Standard*, Meas. Sci. Technol. 14 (2003), pp. 1229–1236.

[12] S. Middelhoek, D.J. Noorlag, *Three-dimensional Representation of Input and Output Transducers, Sensors and Actuators*, 2 (1981), pp. 29–41.

[13] P.K. Stein, *The Unified Approach to the Engineering of Measurement Systems for Test & Evaluation, Part 1: Basic Principles*, Stein Engineering Services, 1998, ISBN 1.881472.00.0.

## Chapter 3

[1] *Guide to the Expression of Uncertainty in Measurement*, International Organization for Standardization, Geneva, Switzerland, 1995, ISBN 92-67-10188-9.

[2] A. Papoulis, *Probability, Random Variables and Stochastic Processes*, Third Edition, McGraw-Hill, New York, 1991.

[3] E. Kreyszig, *Introductory Mathematical Statistics, Principles and Methods*, John Wiley & Sons, New York, 1970.

## Chapter 4

[1] E.A. Faulkner, J.B. Grimbleby, *The Effect of Amplifier Gain-bandwidth Product on the Performance of Active Filters*, The Radio and Electronic Engineer, 43, 9 (Sept. 1973), pp. 547–552.

[2] A. Budak, D.M. Petrela, *Frequency Limitations of Active Filters Using Operational Amplifiers*, IEEE Trans. on Circuit Theory, CT-19, 4 (July 1972, pp. 322–328.

[3] R.P. Sallen, E.L. Key, *A Practical Method of Designing RC Filters*, IRE Trans. on Circuit Theory, CT-2, March 1955, pp. 74–85.

## Chapter 5

[1] J.H. McLellan et al., *Signal Processing First*, Pearson Education, 2003, ISBN 0-13-120265-0.

## Chapter 6

[1] F. Adamo, F. Attivissimo, N. Giaquinto, M. Savino, *Measuring the Static Characteristics of Dithered A/D Converters*, Measurement, 32 (2002), pp. 231–239.

[2] B. Vargha, I. Zoltán, *Calibration Algorithm for Current-output R-2R ladders*, IEEE Trans. Instr. & Measurement, 50, 5 (October 2001), pp. 1216–1220.

[3] P. Carbone, M. Caciotta, *Stochastic-flash Analog-to-digital Conversion*, IEEE Trans. Instr. & Measurement, 47, 1 (February 1998), pp. 65–68.

## Chapter 7

[1] A.A. Bellekom, *Origin of Offset in Conventional and Spinning-current Hall Plates*, PhD thesis Delft University of Technology, 1998, ISBN 90-407-1722-2.

[2] F. Burger, P.-A. Besse, R.S. Popovic, *New Fully Integrated 3-D Silicon Hall Sensor for Precise Angular-position Measurement*; Sensors & Actuators A 67 (1998), pp. 72–76.

[3] O. Dezuari, E. Belloy, S.E. Gilbert, M.A.M. Gijs, *Printed Circuit Board Integrated Fluxgate Sensor*, Sensors & Actuators 81 (2000), pp. 200–203.

[4] K.E. Kuijk, W.J. van Gestel, F.W. Gorter, *IEEE Trans on MAG*, 11 (1975), pp. 1215.

[5] D. Atkinson, P.T. Squire, M.G. Maylin, J. Gore, *An Integrating Magnetic Sensor Based on the Giant Magneto-impedance Effect*, Sensors & Actuators 81 (2000), pp. 82–85.

[6] G. Rieger, K. Ludwig, J. Hauch, W. Clemens, *GMR Sensors for Contactless Position Detection*, Sensors & Actuators A 91 (2001), pp. 7–11.

[7] C. Giebeler, D.J. Adelerhof, A.E.T. Kuiper, J.B.A. van Zon, D. Oelgeschläger, G. Schulz, *Robust GMR Sensors for Detection and Rotation Speed Sensing*, Sensors & Actuators A 91 (2001), pp. 16–20.

[8] H. Wakiwaka, M. Mitamura, *New Magnetostrictive Type Torque Sensor for Steering Shaft*, Sensors & Actuators A 91 (2001), pp. 103–106.

[9] O.J. González, E. Castaño, J.C. Castellano, F.J. Gracia, *Magnetic Position Sensor Based on Nanocrystalline Colossal Magnetoresistances*, Sensors & Actuators A 91 (2001), pp. 137–143.

[10] S. Natarajan, *A Modified Linearized Thermistor Thermometer using an Analog Multiplier*; IEEE Trans. Instrumentation and Measurement, 39, 2 (April 1990), pp. 440–441.

[11] S. Kaliyugavaradan, P. Sankaran, V.G.K. Murti, *A New Compensation Scheme for Thermistors and its Implemantation for Response Linearization Over a Wide Temperature Range*; IEEE Trans. Instrumentation and Measurement, 42, 5 (October 1993), pp. 952–956.

[12] G.C.M. Meijer, A.W. van Herwaarden (ed.), *Thermal Sensors;* IOP Publ., London, 1994, ISBN 0-7503-0220-8.

## Chapter 8

[1] ISA-S37.12, *Specifications and Tests for Potentiometric Displacement Transducers*, Instrument Society of America, Pittsburgh, USA, 1977, ISBN 87664-359-4.

[2] Ph.A. Passeraub, P.-A. Besse, A. Bayadroun, S. Hediger, E. Bernasconi, R.S. Popovic, *First Integrated Inductive Proximity Sensor with On-chip CMOS Readout Circuit and Electrodeposited 1 mm Flat Coil*, Sensors & Actuators 76 (1999), pp. 273–278.

[3] Ph.A. Passeraub, P.A. Besse, S. Hediger, Ch. De Raad, R.S. Popovic, *High-resolution Miniaturized Inductive Proximity Sensor: Characterization and Application for Step-motor Control*, Sensors & Actuators A 68 (1998), pp. 257–262.

[4] H. Fenniri, A. Moineau, G. Delaunay, *Profile Imagery Using a Flat Eddy-current Proximity Sensor*, Sensors & Actuators A, 45 (1994), pp. 183–190P.

[5] Y. Hamasaki, T. Ide, *A Multilayer Eddy Current Microsensor for Nondestructive Inspection of Small Diameter Pipes*, Tansducers '95 - Eurosensors IX, Stockholm, Sweden, June 25–29, 1995, pp. 136–139.

[6] H. Kawai, *The Piezoelectricity of PVDF*, Japanese Journal of Applied Physics, 8 (1969), pp. 975–976.

[7] B. André, J. Clot, E. Partouche, J.J. Simonne, *Thin Film PVDF Sensors Applied to High Acceleration Measurements*, Sensors and Actuators A, 33 (1992), pp. 111–114.

[8] R. Marat-Mendes, C.J. Dias, J.N. Marat-Mendes, *Measurement of the Angular Acceleration Using a PVDF and a Piezo-composite*, Sensors & Actuators 76 (1999), pp. 310–313.

[9] M.G. Silk, *Ultrasonic Transducers for Nondestructive Testing*, Adam Hilger, Bristol, 1984, ISBN 0-85274-436-6.

# Chapter 9

[1] A.J. Bard, L.R. Faulkner, *Electrochemical Methods, Fundamentals and applications*, John Wiley and Sons, New York, 1980.

[2] B.J. Birch, T.E. Edmonds, *Potentiometric Transducers, Chemical Sensors*, in T.E. Edmonds: Blackie and Son, Glasgow and London, 1988.

[3] G.R. Langereis, *An Integrated Sensor System for Monitoring Washing Processes*, PhD-thesis, University of Twente, ISBN 90-365-1272-7, 1999.

[4] W. Olthuis, P. Bergveld, *Reliability and Selectivity at Electrolyte Conductivity Sensing*, Biocybernetics and Biomedical Engineering, 19 (1), 1999.

[5] W. Olthuis, W. Streekstra, P. Bergveld, *Theoretical and Experimental Determination of Cell Constants of Planar-Interdigitated Electrolyte Conductivity Sensors*, Sensors and Actuators B (Chemical), pp. 24–25, 1995.

# Chapter 10

[1] B. Jähne, *Practical Handbook on Image Processing for Scientific and Technical Applications*, Second Edition, CRC Press, 2004.

[2] F. van der Heijden, *Image-based Measurement Systems – Parameter Estimation and Object Recognition*, John Wiley & Sons, Chichester, 1995.

[3] T.S. Curry, J.E. Dowdey, R.C. Murry, *Christensen's Physics of Diagnostic Radiology*, Fourth Edition, Williams & Wilkins, Baltimore, 1990.

[4] P.Z. Peebles, *Radar Principles*, John Wiley & Sons, New York, 1998.

[5] R.O. Nielsen, *Sonar Signal Processing*, Artech House, Boston, 1991.

## Chapter 11

[1] ISA S37.1: *Electrical transducer nomenclature and terminology*, 1975, ISBN 0-87664-113-3.

[2] ISA-S37.12: *Specification and Tests for Potentiometric Displacement Transducers*, 1977, ISBN 0-87664-359-4.

[3] P.P.L. Regtien, K.H. Commissaris, *A New Approach to Practical Training Equipment for Measurement*, Proceedings of the IMEKO TC1 Symposium on Virtual and Real Tools for Education in Measurement, 17–18 September 2001 in Enschede, The Netherlands, ISBN 90.365.1664.1.

[4] H.J. Bergveld, W.S. Kruijt, P. Notten, *Battery Management Systems, Design by Modeling*, Philips Research book series, vol. 1, Kluwer Academic Publishers, Boston, 2002.

# Solutions to Exercises

## Chapter 1

1. a. ordinal; b. nominal; c. absolute; d. ratio; e. interval; f. ratio.

2. MUX: in SDM digital, in TDM analogue (high accuracy required); ADC: in TDM only one converter, fast, in SDM one converter per channel, slower.

3. a. Combined transduction/actuation: from temperature to expansion of liquid to length of the liquid column. b. Transduction from weight via displacement to electric signal; some signal amplification and AD-conversion; possibly a microprocessor for advanced functions (memory), actuation from (digital) to an optical display. c. Transduction from speed to pulse rate; conditioning not really necessary when the pulses have sufficient SNR; a processor performs conversion to speed data by some counting process; digital output from processor via some DA conversion and amplification to a control signal for the throttle.

## Chapter 2

1. a. $[E] \cdot [D] = (V/m) \cdot (C/m^2) = J/m^3$. b. $[V] \cdot [I] = (W/A) \cdot (A) = W \neq J/m^3$. c. $[T] \cdot [S] = (N/m^2) \cdot (1) = J/m^3$. d. $[\Phi] \cdot [H] = (J/A) \cdot (A/m) = J/m \neq J/m^3$. e. $[p] \cdot [V] = (N/m^2) \cdot (m^3) = N \cdot m = J$ (although not $J/m^3$, $p$ and $V$ are nevertheless called conjugated quantities).

2. Direct: photodiode, piezoelectric sensor and thermocouple; modulating: photoresistor, Hall sensor and strain gauge.

3. We see just three equal diamond shaped side planes; the body diagonal makes an angle $\arctan\sqrt{2}$ with the adjacent edges, so the apparent area of the cube amounts $3a \cdot \cos(\arctan\sqrt{2}) \approx 1.73a$.

4. Same procedure, resulting in:

$$T = \left(\frac{\partial^2 U}{\partial S^2}\right)_{\sigma,D} S + \left(\frac{\partial^2 U}{\partial S \partial D}\right)_{\sigma} D + \left(\frac{\partial^2 U}{\partial S \partial \sigma}\right)_{D} \Delta\sigma$$

$$\Delta\Theta = \left(\frac{\partial^2 U}{\partial\sigma\partial S}\right)_D S + \left(\frac{\partial^2 U}{\partial\sigma\partial D}\right)_S D + \left(\frac{\partial^2 U}{\partial D^2}\right)_{S,\sigma} \Delta\sigma$$

$$E = \left(\frac{\partial^2 U}{\partial D\partial S}\right)_\sigma S + \left(\frac{\partial^2 U}{\partial D^2}\right)_{\sigma,E} D + \left(\frac{\partial^2 U}{\partial D\partial\sigma}\right)_E \Delta\sigma$$

5. Heat balance: $\Phi_p = \Phi_c + \frac{1}{2}\Phi_J$, with $\Phi_p$ the Peltier heat withdrawn from the cold side, $\Phi_c$ the conduction heat flowing from the hot side to the cold side and $\Phi_J$ the electrically generated heat in the device. It is supposed that half of the Joule heat flows to the cold plate and the other half to the hot plate.
Further: $\Phi_p = \alpha T_c I_p$, $\Phi_c = (T_h - T_c)K_a$ and $\Phi_J = I_p^2 R_e$. Substitution in the heat balance results in:

$$T_h - T_c = \frac{\alpha T_c I_p - (1/2) I_p^2 R_e}{K_a}$$

which is a quadratic function of the Peltier current. The maximal temperature difference occurs at a Peltier current $I_p(\max) = \alpha T_c / R_e$, and amounts $(T_h - T_c)_{\max} = \frac{1}{2}\alpha^2 T_c^2/(R_e K_a) = \frac{1}{2} z T_c^2$ where $z = \frac{1}{2}\alpha^2/(R_e K_a)$, the figure of merit of the device materials.

## Chapter 3

1. The resolution error of the AD converter has a uniform distribution. The corresponding standard deviation is:

$$\sigma_{ADC} = \frac{3}{\sqrt{12}} \text{ mV}$$

We assume that the noise at the output of the amplifier and the resolution error are independent (a reasonable assumption that only breaks down if the magnitude of the signal is less than about one quantization step). The covariance between noise and round-off error must be zero. Therefore, the variances add up independently:

$$\sigma_{total}^2 = \sigma_{noise}^2 + \sigma_{ADC}^2 = (4 + 9/12) \text{ mV}^2$$
$$\sigma_{total} = \sqrt{4 + 9/12} \approx 2.18 \text{ mV}$$

2. The standard deviation of the result is $\sqrt{1^2 + 1^2 + 1/12}$ mV $\approx 1.44$ mV. Without AD-converter this would be $\sqrt{1^2 + 1^2}$ mV $\approx 1.41$ mV. Apparently, the influence of the AD-converter is small. The resulting error is approximately normally distributed.

3. The uncertainty due to quantization is $\sqrt{1/12} \approx 0.3$ mV. This is much larger than the uncertainty due to the noise and the systematic error of the sensor/amplifier.

Solutions to Exercises    347

Therefore, the quantization error is dominant. If the measurand is static and the noise is negligible, the quantization error is a systematic error. Therefore, repeating the measurement cannot improve the accuracy. The uncertainty remains. 0,3 mV.

4. The volume is calculated as follows: $Volume = length \cdot width \cdot height$. Thus:

$$\left(\frac{\Delta Volume}{Volume}\right)^2 = \left(\frac{\Delta length}{length}\right)^2 + \left(\frac{\Delta width}{width}\right)^2 + \left(\frac{\Delta height}{height}\right)^2 = 3\left(\frac{2}{100}\right)^2$$

The relative uncertainty is $1.73 \cdot 2\% = 3.5\%$.

5. The number of rain drops in the gauge is Poisson distributed. The volume of the collected water is $5 \cdot 2 \text{ cm}^3 = 10 \text{ cm}^3$. The number of rain drops must be: $10/0.1 = 100$. Since the number has a Poisson distribution, the variance of the number equals its mean, i.e. about 100. The standard deviation is 10. The relative uncertainty is 10%. The standard uncertainty of the measured level is 0.2 cm.

6. a. The estimated mean is just the average: $\bar{z} = \sum z_n/N = 7.6$
   b. The sample variance is $S^2 = \sum(z_n - \bar{z})^2/(N-1) = 1.16$
   c. The uncertainty of the mean is $\sigma_z/\sqrt{N}$. Since we do not know $\sigma_z$, we substitute its estimate $S$. Hence: $\hat{\sigma}_{\bar{z}} = S/\sqrt{N} = 0.34$
   d. The standard deviation of $S^2$ is $\sigma_{S^2} = \sqrt{2\sigma_z^2}/\sqrt{N-1}$. Substitution of $S^2$ for $\sigma_z^2$ yields $\hat{\sigma}_{S^2} = 0.54$

## Chapter 4

1. The DM gain of the first stage is $(1 + 2R_b/R_a) = 31$, and that of the second stage is $R_2/R_1 = 15$. The total DM gain is the product, so 465. The CMRR of the second stage is found using equation (4.14), which gives $(1 + 15)/(4 \cdot 2 \cdot 10^{-3}) = 2000$. This value is multiplied by the factor $(1 + 2R_b/R_a) = 31$, resulting in a total CMRR = 62 000.

2. The DM transfer of the oscilloscope is 1 (the value on the screen is the same as at the input). The CM gain amounts to $2 \cdot 10^{-5}/10 = 2 \cdot 10^{-6}$. The CMRR is just the ratio of the DM gain and CM gain, resulting in a CMRR of $5 \cdot 10^5$.

3. The transfer of the configuration can be found by setting up the following equation:

$$V_o = A(V_o - V_i) + \frac{A}{\text{CMRR}} V_i$$

with $A$ the DC voltage gain of the operational amplifier. After rearrangement the transfer is found to be:

$$\frac{V_o}{V_i} = \frac{A}{1+A} \cdot \left(1 + \frac{1}{\text{CMRR}}\right) \approx \left(1 - \frac{1}{A}\right)\left(1 + \frac{1}{\text{CMRR}}\right)$$

The deviation from the ideal transfer due to a non-infinite gain is $1/A = 10^{-5}$, and due to a non-infinite CMRR is $1/\text{CMRR} = 10^{-4}$ (since 80 dB is equivalent to $10^4$).

4. The ratio between the frequencies is just 20; both frequencies are on the declining asymptote of the frequency characteristic. So the ratio of the transfer is $20^n$, with n the order of the filter. Since the transfer at 3 kHz is 1 (asymptotically) the transfer at 150 kHz is $20^2 = 400$ lower, so $2.5 \cdot 10^{-3}$.

5. The question is where to position the cutoff frequency between $f = f_L = 1$ Hz and $f = f_H = 50$ Hz, such that both requirements are met. The amplitude transfer of an $n$th-order LP filter is $|H| = (1 + f^2/f_c^2)^{-n/2}$ with $f_c$ the cutoff frequency. For $f \ll f_c$ this is approximated by $|H| \approx 1 - (n/2)(f^2/f_c^2)$. For $f \gg f_c$ the approximation is $|H| \approx (f_c/f)^n$. With respect to the first requirement ($f = f_L$) this means: $(n/2)(f_L^2/f_c^2) < 1/100$ or $f_c > 10 \cdot f_L \cdot \sqrt{n/2} = 10 \cdot \sqrt{n/2}$. The second requirement (for $f = f_H$) leads to $(f_c/f_H)^n < (1/100)$ or $f_c < f_H/\sqrt[n]{100} = 50/\sqrt[n]{100}$. This is only possible for $n > 4$, so the minimum order is 5.

6. The output voltage of a bridge with just 1 active element is: $V_o = (\Delta R/4R) \cdot V_s$ with $V_s$ the bridge supply voltage. The total amplifier input noise over a bandwidth 4 Hz is $V_n = 20 \cdot \sqrt{4} = 40$ nV. The relative resistance change that yields the same voltage is found by: $(\Delta R/4R) \cdot V_s = V_n$ so $(\Delta R/R)_{\min} = 2 \cdot 10^{-8}$.

## Chapter 5

1. The Nyquist rate is simply half of the sampling frequency. In this case it is 1000 Hz.

2. The sampling frequency is 2000 Hz. Therefore the Nyquist frequency $f_0$ is 1000 Hz. Frequencies $f_0 + \Delta f$ will be folded back to frequencies $f_0 - \Delta f$. If we want the frequency content to be clean below 500 Hz, we will have to take $\Delta f$ maximally 500 Hz, or $f_c = 1500$ Hz.

3. The minimum time required to digitise one sample is $t = t_h + t_{conv}$ where $t_h$ is the acquisition time of the s/h device and $t_{conv}$ is the conversion time of the AD-converter. We ignore the other (short) times mentioned in Table 5.2. Thus we obtain a digitisation time of 20 $\mu$s, or a sample rate $f_s$ equal $1/20 \cdot 10^{-6} = 50$ kHz. The Nyquist criterion states that the maximum frequency left undisturbed is then $\frac{1}{2}{}^*f_s = 25$ kHz. Note that in Section 5.2.3 it is advised to keep a lower bandwith

## Chapter 6

1. Using equation (6.3) and $n = 10$ gives SNR $= 61.78$ dB.

2. Use equation (6.3): $6n + 1.78 > 60$, so $n \geq 10$ (bits).

3. The smallest common multiple of 50 and 60 Hz is 300 Hz. The minimum integration time is 1/300 s; the highest conversion rate at full scale input is 150 conv./s.

4. Direct ADC: $2^{10} = 1024$; cascaded ADC: 10; two-stage flash converter: $2^5 + 2^5 = 64$; SAR converter: 1.

5. The output varies between codes 1011 (6 875 V) and 1100 (7.5 V).

| N  | $E(n)$ (V) | $V_s$ (V) | Code | $V_p$ (V) | $Av(n)$ |
|----|------------|-----------|------|-----------|---------|
| 1  | 7.300      | 7.300     | 1011 | 6.875     |         |
| 2  | 0.425      | 7.725     | 1100 | 7.500     |         |
| 3  | −0.200     | 7.525     | 1100 | 7.500     |         |
| 4  | −0.200     | 3.250     | 0101 | 6.875     |         |
| 5  | 0.425      | 7.725     | 1100 | 7.500     | 7.250   |
| 10 | −0.200     | 7.465     | 1011 | 6.875     | 7.250   |
| 15 | −0.200     | 7.715     | 1100 | 7.500     | 7.292   |
| 20 | −0.200     | 7.340     | 1100 | 6.875     | 7.281   |

## Chapter 7

1. a. $R_{max} = V/I_{max} = 4.5/2.2 = 2\,\Omega$;
b. Voltage measurement: the voltage remains rather constant, so 0.1% of FS applies, so 10 bit resolution will do. Current measurement: the most unfavourable situation occurs at the smallest current, 10 mA, that should be measured with a resolution of 1%, that is 0.1 mA. This current flows through the test resistance (2Ω), so the LSB of the ADC should be less than 0.2 mV. The maximum voltage is 4.5 V, so the ratio is 22500, corresponding to a resolution of at least 15 bit ($2^{15} = 32\,768$).

2. The relative resistance change is $k \cdot x$ which appears to be much less than 1. So the approximate expression for the output voltage can be used: $V_o = \frac{1}{4} V_i \cdot k \cdot x = 10 \cdot 10 \cdot 10^{-5}/4 = 250\,\mu V$.

3. a. Solve $T$ from the second order equation (7.14), with $c = d = \cdots = 0$, for $R(T)/R_0 = 1.80$. This gives $T = 211.3°C$.
b. Solve $T$ from the linear equation ($b = c = \cdots = 0$), which gives $T = 204.7°C$. Apparently, the non-linearity error is about 6 K.
c. Solve $T$ from the third order equation ($d = 0$), and compare the result with the solution of the second order equation. An approximated value is obtained as follows: the 3rd order term is $R_0 c T^3 \, \Omega$, or about $R_0 c T^3 / R_0 a = cT^3/a$ K, or about 13 K.

4. Since $V_o = -V_R R(T)/R_1$ or $R(T) = -V_o R_1 / V_R$ the maximum total relative uncertainty in $R(T)$ is the sum of the individual relative uncertainties, that is 1.5%. At 100°C $R(T) = 139 \, \Omega$. The error in $R(T)$ amounts 1.5% of this value or 2.085 $\Omega$, which is equivalent to $2.085/0.39 = 0.5$ K.

5. $V_o = -V_R R(T)/R_1$, so $\Delta V_o = V_R R_0 a \Delta T / R_1 = 2$ mV.

6. Since the junctions with the connecting wires have the same temperature, these junctions do not contribute to the output voltage. A K-couple has a sensitivity of $\alpha_{bc} = 39 \, \mu V/K$ (Table 7.1), so $V_o = \alpha_{bc}(T_x - T_y) = 39(20 - 100) \, \mu V = -3.12$ mV.

7. According to equation (7.32) $y = 2x/L$, so $x = yL/2 = -0.6 \cdot 20 = -12$ mm.

## Chapter 8

1. When stressed, the volume of a body will increase somewhat. This means:

$$\frac{dV}{V} = \frac{dA}{A} + \frac{dl}{l} > 0 \rightarrow 2\frac{dr}{r} + \frac{dl}{l} > 0 \rightarrow 2\frac{dr}{r} > -\frac{dl}{l} \rightarrow \frac{dr/r}{dl/l} > -\frac{1}{2} \rightarrow \mu$$
$$= -\frac{dr/r}{dl/l} < \frac{1}{2}$$

2. At $x = +4$ mm the ration of the upper and lower part of the potentiometer is $16:24 = 2:3$, which is also the ratio of the resistances (linearity presumed). $V_o = (\frac{1}{2} - \frac{3}{5})V_i = -0.8$ V.

3. The transfer above the cut-off frequency is $H = C_e/(C_e + C_c) = 1/(1 + C_c/C_e) \approx 1 - (C_c/C_e)$, so neglecting the last term gives a relative error of $-C_c/C_e = -80/2000 = -4\%$.

4. $C_1 = \varepsilon_0 A/(d_1) = \varepsilon_0 A/(d - \Delta d)$ and $C_2 = (\varepsilon_0 A/d_2) = \varepsilon_0 A/(d + \Delta d)$ so

$$\Delta C = \frac{\varepsilon_0 A}{d - \Delta d} - \frac{\varepsilon_0 A}{d + \Delta d} \approx \frac{2\varepsilon_0 A \Delta d}{d^2} = \frac{2 \cdot 8.8 \cdot 10^{-12} \cdot 1.0 \cdot 10^{-4} \cdot 1.0 \cdot 10^{-6}}{1.0 \cdot 10^{-6}}$$
$$= 17.6 \cdot 10^{-16} F$$

5. It is assumed that the value of $R_f$ is such that the cut-off frequency is below 15 kHz. In that case:

$$V_o = \frac{C_1 V_i - C_2 V_i}{C_f} = \frac{2 \cdot \Delta C}{C_f} V_i = \frac{2 \cdot 17.6 \cdot 10^{-16}}{10^{-12}} \cdot 5 = 17.6 \text{ mV}$$

6. An empty tank has capacitance $C_{emp} = \varepsilon_0 hw/d$ with $h$ the plate height, $w$ the plate width and $d$ the plate distance.

Filled up to a height $x$ the capacitance is the sum of the empty and the filled parts:

$$C_x = \frac{\varepsilon_0 \varepsilon_r x w}{d} + \frac{\varepsilon_0 (h - x) w}{d}$$

The ratio is:

$$\frac{C_{emp}}{C_x} = \frac{\varepsilon_r x + h - x}{h} = \frac{x}{h}(\varepsilon_r - 1) + 1 \quad \text{or} \quad \frac{x}{h} = \frac{C_{emp}}{C_x} \frac{1}{\varepsilon_r - 1}$$

For $\varepsilon_r = 5$ and $C_{emp}/C_x = 2$ this results in $x/h = \frac{1}{4} = 25\%$.

## Chapter 9

1. $[R] = N \cdot m/(mol \cdot K)$, $[T] = K$ and $[F] = A \cdot s/mol$. Thus $[RT/F] = N \cdot m/(A \cdot s) = $ Volt (!). At $T = 298K$, $RT/F = 25.7$ mV.

2. Generally, both the working electrode and the reference electrode have a relatively high electrode-solution impedance. Measuring with a low-ohmic voltmeter would load the electrochemical cell to such an extent that the reading does no longer truly represent the cell potential. Moreover, a current through the electrochemical cell changes the composition of the working electrode and might damage the reference electrode.

3. $E^0_{Ag/Ag+} = 0.80$ V. Substituting $50 \cdot 10^{-3}$ mol/dm$^3$ Ag$^+$ in Nernst law yields: 0.723 V. $E^0_{Al/Al3+} = -1.67$ V. Substituting $20 \cdot 10^{-3}$ mol/dm$^3$ Al$^{3+}$ in Nernst Law yields: $-1.703$ V. Therefore the measured potential difference will be 2.426 V.

4. Surface area of a sphere with radius 1 m: $4\pi$ m$^2$ · $1\mu A = 1 \cdot 10^{-6}$ C/s = $6.24 \cdot 10^{12}$ elektronen/s. Thus the flux is $4.97 \cdot 10^{11}$ elektronen/(m$^2$s).

5. Using equation (9.19) simply yields: $D_{o2} = 4.15 \cdot 10^{-11}$ m$^2$/s.

6. The Ag-flux is $1.04 \cdot 10^{-4}$ mol/m$^2$s = $1.12 \cdot 10^{-5}$ kg/m$^2$s = $1.42 \cdot 10^{-9}$ m/s = 1.42 nm/s.

7. The viscosity of the medium. The viscosity decreases with increasing temperature and thus the conductivity increases with increasing temperature.

8. $C = \varepsilon_r \varepsilon_0$/cell-constant. $R = \rho \cdot$ cell-constant. $RC = \varepsilon_r \varepsilon_0 \rho$. $[RC] = $ s. Check this with $[\varepsilon_r \varepsilon_0 \rho]$!

9. Using equation (9.36) yields: G = 1.17 mS (R = 855$\Omega$).

## Chapter 10

1. The distance between the points in pixel coordinates is:

$$\sqrt{(130 - 100)^2 + (170 - 130)^2} = 50$$

This corresponds to a distance of $50 \cdot 0.006$ mm at the image plane. The distance between object plane and pinhole is 490 mm. The distance between image plane and pinhole is 10 mm. Therefore, the geometric magnification is $\frac{1}{49}$. The length of the line segment is $50 \cdot 0.006 \cdot 49 = 14.7$ mm.

2. The geometric relation of a pinhole camera is (see equation (10.8)):

$$x = \frac{-f}{Z-f}X \quad y = \frac{-f}{Z-f}Y$$

The geometric set-up of X-ray imaging is (see also Figure 10.14):

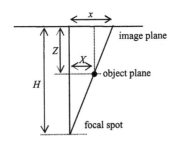

If Z is defined as the distance between object and image plane, and H is the distance between the focal spot (X-ray point source), then a point with coordinates (X, Y, Z)

is imaged at:
$$x = \frac{-H}{Z-H}X \quad y = \frac{-H}{Z-H}Y$$
Hence, X-ray imaging follows the pinhole model.

3. b is not correct. Doppler is suitable for velocity measurement. The range is obtained via the time-of-flight.

4. The welding process involves body surfaces with large temperature gradients. According to Planck's law the dynamic range of the emittance of a surface with these varying temperatures will be very large. We need a device that measures this emittance (this excludes X-ray imaging and the flying spot camera). The CMOS is preferred above the CCD because the former has a larger dynamic range.

# Index

AC voltage, measurement of 166
accuracy 44, 72
acoustic arrays 243
acoustic distance measurement 236
acoustic impedance 237
acoustic radiation 278, 279
acoustic transducers 239
actuator function 9
AD conversion 8
   direct 152
   integrating 154
   parallel 149
   round off error 321
   sigma-delta 157
AM signals, frequency spectrum 105
amplification noise 67
amplifiers
   analogue 87
   chopper 112
   differential 95
   inverting 93
   lock-in 111
   non-inverting 91
   operational 87
analogue to digital conversion 141 et seq
   conversion errors 141
analogue filters 97
   first order lowpass 99
analogue signal
   conditioning 87 et seq
   reconstruction 124

bandwidth 72
battery, SoC definitions 324
   system structure 325
Beaufort scale 332
binomial distribution 48

chemical quantities, measurement of 251 et seq
   amperometry 257
      oxygen sensor 261
   electrolyte conductivity 267
      sensing 268, 271
   potentiometry 252
   Pt black electrodes 273

colour vision 293
   digital representation 294
combinational operations 134
common mode rejection ratio 73, 96
compensation 79
   balanced 80
   method and feedback 82
computer vision 301
conventional true value 44
current to voltage converter 90

DA converters
   parallel 145
   serial 148
data acquisition, distribution, processing 8
DC voltage, measurement of 164
decoding 133
demodulation 109
density functions 47
derived quantities 17
design automation 312
detection 105
   synchronous 111
detectivity 32
differential non-linearity 144
differentiator 101
digital to analogue conversion 141 et seq
digital functions 133
digital image processing 301
digital signal
   binary 117
   conditioning 117
      hardware 135
   processor (DSP) 137
Dirac function 46
displacement quantities 27
displacement sensor
   capacitive 205
   eddy current 213
   inductive proximity and displacement 211
   magnetic and inductive 210
   magnetic proximity 210
   optical 217
      by intensity 217
      by triangulation 219
   transformer type 214
distance measurements 244
distribution functions 48

356  Index

electrical current, measurement of 169
electrical quantities 23
  definitions 23
  measurement of 163 et seq
encoding 133
error
  absolute 44
  budget 61
  drift 71
  dynamic 70
  environmental 70
  relative 44
  measurement 44
    and uncertainty 43 et seq
  noise 71
  and probability theory 44
  propagation 58
  random 63, 70
  reduction techniques 74
  resolution 70
  sampling 119
    information losses by 121
    theorem 123
  sources 68
    elimination 75
  systematic 63
  thermal 71
expectation 49

Faradaic, non-Faradaic processes 267
filter,
  Butterworth 104
  first order highpass 101
  over-sampling 127
  pre-sampling 125
  second order bandpass 102
force quantities 27
Fourier transform 122

glitch errors 144
guarding and grounding 77
  active 170

histogram, 51
  normalized 50
histogramming 50
Hooke's Law 27

illumination techniques 287
image segmentation 303
imaging
  device 286
    advanced 295
  digital 301
  instruments 277 et seq

NMR 300
optical 286
range 298
X-ray 296
impedance, measurement of 171
indication time 72
instruments, measurement, requirements
  for 6
integral non-linearity 143
integrator 99
interferometer
  Fabry-Perot 230
  Mach-Zehnder 229
  Michelson 229
interferometric displacement sensing 227
irradiance 29

Josephson effect 19

Ladder network 145
Lambert's cosine law 30
light
  absorption 284
  reflection and refraction 284
  scattering 284
light dependent resistor 188
light-material interactions 283
lumped element 33

magnetic circuits 24
magnetic quantities 23
  field strength 23
  flux 23
  induction 23
  measurand 44
magnetic quantities, measurement of 174
  by coil 175
  fluxgate sensor 177
  Hall sensor 175
  magnetoresistive sensor 178
mean deviation 53
measurement, definition 1
  DIN 2
  errors 43 et seq
  Finkelstein, L 2
  history 1
  properties, transformation from
    system,
    computer-based 314
    design of 309 et seq
    general architecture of 8
    linear relationship 58
    nonlinear relationship 60
  units 2
mechanical quantities, measurement of 197

Index   357

microcontrollers 136
modulation 105
Moh's scale of hardness 331
monotonicity 144
multiplexing 9, 128
   analogue 131
   digital 131

noise
   equivalent power 32
   Johnson 71
   Poisson 71
   quantization 141
   shot 71
   thermal 71
Nyquist rate 123

object recognition 305
optical encoder 222
optical quantities, measurement of 163 et seq, 188
optical scanning and sensing devices 289 et seq
   CCD 289
   CMOS camera 291
   multispectral imaging and colour vision 291

particle radiation 278, 279
PC in measurement system 314
perspective projection 288
photodiode 190
photometric quantities 28
physical effects
   symbols, property names, units 35
piezoelectric effect 231
   materials 232
   sensors, accelerometers 231, 233
      interfacing 234
Poisson ratio 201
position sensitive diode 191
postfiltering 76
potentiometric sensors 197
prefixes, decimal and binary multiples 335
probability density function 46, 52
programmable logic devices 136
   read only memory 136
Pt 100 temperature sensors 180

quantities
   conjugate 33
   and properties 21
      framework based on energy considerations 33
quantum Hall effect 20

radian 22
radiant intensity 29
radiation thermometer 186
radiometric quantities 28
rain drops 321
rain gauge, uncertainty 320
random effects 70
random variable
   single, expectation of 335
   two, joint and marginal density 336
      expectation, variance of sum 337
      function expectation 337
regional description 304
repeatability 72
reproducibilty 72
resistive temperature sensors 180
resolver 215
resolving power 279
response time 73
Richter and Mercalli scales 333

sampling
   devices and multiplexers 128
      and hold devices 129
   filter
      over- 127
      pre- 125
scales
   absolute 5
   interval 4
   ordinal 4
   ratio 5
Seebeck effect 183
sensors
   ISA standards 313
   modulating 36
   piezoelectric 39
   selection of level 318
sequential operations 134
shielding 78
slew rate 72
sound intensity 237
sound propagation speed 237
spectral EM radiation
   discrete nature 282
standard
   deviation 49, 53
   electrical impedance 20
   meter 18
   potential difference 19
standards 5, 18
   secondary 18
statistical inference 50
steradian 22
stochastic experiment 44
strain gauges 200

tachometer, optical 221
temperature sensors, integrated 185
thermal quantities 25
   measurement of 163 et seq, 179
thermocouples, thermopiles 184
thermoelectric sensors 182
time domain
   signal characterization 64
time of flight measurements 244
   CW, FM-CW 246
time and time related quantities,
      measurement of 173
tomography, computer 297
transducers 36
   acoustic 239
   electric and piezoelectric 241
   ultrasonic 242
transduction 8
triangulation 219

uncertainty 44
   analysis 55
   combined 58, 322
   and measurement error 43 et seq
   type A evaluation 56
   type B evaluation 57
units,
   derived 334
   seven standard 16
   SI derived 17
   systems of 15

variables, dependent and independent 21
   generalized I and V 33
   through and across 22, 33
variance 49
virtual instrumentation 7, 316
   design 323

Wheatstone bridge 81, 108